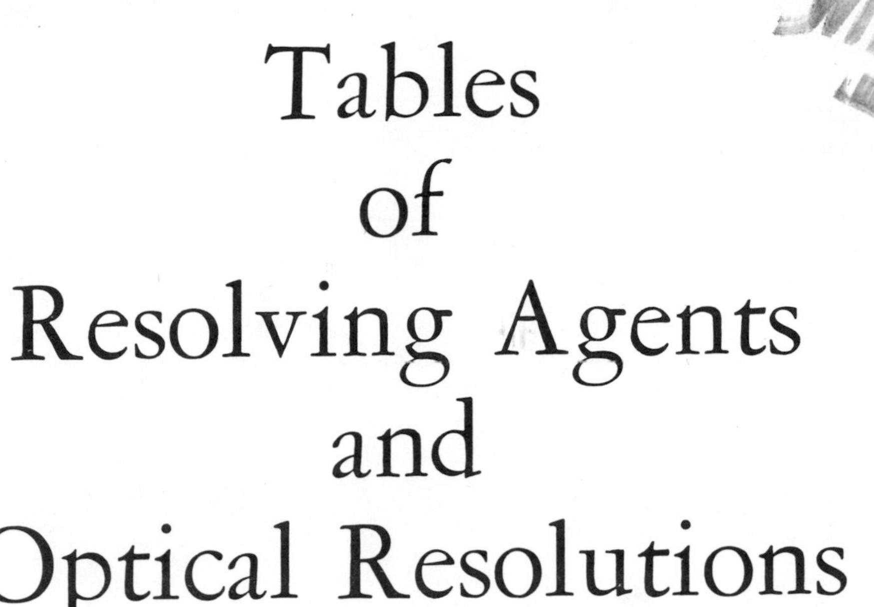

# Tables of Resolving Agents and Optical Resolutions

SAMUEL H. WILEN
Department of Chemistry
The City College of the
City University of New York

Edited by
ERNEST L. ELIEL
Department of Chemistry
University of Notre Dame

UNIVERSITY OF NOTRE DAME PRESS • NOTRE DAME – LONDON

International Standard Book Number 0-268-00467-6

Library of Congress Catalog Card Number: 79-186520

Printed in the United States of America by
NAPCO Graphic Arts, Inc., Milwaukee, Wisconsin

## CONTENTS

# PREFACE

This little volume is dedicated to those scientists who, along with their students and associates, were in large measure responsible for developing methods and techniques of resolution and for discovering useful resolving agents. Included among these scientists are Louis Pasteur, Sir William J. Pope, Joseph Kenyon and Arne Fredga on the European continent and A. W. Ingersoll on the American continent.

The review of resolution methods and the search for resolving agents and resolutions of which this volume forms an integral part was initiated at the University of Groningen, the Netherlands, during the tenure of my sabbatical leave (1968-69). I gratefully acknowledge the hospitality and cooperation of Professor Hans Wynberg and of his associates there. My thanks go to Mr. Gaatse J. Douma and Mr. Jan Schut who commenced the search in Groningen and to Mr. Robert Gallucci and Miss Niza J. Leichtman who assisted in the tabulation in New York. Professor Arne Fredga provided numerous data for these Tables, some of them unpublished, as well as encouragement for which I am very appreciative.

I particularly thank my wife, Rosamond Lewis Wilen, who helped me to keep track of hundreds of resolutions and performed a variety of editorial chores and Professor Ernest L. Eliel who encouraged the work from beginning to end and found a way to have it published.

The camera-ready manuscript for these Tables was prepared by Juris Verlag, Zurich, Switzerland with the help of a grant-in-aid from Firmenich & Cie., Geneva. I am very grateful to Firmenich & Cie. and to their Director of Research, Dr. G. Ohloff for this assistance without which the Tables could not have been produced in their entirety.

December 1971 Samuel H. Wilen

## Chapter I

# Introduction

---

A review of the literature on optical resolutions of interest to organic chemists was undertaken in 1968 since there were no recent extensive reviews in existence. The two most recent reviews, those of Eliel (1) and of Theilacker (2), had appeared in 1962 and 1955, respectively. The former was not meant to be a compendium of resolutions, while the latter was; however, neither review cites a large number of compounds resolved.

The present review is appearing in two parts, the discussion as a chapter in Volume 6 of Topics in Stereochemistry (3) and the data upon which the review is based in the form of these Tables of Resolving Agents and Resolutions. These Tables may thus be considered as a supplement to Topics in Stereochemistry, Volume 6.

The reason for the publication of the review in two parts is that together the two would have comprised an entire volume in the Topics in Stereochemistry series. This was not felt to be economically justifiable. And yet, it seemed worthwhile to make the tabulated data on resolutions available since no comparable listing exists. To accomodate the tables, a separate publication of modest size, and hopefully of reasonable cost, was planned and culminated in this volume.

Some comments on the literature search associated with the review and tables may be of interest. The principal approach utilized was to search Chemical Abstracts through its subject indexes. The very character of this journal, i.e., its emphasis on abstracting of information on specific chemical compounds and its lack of emphasis on processes and concepts, conspire to make systematic and reliable retrieval of articles describing resolutions difficult. The search was carried out by examining index entries under the terms resolution, optical resolution, stereochemistry, isomer, enantiomer, and chiral. Resolutions were located under all these entries, in particular under the entry resolution, yet, disconcertingly, many more were found through page by page scanning of abstract issues.

It soon became evident that an exhaustive search would be impracticable, at least for one individual to undertake in a reasonable period of time. Indeed, it seems questionable whether an exhaustive survey and tabulation of resolutions could be successfully achieved at all given the difficulty of locating them. Even machine retrieval of resolutions does not appear possible; what is not put into the abstract or the indexes cannot, of course, be retrieved.

Thus, a more limited, though substantial and comprehensive tabulation became one of the goals of this work. This goal was to list a large and representative number of resolutions of several classes of organic compounds resolved principally by conventional, i.e., non-chromatographic and non-biological methods. The specific limitations are stated in Section 2 of reference (3) where resolution methods and resolving agents are discussed in detail. The search covered the literature over a period of approximately 20 years, to the beginning of 1971.

In addition to the search in Chemical Abstracts, Annual Reports of the Chemical Society, Organic Syntheses, and several primary journals were searched over varying periods of time. The latter included extensive issue by issue and even page by page searching of appropriate sections of the Journal of the American Chemical Society and of Arkiv Kemi. These two journals in particular appear to contain more resolutions (in the 1950-1970 period) than any others known to this author.

In this manner, a wide variety of resolutions were uncovered and it is felt that few, if any, major resolution methods were missed and are not illustrated in the tables.

The search for resolution methods, for resolving agents and for resolutions yielded the bonus of optical rotation data on the compounds resolved. No extensive collection of resolutions and rotation data on the corresponding resolved compounds appears to exist. For example, the often quoted review of Velluz (4) includes a list of approximately 125 resolutions. The article of Ingersoll (5) reviewing resolutions of alcohols lists some 150 resolutions. Neither of these lists includes any detail and in fact only states the name of the compound resolved and the resolving agents used. It seemed worthwhile, therefore, to arrange the collected data in tabular form and to make them available to other chemists. In fact, not even all the data collected were incorporated in these Tables in order to keep their size and hence the cost of the book reasonable.

The search also uncovered many references on the preparation and properties of resolving agents themselves and it seemed appropriate to include this information here as well (Chapter 3).

In addition to providing up-to-date information on resolving agents and resolutions, it is hoped that these Tables may serve as a key reference for those interested in the improvement of classical resolution procedures in general and as a starting point for those interested in the development of new resolving agents in particular.

Added in Proof. A new review of methods of optical resolution has appeared in 1971 [P. H. Boyle, Quart. Rev. (London), 25, 323 (1971)] and a review of resolution methods in the Russian language has just come to the attention of this author [ A. P. Terent'ev and V. M. Potapov, Usp. Khim., 26, 1152 (1957)] .

The latter review appears to be little known to scientists in Western Europe or on the American continent. An English translation does not seem to be available and no references to it have been found in the literature search described in Chapter I.

Neither of these reviews contains an extensive tabulation of compounds resolved.

## References to Chapter I

(1) E. L. Eliel, Stereochemistry of Carbon Compounds, McGraw-Hill, New York, 1962, Chap. 4.

(2) W. Theilacker, in Methoden der Organischen Chemie (Houben-Weyl), 4th ed., Vol. 4, Part 2, E. Müller, ed., Georg Thieme Verlag, Stuttgart, 1955, Chap. 7.

(3) S. H. Wilen, "Resolving Agents and Resolutions in Organic Chemistry", in Topics in Stereochemistry, Vol. 6, N. L. Allinger and E. L. Eliel, eds., Wiley-Interscience, New York, 1971.

(4) L. Velluz, ed., Substances Naturelles de Synthèse, Vol. IX, Masson, Paris, 1954, pp. 119-174.

(5) A. W. Ingersoll, in Organic Reactions, R. Adams, ed., Vol. 2, Wiley, New York, 1944, Chap. 9.

## Chapter II

## Addendum to "Survey of Resolving Agents and Resolutions"

---

Since these Tables are going to press slightly later than the Chapter "Resolving Agents and Resolutions in Organic Chemistry," (1) an opportunity provided itself to include this brief addendum in which recent developments and other items relating to resolutions which could not be included in the chapter are described.

Acids and their derivatives. Resolution of a carboxylic methyl ester through diastereomeric ester synthesis *via* ester interchange has been reported (2). (+)-Menthol was the alcohol resolving agent used. After recrystallization of the mentholate esters, the methyl ester was recovered by reaction with methanol at 110-120$^{o}$ in an autoclave.

T. Svensson has briefly outlined some negative aspects of the use of α-(2-naphthyl)-ethylamine in the resolution of acids (3). In addition to the need of preparing and resolving this resolving agent prior to use, he reports a resolution in which much solvent and many recrystallizations were required to attain maximum rotation of the diastereomeric salts.

Amines. A water insoluble acid, (-)-diisopropylidene-2-oxo-L-gulonic acid [1] which is an intermediate in the commercial synthesis of ascorbic acid [Hoffmann-La Roche Inc.] is recommended as an easily recoverable resolving agent for amines (4). α-Methylbenzylamine, (and

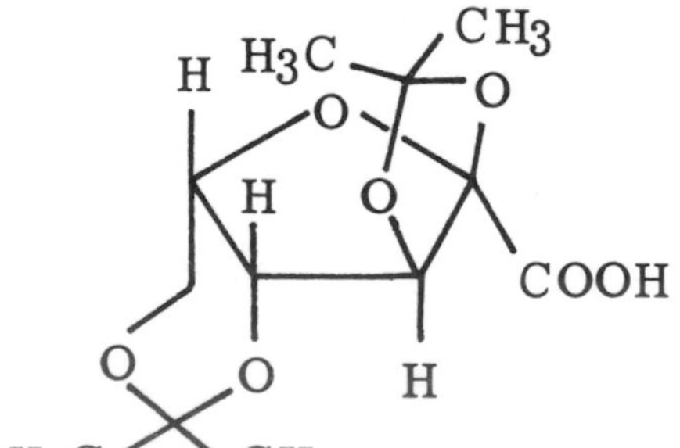

[1]

other phenylethylamines; ref. (5)), threo-2-amino-1-(p-nitrophenyl)-1,3-propanediol, α-(1-naphthyl)-ethylamine and heterocyclic amines such as 3-methoxymorphinan are readily resolved with [1].

L-(+)-Aspartic acid has been utilized in the resolution of the base 1-phenyl-2-(4-tolyl)-ethylamine (6).

Alcohols. A few alcohols have been resolved through their covalent tartranilates formed by acid-catalyzed reaction of the alcohol with (+)-tartranil, [2] (7).

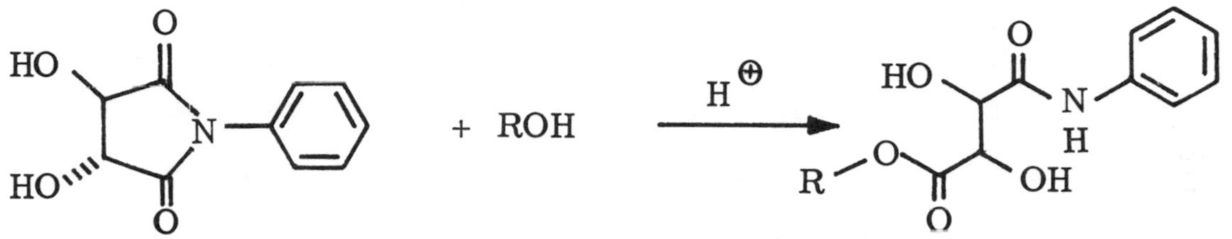

HO, HO, O, O, N + ROH $\xrightarrow{H^{\oplus}}$ HO, O, N, H, R, O, O, OH

[2]

(+)-Tartranil is now commercially available. Resolutions via this method may be attempted and, if separation of the tartranilates by crystallization is unavailing, may be coupled with chromatographic separation of the covalent diastereomers.

The preparative gas chromatographic separation of alcohols through separation of the N-trifluoroacetyl-L-alaninate esters has been reported (8). In this way, both enantiomers of, e.g., 3,3-dimethyl-2-butanol have been obtained in greater than 98% optical purity. The method has so far been applied to quantities of alcohol no greater than 1 ml.

Aldehydes and ketones. 5-Methyl-5-(α-methyl-β-phenylethyl)-semi-oxamazide, [3] (Hosemiazide), prepared from deoxyephedrine,

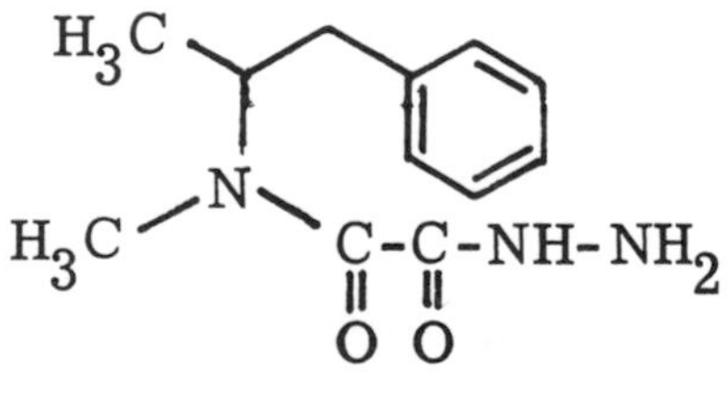

$H_3C$, N, $H_3C$, C-C-NH-$NH_2$, O O

[3]

has been suggested as a reagent for the resolutions of aldehydes and ketones (9). 2-Flavanone has been resolved with [3].

The partial resolution of trans-2,3-di-tert-butylcyclopropanone by asymmetric destruction with amphetamine has been reported (10).

Amino acids. B. Sjöberg has pointed out that α-azido acids are fairly strong and consequently can be resolved conventionally with basic resolving agents. They are then easily reduced to amino acids (11a). For example, the resolution of α-azido-m-fluorophenylacetic acid with pseudoephedrine has been reported (11b).

Ephedrine has been employed in the resolution of nine amino acids, e.g., leucine, through the N-benzyloxycarbonylamino derivatives (12). In these resolutions, use of one half equivalent of resolving agent is sufficient.

Miscellaneous. A new method for the resolution of thiols via thioglycolic acids has been described. Phenylbiphenylyl-α-naphthylmethyl mercaptan has been so resolved (13).

Attempts at the resolution of carbodiimides have been discussed and the results considered questionable (14).

The spontaneous resolution of 1,1'-binaphthyl has been reported in the solid state upon heating below the melting point of the compound or upon solidification of melted samples. Pincock and Wilson suggest (15) that spontaneous resolution by crystallization from a melt or solution "may be more general than commonly supposed" for compounds in which enantiomers can rapidly interconvert. The helicene resolutions of Martin and Wynberg and their co-workers immediately come to mind in support of this view (16,17).

Whitesides and Lewis (18) have recently described a nmr shift reagent, tris[3-(tert-butylhydroxymethylene)-d-camphorato]europium (III) [4] that may greatly simplify the determination of enantiomeric purity

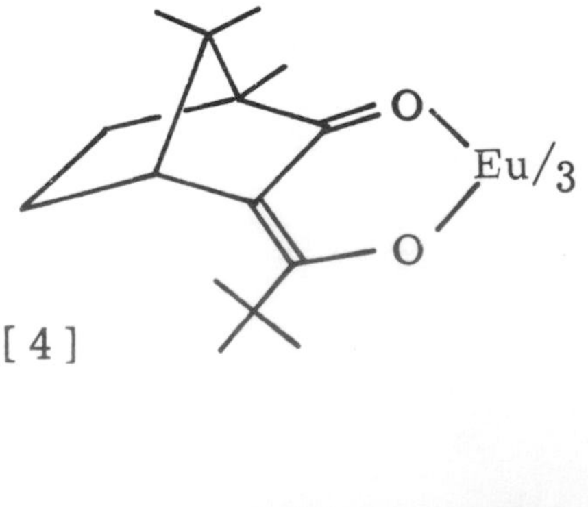

O Eu/3 O

[4]

of amines by the nmr method.

A simple method for the partial resolution of a variety of chiral sulfoxides has been described which involves stereospecific inclusion into β-cyclodextrin (19).

## Added in Proof

New chiral nmr shift reagents, e.g., [5] (20), [6] (21), and [7] (22), have been described which considerably increase the possibility of carrying out enantiomeric purity determinations by the nmr method. Determination of enantiomeric purity of epoxides, alcohols, sulfoxides, aldehydes, ketones and esters is now possible with these reagents, in some cases even at 60 MHz.

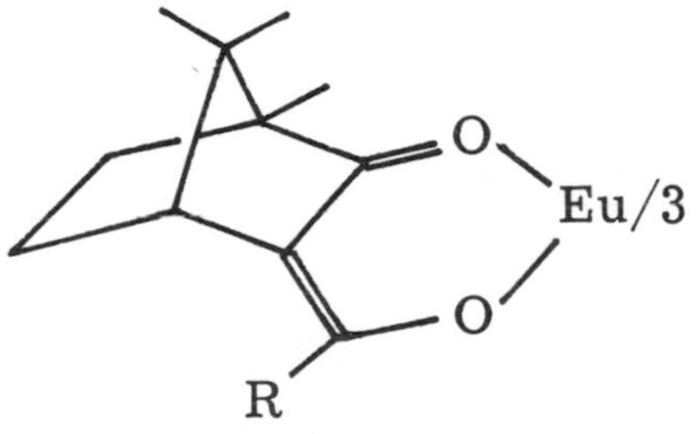

Eu/3

R

[5] R = $CF_3$

[6] R = $CF_3CF_2CF_2$

[7] R = $-CH_2$ ... $CH_3$

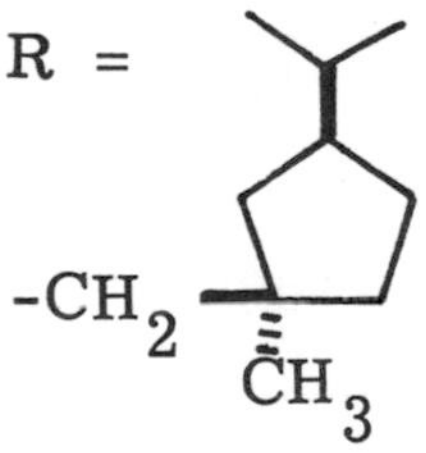

The first resolution of a chiral tetrahedral sulfur compound, ethylmethylphenyloxosulfonium ion, has been reported. Separation of the enantiomeric ions is achieved by reaction of the mercuritri-iodide salt with silver (+)-camphor-10-sulfonate and recrystallization of the resulting diastereomeric salts (23).

## References to Chapter II


(1) S. H. Wilen, "Resolving Agents and Resolutions in Organic Chemistry," in Topics in Stereochemistry, Vol. 6, N. L. Allinger and E. L. Eliel, Eds., Wiley-Interscience, New York, 1971.

(2) E. J. Eisenbraun, G. H. Adolphen, K. S. Schorno, and R. N. Morris, J. Org. Chem., 36, 414 (1971).

(3) T. Svensson, Ark. Kemi, 26, 27 (1967).

(4) C. W. Den Hollander, W. Leimgruber and E. Mohacsi, Belgian Patent 745,032, Jan. 28, 1970 (to F. Hoffmann-La Roche and Co.); Ger. Offen. 2,003,486, August 13, 1970 [Chem. Abstr., 73, 77081s (1970)] and private communication from Dr. E. P. Oliveto.

(5) A. Brossi and S. Teitel, J. Org. Chem., 35, 3559 (1970).

(6) Y. Suzuki, Ger. Offen. 2,023,426, Nov. 19, 1970 (Sumimoto Chemical Co., Ltd.) [Chem. Abstr., 74, 42121y (1971)].

(7) F. Barrow and R. G. Atkinson, J. Chem. Soc., 1939, 638; see also the discussion by A. W. Ingersoll in Organic Reactions, R. Adams, Ed., Vol. 2, Wiley, New York, 1944, p. 383.

(8) G. S. Ayers, J. H. Mossholder, and R. E. Monroe, J. Chromatogr., 51, 407 (1970).

(9) M. Kotake and G. Nakaminami, Proc. Japan Academy, 29, 56 (1953).

(10) D. B. Sclove, J. F. Pazos, R. L. Camp, and F. D. Greene, J. Amer. Chem. Soc., 92, 7488 (1970).

(11) (a) A. Fredga, personal communication; (b) B. A. Ekstrom and B. Sjöberg, Ger. Offen. 1,912, 904, Oct. 9, 1969 [Chem. Abstr., 72, 3484h (1970)].

(12) K. Oki, K. Suzuki, S. Tuchida, T. Saito and H. Kotake, Bull. Chem. Soc. Japan, 43, 2554 (1970).

(13) L. W. Feller, Ph.D. Dissertation, Johns Hopkins University, 1969; Diss. Abstr. Int. B., 30, 557 (1969).

(14) F. A. L. Anet, J. C. Jochims and C. H. Bradley, J. Amer. Chem. Soc., 92, 2557 (1970).

(15) R. E. Pincock and K. R. Wilson, J. Amer. Chem. Soc., 93, 1291 (1971).

(16) R. H. Martin, M. Flammang-Barbieux, J. P. Cosyn and M. Gelbcke, Tetrahedron Letters, 1968, 3507.

(17) H. Wynberg and M. B. Groen, J. Amer. Chem. Soc., 90, 5339 (1968); idem, ibid., 92, 6664 (1970).

(18) G. M. Whitesides and D. W. Lewis, J. Amer. Chem. Soc., 92, 6979 (1970).

(19) M. Mikolajczyk, J. Drabowicz and F. Cramer, Chem. Comm., 1971, 317.

(20) H.L. Goering, J.N. Eikenberry and G.S. Koermer, J. Amer. Chem. Soc., 93, 5913 (1971).

(21) R.W. Fraser, M.A. Petit and J.K. Saunders, Chem. Commun., 1971, 1450.

(22) G.M. Whitesides and D.W. Lewis, J. Amer. Chem. Soc., 93, 5914 (1971).

(23) M. Kobayashi, K. Kamiyama, H. Minato, Y. Oishi, Y. Takada and Y. Hattori, Chem. Commun., 1971, 1577.

# Chapter III

## Tables of Resolving Agents

A large number of resolving agents are now available commercially. Since some of the suppliers may not be well known, the names of the principal ones in the United States and Western Europe are listed in Table 1.

The principal resolving agents and chiral adsorbents of modern usage are listed in Tables 2 through 7. Molecular formulas of most of these substances are identified by bracketed numbers in column 2 of these tables. These numbers refer to the bold-face structure numbers in the review in _Topics in Stereochemistry_, Volume 6. Where known, the absolute configuration of the resolving agent is given in column 5.

The specific rotations of resolving agents are often required to assess the purity of these substances. Where such values have been found in the literature, or given by commercial suppliers, they have been included in the tables. Since most often critical evaluation of the literature values is lacking, they should be used with caution. Solvent names have been abbreviated as follows: benzene, $C_6H_6$; chloroform, $CHCl_3$; dichloromethane, $CH_2Cl_2$; ethanol, EtOH; methanol, MeOH; pyridine, py; water, $H_2O$; no solvent, neat.

The specific commercial sources for each resolving agent, in optically active (+/-) and/or in racemic form (rac), are tabulated. Some of these resolving agents are also available, or only available, as salts. This information comes from the latest catalogues available (1969-1971). The suppliers do not always use the tabulated names in their catalogs, however. Some synonyms are given in these Tables and also in _Topics in Stereochemistry_, Vol. 6. The symbols utilized for the suppliers are given in Table 1.

References to the preparation (Prep) and resolution (Res) of resolving agents are given in column 7 of Tables 2 to 7.

Some of the compounds listed naturally belong in more than one of these tables. For example, some acidic resolving agents can be used to resolve alcohols as well as amines, the former through diastereomeric ester formation. Similarly, some amines may be used to resolve either acids or alcohols when the latter are first converted to derivatives such as alkyl hydrogen phthalates.

Data on some synthetic resolving agents also appear in Tables 8-14. These entries contain additional useful information and references; they should be consulted along with Tables 2-7.

## TABLE 1 COMMERCIAL SUPPLIERS OF RESOLVING AGENTS *

| | (Code) |
|---|---|
| Aldrich Chemical Company | A |
| Fluka AG ** | F |
| Frinton Laboratories | FR |
| K and K Laboratories | KK |
| Koch-Light Laboratories | KL |
| Norse Laboratories + | N |
| Pfaltz and Bauer Inc. | PB |
| Sigma Chemical Company | S |

* Addresses of these suppliers are given in L. Fieser and M. Fieser, Reagents for Organic Synthesis, Wiley-Interscience, N.Y., Vol. 1, p. 1297 (1967) and Vol. 2, p. 469 (1969).

** Columbia Organic Chemicals Co., Inc., Columbia, South Carolina 29205 and Norell Chemical Co., Inc., Landing, N.J. 07850 are distributors of Fluka chemicals in the United States.

\+ Norse Laboratories, Inc., Santa Barbara, California 93103 also offers a custom resolution service.

## TABLE 2 RESOLVING AGENTS FOR ACIDS*

| NAME | Molecular Formula | MW | Structural Formula | Configuration | $[\alpha]_D$[+] | References | Suppliers |
|---|---|---|---|---|---|---|---|
| 2-Amino-1-butanol | $C_4H_{11}NO$ | 89.14 | | S-(+) | $+10.11^o$ (neat) ($t = 20^o$)<br>$-10.10^o$ (neat) ($t = 20^o$) | Res 1 | rac: A, F, KK, KL, N, PB<br>+/- : A, KK, N, PB |
| threo-2-Amino-1-phenyl-1,3-propanediol | $C_9H_{13}NO_2$ | 167.21 | [28] | L-(+) | $+18^o$ (c 2 %, $H_2O$) ($t = 27^o$)<br>$+18^o$ (c 2 %, $H_2O$) ($t = 27^o$) | Prep 2 a,b<br>Res 2 b,c | (+) : A, FR, KK |
| threo-2-Amino-1-(p-nitrophenyl)-1,3-propanediol | $C_9H_{12}N_2O_4$ | 212.21 | [27] | 1S,2S-(+) | $+22.6^o$ (c 1.5, MeOH) ($t = 20^o$)<br>$-22.6^o$ (c 1.5, MeOH) ($t = 20^o$) | Res 3<br>Also, Chapter II, ref. 4. | (-): A, PB |
| Amphetamine[a] | $C_9H_{13}N$ | 135.21 | [32] | S-(+) | $+34.5^o$ (neat) ($t = 24^o$)<br>$+35.6^o$ (neat) ($t = 15^o$)<br>$+38^o$ ($C_6H_6$) | Prep 4b<br>Res 4a | rac: KK; HCl salt: S<br>+/-: A, KK; S (sulfate) |
| Arginine $HN{=}C(NH_2)-NH-CH_2CH_2CH_2-CH(NH_2)-COOH$ | $C_6H_{14}N_4O_2$ | 174.21 | | L-(+) | $+12.5^o$ (c 3.5, $H_2O$) ($t = 20^o$) | | rac: A (HCl salt); F, KK, PB, S<br>(+): A, F, FR, KK, KL, PB, S |
| Brucine[b, c] | $C_{23}H_{26}N_2O_4$ | 394.44 | [20] | | $-127^o$ ($CHCl_3$) | | (-): A, F, KK, KL, N, PB, S |
| Cinchonidine | $C_{19}H_{22}N_2O$ | 294.40 | [24] | | $-109.2^o$ (EtOH) ($t = 20^o$) | | (-): A, F, KK, KL, N, PB, S |
| Cinchonine | $C_{19}H_{22}N_2O$ | 294.40 | [25] | | $+229^o$ (EtOH) ($t = 20^o$) | | (+): F, KK, KL, N, PB, S |

* References to Table 2 are on p. 22.

\+ Rotation values for common resolving agents given in the Merck Index (8th Ed.) are unreferenced.

a 1-Phenyl-2-aminopropane; (+)-Dexedrine.

b Recovery of brucine from salts is described in A.I. Vogel, Practical Organic Chemistry, 3rd. ed., Longmans, London, 1957, p. 507; also, F. Saunders, J. Amer. Chem. Soc. 50, 1231 (1928).

c Purification of brucine: C.H. De Puy, F.W. Breitbeil and K.R. Debruin, J. Amer. Chem. Soc., 88, 3347 (1966).

TABLE 2 (continued)

| NAME | Molecular Formula | MW | Structural Formula | Configuration | $[\alpha]_D$ | References | Suppliers |
|---|---|---|---|---|---|---|---|
| Cinchonine methoxydroxide | $C_{20}H_{26}N_2O_2$ | 326.44 | | | | Prep 5 | |
| Dehydroabietylamine[a,e] | $C_{20}H_{31}N$ | 285.46 | [19] | | $+46^o$ (MeOH)<br>$+56.1^o$ (c 2.4, py) (t = $20^o$) | Prep 6, 7 | (+): A, KK, N, PB |
| Deoxyephedrine[b,c] | $C_{10}H_{15}N$ | 149.24 | [33] | S-(+) | $+17.91^o$ ($H_2O$) (t = $26^o$)<br>on the hydrochloride | Prep 8c<br>Res 8a, b | rac: F (HCl salt), KK, FB<br>+/-: A, KK, PB; S (HCl salt) |
| threo-2-(N, N-Dimethylamino)-1-(p-nitrophenyl)-1, 3-propanediol | $C_{11}H_{16}N_2O_4$ | 240.26 | [29] | D-(-) | $+26 \pm 1^o$ (c 1, EtOH)<br>$-26 \pm 1^o$ (c 1, EtOH) | Prep 9 | |
| Ephedrine[d] | $C_{10}H_{15}NO$ | 165.24 | erythro [26] | 1R, 2S-(-) | $-6.3^o$ (EtOH) (t = $20^o$)<br>$+34.59^o$ (c 4, $H_2O$) (t = $25^o$)<br>on the hydrochloride | Prep 10a<br>Res 10b | rac, HCl salt: A, F, KK, PB, S<br>+/-: A, F, KK, KL, PB, S |
| α-Fenchylamine | $C_{10}H_{19}N$ | 153.27 | [36] | | $+25.5^o$ (c 5, 95% EtOH)<br>$-25.4^o$ (c 4, 95% EtOH) (t=$25^o$) | Prep } 11<br>Res } | |
| Menthol (OH) | $C_{10}H_{20}O$ | 156.27 | | 1R-(-) | $+48.6^o$ (c 2%, $CHCl_3$) (t = $20^o$)<br>$-50^o$ (c 10%, EtOH) (t = $18^o$) | Res 12 | rac: A, F, N, PB<br>(-): A, F, KK, N, PB |

a Isolation from Hercules Powder Co., Amine D, reference 7.

b Desoxyephedrine; metamphetamine.

c Conversion of amine hydrochloride to amine: J. Jacobus and T. B. Jones, J. Amer. Chem. Soc., 92, 4583 (1970).

d Conversion of amine hydrochloride to amine: K. Oki, et al., Bull. Chem. Soc. Japan, 43, 2554 (1970).

e Dehydroabietylamine hydrochloride is insoluble in water. Decomposition of dehydroabietylamine salts is best effected with base: R. L. Clarke and S. J. Daum, J. Med. Chem., 13, 320 (1970).

TABLE 2 (continued)

| NAME | Molecular Formula | MW | Structural Formula | Configuration | $[\alpha]_D$ | References | Suppliers |
|---|---|---|---|---|---|---|---|
| α-Methylbenzylamine[a] | $C_8H_{11}N$ | 121.18 | | S-(-) | +39.7° (neat) (t = 29°)<br>+40.1° (neat) (t = 25°)<br>-40.1° (neat) (t = 25°)<br>-40.3° (neat) (t = 20°) | Res 13<br>Also, Chapter II, ref. 4 | rac: A, F, KK, KL, N<br>+/-: A, KK, KL, N, PB |
| α-Methylbenzylisothiouronium acetate | $C_{11}H_{16}N_2O_2S$ | 240.32 | [38] | | +160° (c 2.5, 96% EtOH) (t = 20°)<br>-158° (EtOH) | Prep 14 | |
| Morphine | $C_{17}H_{19}NO_3$ | 285.35 | [30] | | -132° (c 1, MeOH) (t = 25°)<br>on the monohydrate | | |
| α-(1-Naphthyl)-ethylamine[b] | $C_{12}H_{13}N$ | 171.24 | [34] | R-(+) | +82.8° (neat) (t = 17°)<br>+81.7° (neat) (t = 23°)<br>-80.8° (neat) (t = 25°) | Prep 15b<br>Res 15a[c] | rac: KK, N<br>+/-: A, KK, N, PB |
| α-(2-Naphthyl)-ethylamine | $C_{12}H_{13}N$ | 171.24 | [35] | R-(+) | +19.4° (EtOH) (t = 19°)<br>-18.9° (EtOH) (t = 20°) | Prep 15a, 16<br>Res 15a, 17[c] | |
| Norepinephrine[d]<br>HO, HO, OH, $NH_2$ | $C_8H_{11}NO_3$ | 169.18 | | | +39° ($H_2O$) (t = 27°)<br>-40° ($H_2O$) (t = 25°)<br>on the hydrochloride | Res 18[e] | rac, HCl salt: KK, S<br>+/-: S |

a α-Phenylethylamine. For the effect of storage and $CO_2$ on this amine see N. Kornblum, et al., J. Amer. Chem. Soc., 82, 3099 (1960).

b Crestwood Chemicals, Newport, Tenn.: "Resoline A"

c See also Table 9a, p. 146.

d Arterenol

e See also Table 9b, p. 150.

TABLE 2 (continued)

| NAME | Molecular Formula | MW | Structural Formula | Configuration | $[\alpha]_D$ | References | Suppliers |
|---|---|---|---|---|---|---|---|
| Pseudoephedrine | $C_{10}H_{15}NO$ | 165.24 | threo [26] | 1S, 2S-(+) | +52.9° (c 4.0, EtOH) (t = 22.5°)<br>-52.50° (c 4.0, EtOH) (t = 22.5°) | Prep 10a<br>Res 19 | +/-: S |
| Quinidine | $C_{20}H_{24}N_2O_2$ | 324.43 | [23] | | +230° (c 1.8, $CHCl_3$) (t = 15°) | | (+): A, F, KK, N, PB, S (HCl salt) |
| Quinine | $C_{20}H_{24}N_2O_2$ | 324.43 | [22] | | -117° (c 1.5, $CHCl_3$) (t = 15°) | | (-): A, F, KK, N, PB, S |
| Quinidine methohydroxide | $C_{21}H_{28}N_2O_3$ | 356.46 | | | | Prep 5 | |
| Quinine methohydroxide | $C_{21}H_{28}N_2O_3$ | 356.46 | | | | Prep 5, 20 | |
| Stilbenediamine | $C_{14}H_{16}N_2$ | 212.30 | [37] | | -108° (MeOH) (t = 20°) | Prep } 21<br>Res } | |
| Strychnine | $C_{21}H_{22}N_2O_2$ | 334.42 | [21] | | -139.3° (c 1, $CHCl_3$) (t = 18°) | | (-): A, KK (HCl salt), N, S |
| Yohimbine | $C_{21}H_{26}N_2O_3$ | 354.45 | [31] | | +108° (c 1, py) (t = 20°) | | (+): KK; HCl salt: A, S |

## References to Table 2 - Resolving Agents for Acids


(1) (a) F. H. Radke, R. B. Fearing and S. W. Fox, J. Amer. Chem. Soc., 76, 2801 (1954); (b) G. Zoja, French Patent 1,439,850, May 27, 1966 [Chem. Abstr., 66, 55027g (1967)]; (c) D. Pitrè and E. B. Grabitz, Chimia, 23, 399 (1969) [Chem. Abstr., 72, 54445d (1970)].

(2) (a) H. Berger and E. Haack, German Patent 1,052,409, Mar. 12, 1959 [Chem. Abstr., 55, 6442b (1961)]; British Patent 820,260, Sept. 16, 1959 [Chem. Abstr., 54, 8735b (1960)]; (b) M. Nagawa and Y. Murase, Takamine Kenkyujo Nempo, 8, 15 (1956) [Chem. Abstr., 52, 308d (1958)]; (c) V. D'Amato and R. Pagani, British Patent 738,064, Oct. 5, 1965 [Chem. Abstr., 50, 10773g (1956)].

(3) (a) J. Controulis, M. C. Rebstock and H. M. Crooks, J. Amer. Chem. Soc., 71, 2463 (1949); (b) H. Ikeda and H. Ikeda, J. Sci. Res. Inst. Tokyo, 45, No. 1237, 8 (1951).

(4) (a) F. P. Nabenhauer, British Patent 550,916, Feb. 1, 1943 [Chem. Abstr., 38, 1754[2] (1944)]; U. S. Patents 2,276,508 and 2,276,509, Mar. 17, 1942 [Chem. Abstr., 36, 4522[9] (1942)]; (b) R. E. Ernst, M. J. O'Connor and R. H. Holm, J. Amer. Chem. Soc., 90, 5735 (1968).

(5) R. T. Major and J. Finkelstein, J. Amer. Chem. Soc., 63, 1368 (1941).

(6) L. C. Cheney, U. S. Patent 2,787,637, Apr. 2, 1957 [Chem. Abstr., 51, 13926a (1957)].

(7) W. J. Gottstein and L. C. Cheney, J. Org. Chem., 30, 2072 (1965).

(8) (a) H. Emde, Helv. Chim. Acta, 12, 365 (1929); (b) G. Bellucci, G. Berti, A. Borraccini and F. Macchia, Tetrahedron, 25, 2979 (1969); (c) J. Jacobus and T. B. Jones, J. Amer. Chem. Soc., 92, 4583 (1970).

(9) G. Müller, A. Gaston, A. Poittevin and T. Vesperto, French Patent 1,481,978, May 26, 1967 (to Roussel-UCLAF) [Chem. Abstr., 69, 2698g (1968)]; G. Nominé, G. Amiard, and V. Torelli, Bull. Soc. Chim. Fr., 1968, 3664.

(10) (a) J. B. La Pidus, A. Tye, P. Patil and B. A. Modi, J. Med. Chem., 6, 76 (1963); (b) R. H. Manske and T. B. Johnson, J. Amer. Chem. Soc., 51, 1906 (1929); L. R. Overby and A. W. Ingersoll, ibid., 82, 2067 (1960).

(11) A. W. Ingersoll and H. D. De Witt, J. Amer. Chem. Soc., 73, 3360 (1951).

(12) (a) A. W. Ingersoll, in Organic Reactions, R. Adams, ed., Vol. 2, Wiley, New York, 1944, p. 399; (b) B. A. Everard and J. A. Mills, J. Chem. Soc., 1950, 3386; (c) L. Velluz, Substances Naturelles de Synthèse, Vol. IX, Masson, Paris, 1954, p. 170.

(13) A. Ault, Org. Syn., 49, 93 (1969).

(14) W. Klötzer, Monatsh. Chem., 87, 346 (1956).

(15) (a) E. Samuelsson, Svensk. Kem. Tidskr., 34, 7 (1922); idem, Thesis, University of Lund, 1923; (b) J. Jacobus, M. Raban and K. Mislow, J. Org. Chem., 33, 1142 (1968).

(16) M. L. Moore, in Organic Reactions, R. Adams, ed., Vol. 5, Wiley, New York, 1949, p. 320.

(17) A. Fredga, B. Sjöberg and R. Sandberg, Acta Chem. Scand., 11, 1609 (1957).

(18) (a) B. F. Tullar, J. Amer. Chem. Soc., 70, 2067 (1948); (b) B. F. Tullar, U. S. Patent 2,774,789, Dec. 18, 1956 [Chem. Abstr., 51, 10576c (1957)].

(19) E. Späth and R. Göhring, Monatsh. Chem., 41, 319 (1920).

(20) J. Knable and R. Kräuter, Arch. Pharm., 298, 1 (1965).

(21) I. Lifschitz and J. G. Bos, Rec. Trav. Chim. Pays-Bas, 59, 173 (1940).

TABLE 3 RESOLVING AGENTS FOR AMINES AND OTHER BASES[*]

| NAME | Molecular Formula | MW | Structural Formula | Configuration | $[\alpha]_D$ | References | Suppliers |
|---|---|---|---|---|---|---|---|
| N-Acetyl-3,5-dibromotyrosine | $C_{11}H_{11}NO_4Br_2$ | 381.04 | [53] | L-(+) | $+34.5^\circ$ (c 4, MeOH) ($t = 25^\circ$) | Prep 1 | |
| N-Acetylleucine | $C_8H_{15}NO_3$ | 173.21 | [52] | L-(-) | $-24.1^\circ$ (c 4, MeOH) ($t = 25^\circ$) | Prep 2 | rac: F, KK, KL, PB, S<br>+/-: F, KK, PB, S |
| Arabonic acid<br>HO, OH, HO, COOH, OH<br>(Arabinonic acid) | $C_5H_{10}O_6$ | 166.13 | | D-(+) | $+45^\circ$ ($c = 4.6\%$, $H_2O$, $t = 26^\circ$)<br>on the Ba salt[a] | | K salt: KK<br>Ca salt: KL, PB |
| Atrolactic acid<br>$CH_3$, C-COOH, OH | $C_9H_{10}O_3$ | 166.18 | | R-(-) | $-31^\circ$ (95 % EtOH) ($t = 25^\circ$)<br>[82% opt. pure] | Prep 3<br>Res 4 | rac: A, F, KK, KL, PB |
| 3-Bromocamphor-9-sulfonic acid[b, c] | $C_{10}H_{15}O_4SBr$ | 311.21 | [48] | | $+85.8^\circ$ ($H_2O$) ($t = 25^\circ$)<br>$-86.6^\circ$ ($H_2O$) ($t = 25^\circ$)<br>On the ammonium salt:<br>$+84.5^\circ$ ($H_2O$) ($t = 20^\circ$) | Prep } 5<br>Res } | +/-, $NH_4^+$ salt: A, KK, PB, N |
| Camphoric acid | $C_{10}H_{16}O_4$ | 200.23 | [50] | 1S,3S-(+) | $+48.45^\circ$ (EtOH) ($t = 20^\circ$)<br>$-48.47^\circ$ (EtOH) ($t = 20^\circ$) | Prep 6 | rac: F, KK, N, PB, S<br>+/-: A, F, KK, N, PB, S |

* References to Table 3 are on p. 29.

a J.F. Haskins and M.J. Hogsed, J. Org. Chem., 15, 1264 (1950).

b $\alpha$-Bromocamphor-$\pi$-sulfonic acid.

c Isolation of acid from the ammonium salt by means of an ion-exchange resin: S. Gronowitz, I. Sjögren, L. Wernstedt and B. Sjöberg, Ark. Kemi, 23, 129 (1965).

TABLE 3 (continued)

| NAME | Molecular Formula | MW | Structural Formula | Configuration | $[\alpha]_D$ | References | Suppliers |
|---|---|---|---|---|---|---|---|
| Camphor-3-sulfonic acid[a] | $C_{10}H_{16}O_4S$ | 232.30 | [49] | | $+103^o$ (c 4, $H_2O$) (546 nm) | Prep 7 | |
| Camphor-10-sulfonic acid[b] | $C_{10}H_{16}O_4S$ | 232.30 | [47] | 1R-(+) | $+43.5^o$ (EtOH) ($t = 20^o$)<br>$-18.52^c$ (c 5.40, $H_2O$) ($t = 24^o$) | Prep 8, 9<br>Res 38 | rac: A, F, KK, PB<br>(+): A, F, N, KK, PB |
| Camphor-10-sulfonyl chloride | $C_{10}H_{15}O_3SCl$ | 250.74 | | 1R-(+) | $+36.5^o$ (c 7.7, $CHCl_3$) ($t = 29^o$) | Prep 10, 11 | |
| Cellulose acetate | see Table 7 | | | | | | |
| 4-Chlorotartranilic acid<br>Cl-$C_6H_4$-N(H)-C(=O)-CH(OH)-CH(OH)-C(=O)-OH | $C_{10}H_{10}NO_5Cl$ | 259.65 | | 2R, 3R-(+) | $+108.9^o$ (c 1.6, 95% EtOH) ($t = 25^o$) | Prep 12 | (+): A |
| trans-Chrysanthemic acid | $C_{10}H_{16}O_2$ | 168.23 | [structure: cyclopropane ring with two $CH_3$, H, COOH, and $C=C(CH_3)_2$] | | $+14.4^o$ (EtOH) ($t = 23^o$)<br>$-14.4^o$ (EtOH) ($t = 20^o$) | Prep } 13<br>Res } | rac, Ethyl ester: F, KK, KL, PB, S |
| O, O'-Dibenzoyltartaric acid[c, d] | $C_{18}H_{14}O_8$ | 358.31 | [39] | 2R, 3R-(-) | $+116.0^o$ (c 4, EtOH) ($t = 25^o$)<br>$-118^o$ (c 4, EtOH) ($t = 25^o$)<br>$^+/- 132^o$ (EtOH) ($t = 20^o$) (546 nm) | Prep } 14<br>Res } | $^+/-$: A, F, FR, KK, KL, N, PB, S |

a $\alpha$-Camphorsulfonic acid.

b Reychler's acid; also known as $\omega$- or $\beta$-camphorsulfonic acid.

c (-)-Anhydride available (KK).

d For the recovery of this acid from resolutions see Italian Patent 623, 011, July 7, 1961; Chem. Abstr., 57, 4601 i (1962).

TABLE 3 (continued)

| NAME | Molecular Formula | MW | Structural Formula | Configuration | $[\alpha]_D$ | References | Suppliers |
|---|---|---|---|---|---|---|---|
| Di-O-isopropylidene-2-oxo-L-gulonic acid | $C_{12}H_{18}O_7$ | 274.27 | | L-(-) | -20.37° (c 2, MeOH) (t = 25°) (personal communication from Dr. E. P. Oliveto) | Prep 15 | |
| 6,6'-Dinitro-2,2'-diphenic acid | $C_{14}H_8N_2O_8$ | 332.23 | | S-(-) | +127° (c 2.4, MeOH) (t = 25°)<br>-127° (c 1.6, MeOH) (t = 29°) | Prep, Res 16<br>See also Table 8d | |
| O,O'-Di-p-toluoyltartaric acid | $C_{20}H_{18}O_8$ | 386.36 | [40] | 2S,3S-(+) | +140° (EtOH) (t = 25°)<br>-140° (EtOH) (t = 20°) | Prep 17<br>Res 18 | +/-: A, F, FR, KK, N, PB |
| Glutamic acid | $C_5H_9NO_4$ | 147.13 | [54] | 2S-(+) | +31.4° (6 N HCl) (t = 22.4°)<br>-30.5° (6 N HCl) (t = 20°) | Res 19 | rac: A, KK, KL, N, PB, S<br>+/-: A, F, KK, KL, N, PB, S |
| Hydratropic acid | $C_9H_{10}O_2$ | 150.18 | | S-(+) | +98.8° (neat) (t = 21°)<br>+78.9° (EtOH) (t = 25°)<br>-78.4° (EtOH) (t = 25°) | Prep, Res 20 | rac: KK, KL, PB |
| Lactic acid | $C_3H_6O_3$ | 90.08 | | S-(+) | +2.6° ($H_2O$) (t = 21-22°)<br>-2.6° ($H_2O$) (t = 21.5°)<br>both at 546 min | Res 21 | rac: A, PB, S<br>+/-: PB, S; aqueous soln: A, F, KK, KL |

TABLE 3 (continued)

| NAME | Molecular Formula | MW | Structural Formula | Configuration | $[\alpha]_D$ | References | Suppliers |
|---|---|---|---|---|---|---|---|
| Lactose [a] | $C_{12}H_{22}O_{11}$ | 342.31 | | D-(+) | $+52.3 \pm 1^{o}$ ($H_2O$) ($t = 25^{o}$)[b] | Prep 22 | A, F, KL, PB, S |
| Malic acid | $C_4H_6O_5$ | 134.09 | [44] | 2S-(-) | $-2.3^{o}$ ($H_2O$) | Prep, Res 23 | rac: A, F, KL, PB, S<br>+/-: A, F, KK, KL, N, PB, S |
| Mandelic acid | $C_8H_8O_3$ | 152.15 | [45] | R-(-) | $+156.6^{o}$ ($H_2O$) ($t = 20^{o}$)<br>$-156.9^{o}$ ($H_2O$) ($t = 20^{o}$) | Res 24 | rac: A, F, FR, KK, KL, PB, S<br>+/-: A, F, KK, KL, N, PB, S |
| Menthoxyacetic acid | $C_{12}H_{22}O_3$ | 214.31 | [51] | 1R, 2R, 5R-(-) | $-92.5^{o}$ (EtOH) ($t = 25^{o}$) | Prep 25, 26[c] | (-): A, KK, N, PB |
| Menthoxyacetyl chloride | $C_{12}H_{21}O_2Cl$ | 232.75 | | | $-89.6^{o}$ (neat ?) ($t = 25^{o}$) | Prep 25, 27 | |
| Menthyl chloroformate<br>[structure: menthyl ring with OCCl, C=O] | $C_{11}H_{19}O_2Cl$ | 218.77 | | | | Prep 28 | |
| Menthyl isocyanate | $C_{11}H_{19}NO$ | 181.28 | | | $-64.2^{o}$ (c 5, $C_6H_6$) ($t = 22^{o}$) | Prep 29 | |

a $\alpha$-Lactose monohydrate appears to be the common form.

b Mutarotates; value given is the final value.

c An improved procedure is given by T.G. Cochran and A.C. Huitric, J. Org. Chem., 36, 3046 (1971).

TABLE 3 (continued)

| NAME | Molecular Formula | MW | Structural Formula | Configuration | $[\alpha]_D$ | References | Suppliers |
|---|---|---|---|---|---|---|---|
| α-Methylbenzyl isocyanate (C₆H₅-CH(CH₃)-NCO) | $C_9H_9NO$ | 147.18 | | S-(-) | $[M]_D^{20}$ -14.5° (neat)<br>-2° (c 3.541, $C_6H_6$) (t = 24°) | Prep 30 | +/-: F |
| N-(α-Methylbenzyl)-phthalamic acid | $C_{16}H_{15}NO_3$ | 269.31 | [43] | R-(+) | +47.60° (c 1, EtOH) (t = 20°)<br>-47.10° (c 2, EtOH) (t = 20°) | Prep 31 | +/-: N, KK |
| N-(α-Methylbenzyl)-succinamic acid | $C_{12}H_{15}NO_3$ | 221.26 | [42] | R-(+) | +112.05° (c 2, EtOH) (t = 20°)<br>-111.90° (c 2, EtOH) (t = 20°) | Prep 31 | |
| α-Methylphenylmethanesulfonic acid [a] | $C_8H_{10}O_3S$ | 186.23 | [103] | | -11.8° (c 1.90%, $H_2O$) (t = 18°) on the potassium salt | Prep, Res 32 | +/-: KK |
| 2-Nitrotartranilic acid | $C_{10}H_{10}N_2O_7$ | 270.20 | [41] | 2R,3R-(+) | +90.0° (c 0.8, $H_2O$) (t = 25°) | Prep 12 | |
| 2-Phenoxypropionic acid | $C_9H_{10}O_3$ | 166.18 | [46] | D-(+) | +40.0° (EtOH) (t = 23°) | Res 33 | rac: KK, KL, PB |
| 2-Pyrrolidone-5-carboxylic acid [b] | $C_5H_7NO_3$ | 129.12 | [56] | S-(-) | -11.9° (c 2, $H_2O$) (t = 20°) | Prep 34 | rac: F, KK, KL, PB, S<br>(-): A, F, KK, KL, N, PB, S |
| Tartaric acid | $C_4H_6O_6$ | 150.09 | | 2R,3R-(+) | +12.0° ($H_2O$) (t = 20°)<br>-12.0° ($H_2O$) (t = 20°) | Res 35<br>Prep 36 | rac: A, F, KK, KL, N, PB<br>+/-: A, F, KK, KL, PB, S |
| N-(p-Toluenesulfonyl)-glutamic-acid | $C_{12}H_{15}NO_6S$ | 301.32 | | L-(+) | +14.3° (MeOH) (t = 20°) | Prep 37 | (+): KK, N, PB |

a 1-Phenylethanesulfonic acid.
b Pyroglutamic acid.

## References to Table 3 - Resolving Agents for Amines and Other Bases

(1) H. D. De Witt and A. W. Ingersoll, J. Amer. Chem. Soc., 73, 5782 (1951).

(2) H. D. De Witt and A. W. Ingersoll, J. Amer. Chem. Soc., 73, 3359 (1951).

(3) E. L. Eliel and J. P. Freeman, Organic Syntheses, Coll. Vol. IV, 1963, p. 58.

(4) L. Smith, J. Prakt. Chem., 84, 731 (1911).

(5) (a) A. W. Ingersoll and S. H. Babcock, J. Amer. Chem. Soc., 55, 341 (1933); (b) H. Regler and F. Hein, J. Prakt. Chem., 148, 1 (1937); (c) L. Velluz, Substances Naturelles de Synthèse, Vol. IX, Masson, Paris, 1954, p. 162.

(6) (a) W. A. Noyes, Am. Chem. J., 16, 501 (1894); (b) Purification: O. Aschan, Ber., 27, 2003 (1894); (c) C. P. Berg, J. Biol. Chem., 115, 9 (1936).

(7) M. Frèrejacque, Compt. rend., 183, 607 (1926); Ann. Chim. (Paris) [10] 14, 147 (1930) (see p. 165).

(8) P. D. Bartlett and L. H. Knox, Org. Syn., 45, 12 (1965).

(9) L. Velluz, Substances Naturelles de Synthèse, Vol. IX, Masson, Paris, 1954, p. 164.

(10) (a) P. D. Bartlett and L. H. Knox, Org. Syn., 45, 14 (1965); (b) N. Fischer and G. Opitz, ibid., 48, 106 (1968).

(11) L. Velluz, Substances Naturelles de Synthèse, Vol. IX, Masson, Paris, 1954, p. 169.

(12) T. A. Montzka, T. L. Pindell and J. D. Matiskella, J. Org. Chem., 33, 3993 (1968).

(13) (a) I. G. M. Campbell and S. H. Harper, J. Chem. Soc., 1945, 283; (b) idem, J. Sci. Food Agr., 3, 189 (1952).

(14) (a) C. L. Butler and L. H. Cretcher, J. Amer. Chem. Soc., 55, 2605 (1933); (b) G. Losse, Chem. Ber., 87, 1279 (1954);

(c) D. C. Iffland and H. Siegel, J. Amer. Chem. Soc., 80, 1947 (1958); (d) L. Velluz, Substances Naturelles de Synthèse, Vol. IX, Masson, Paris, 1954, p. 165.

(15) T. Reichstein and A. Grüssner, Helv. Chim. Acta, 17, 311 (1934).

(16) (a) A. W. Ingersoll and J. R. Little, J. Amer. Chem. Soc., 56, 2123 (1934); (b) M. Siegel and K. Mislow, ibid., 80, 473 (1958).

(17) (a) A. Stoll and A. Hofmann, Helv. Chim. Acta, 26, 922 (1943); (b) D. A. A. Kidd, J. Chem. Soc., 1961, 4675.

(18) J. H. Hunt, J. Chem. Soc., 1957, 1926.

(19) Resolution methods are reviewed in C. W. Huffman and W. G. Skelly, Chem. Rev., 63, 625 (1963).

(20) (a) S. I. Goldberg and F.-L. Lam, J. Org. Chem., 31, 2336 (1966); (b) K. Petterson, Arkiv Kemi, 10, 283 (1956); (c) C. L. Arcus and J. Kenyon, J. Chem. Soc., 1939, 916.

(21) T. Purdie and J. W. Walker, J. Chem. Soc., 61, 754 (1892); H. Borsook, H. M. Huffman and Y.-P. Liu, J. Biol. Chem., 102, 449 (1933).

(22) V. Prelog and P. Wieland, Helv. Chim. Acta, 27, 1127 (1944); N. J. Leonard and W. J. Middleton, J. Amer. Chem. Soc., 74, 5117 (1952).

(23) A. McKenzie, J. Plenderleith and N. Walker, J. Chem. Soc., 123, 2875 (1923).

(24) (a) R. Roger, J. Chem. Soc., 1935, 1544; (b) R. H. F. Manske, and T. B. Johnson, J. Amer. Chem. Soc., 51, 1906 (1929); (c) J. L. Norula and J. Kenyon, Current Sci. (India), 32, 260 (1963) [Chem. Abstr., 59, 8643g (1963)]; (d) L. Velluz, Substances Naturelles de Synthèse, Vol. IX, Masson, Paris, 1954, p. 166.

(25) A. W. Ingersoll, in Organic Reactions, R. Adams, ed., Vol. 2, Wiley, New York, 1944, p. 398.

(26) M. T. Leffler and A. E. Calkins, Organic Syntheses, Coll. Vol. III, 1955, p. 544.

(27) M. T. Leffler and A. E. Calkins, Organic Syntheses, Coll. Vol. III, 1955, p. 547.

(28) J. W. Westley and B. Halpern, J. Org. Chem., 33, 3978 (1968).

(29) (a) G. A. Haggis and L. N. Owen, J. Chem. Soc., 1953, 389; (b) L. Velluz, Substances Naturelles de Synthèse, Vol. IX, Masson, Paris, 1954, p. 170.

(30) A. P. Terent'ev, R. A. Gracheva, V. T. Bezruchko and O. G. Poltava, Zh. Obshch. Khim., 37, 1473 (1967); Engl. transl., p. 1397; T. L. Cairns, J. Amer. Chem. Soc., 63, 871 (1941).

(31) E. Felder, D. Pitrè and S. Boveri, Helv. Chim. Acta, 52, 329 (1969).

(32) E. B. Evans, E. E. Mabbott and E. E. Turner, J. Chem. Soc., 1927, 1159.

(33) (a) B. Sjöberg and S. Sjöberg, Arkiv Kemi, 22, 447 (1964); (b) E. Fourneau and G. Sandulesco, Bull. Soc. Chim. Fr. [4] 31, 988 (1922); (c) Y. G. Perron, W. F. Minor, C. T. Holdrege, W. J. Gottstein, J. C. Godfrey, L. B. Crast, R. B. Babel and L. C. Cheney, J. Amer. Chem. Soc., 82, 3934 (1960); (d) W. J. Gottstein, U. S. Patent 3,454,626, July 8, 1969 [Chem. Abstr., 72, 12917d (1970)]; W. J. Gottstein and L. C. Cheney, J. Org. Chem., 30, 2072 (1965).

(34) R. J. Dearborn and J. A. Stekol, U. S. Patent 2,528,267, Oct. 31, 1950 [Chem. Abstr., 45, 2984f (1951)].

(35) E. Walton, J. Soc. Chem. Ind., 64, 219 (1945).

(36) L. Velluz, Substances Naturelles de Synthèse, Vol. IX, Masson, Paris, 1954, p. 166.

(37) British Patent 1,169,310, November 5, 1969 [Chem. Abstr., 72, 55453s (1970)].

(38) K. L. Marsi, C. A. Van der Werf and W. E. McEwen, J. Amer. Chem. Soc., 78, 3063 (1956).

## TABLE 4 RESOLVING AGENTS FOR ALCOHOLS*

| NAME | Molecular Formula | MW | Structural Formula | Configuration | $[\alpha]_D$ | References | Suppliers |
|---|---|---|---|---|---|---|---|
| 3β-Acetoxy-$\Delta^5$-etienic acid[a] | $C_{22}H_{32}O_4$ | 360.48 | [60] | | | Prep 1 | PB[b,c] |
| 3β-Acetoxy-$\Delta^5$-etienic acid chloride | $C_{22}H_{31}O_3Cl$ | 378.95 | | | | Prep 2 | |
| ω-Camphanic acid | $C_{10}H_{14}O_4$ | 198.22 | | | $-9.3^o$<br>$[\alpha]_1$ $+7.0^o$ (EtOH)[d]<br>$[\alpha]_1$ $-7.14^o$ (EtOH)[d] | Prep 3 | |
| ω-Camphanic acid chloride | $C_{10}H_{13}O_3Cl$ | 216.65 | [63] | | $-4^o$ (c 1.1, $C_6H_6$) | Prep 4 | |
| Camphor-10-sulfonyl chloride | | | see Table 3 | | | | |
| O,O'-Diacetyltartaric acid monomethyl ester chloride[e] | $C_9H_{11}O_7Cl$ | 282.63 | [62] | 2R,3R-(+) | | Prep 5 | |
| Lactic acid | | | see Table 3 | | | | |
| Menthoxyacetic acid | | | see Table 3 | | | | |
| Menthoxyacetyl chloride | | | see Table 3 | | | | |
| Menthyl isocyanate | | | see Table 3 | | | | |
| α-Methylbenzyl isocyanate | | | see Table 3 | | | | |
| Tartranil | $C_{10}H_9NO_4$ | 207.19 | | 2R,3R-(+) | $+130^o$ (c 1, MeOH) (t = $25^o$) | Prep 6 | (+): KK, N |
| Valine | $C_5H_{11}NO_2$ | 117.15 | | L-(+) | $+22.9^o$ (c 0.8, 20% HCl) (t = $23^o$) | | rac: KK, PB, S<br>+/-: A, KK, PB, S |

* References to Table 4 are on p. 33.

a 5-Androsten-3β-ol-17β-carboxylic acid 3-acetate; also known as 3β-acetoxyetiocholenic acid.

b Available from Steraloids, Inc., P.O. Box 127, Pawling, N.Y., 12564 and from Norell Chemical Co., Inc., Landing, N.J., 07850.

c The precursor to this compound, pregnenolone acetate, is widely available (e.g., from A, KK, and PB).

d Beilstein 18, 401 states $[\alpha]^{19.5}_{weiss}$.

e A precursor is 2R,3R-(+)-O,O'-diacetyltartaric acid anhydride; prep.: R.L. Shriner and C.L. Furrow, Jr., Organic Syntheses, Coll. Vol. IV, 1963, p. 242; $[\alpha]^{20}_D$ + 97.2° (c 0.47, pyridine).

References to Table 4 - Resolving Agents for Alcohols

(1) J. Staunton and E. J. Eisenbraun, Org. Syn., 42, 4 (1962); C. Djerassi and J. Staunton, J. Amer. Chem. Soc., 83, 736 (1961).

(2) C. Djerassi, J. Burakevich, J. W. Chamberlin, D. Elad, T. Toda and G. Stork, J. Amer. Chem. Soc., 86, 465 (1964).

(3) (a) O. Aschan, Ber., 27, 3504 (1894); idem, ibid., 28B, 922 (1895); (b) N. Zelinsky and N. Lepeschkin, Justus Liebigs Ann. Chem., 319, 303 (1901) (see p. 311).

(4) H. Gerlach, Helv. Chim. Acta, 51, 1587 (1968).

(5) H. J. Lucas and W. Baumgarten, J. Amer. Chem. Soc., 63, 1653 (1941).

(6) F. Barrow and R. G. Atkinson, J. Chem. Soc., 1939, 638.

TABLE 5 RESOLVING AGENTS FOR ALDEHYDES AND KETONES*

| NAME | Molecular Formula | MW | Structural Formula | Config-uration | $[\alpha]_D$ | References | Suppliers |
|---|---|---|---|---|---|---|---|
| 2,3-Butanediol | $C_4H_{10}O_2$ | 90.12 | | 2R,3R-(-) | +12.85° (neat) (t = 20°)<br>-13.34° (neat) (t = 26°) | Prep 1 | rac: A, F, KK, PB[a] |
| 2,3-Butanedithiol | $C_4H_{10}S_2$ | 122.26 | | 2S,3S-(+) | +14.2° ($CHCl_3$) (t = 25°) | Prep 2 | rac: PB |
| Mandelic acid hydrazide | $C_8H_{10}N_2O_2$ | 166.18 | [85] | | | Prep 3 | rac: FR |
| Menthyl N-aminocarbamate[b] | $C_{11}H_{22}N_2O_2$ | 214.31 | [81] | 2R-(-) | -80° (c 2%, EtOH) (t = 25°)<br>$[\alpha]_D^{25}$ -3.13° (c 1.96, 12, EtOH) | Prep 4 | (-): FR, KK, PB |
| Menthylhydrazine | $C_{10}H_{22}N_2$ | 170.30 | [79] | | -46.05° ($H_2O$)<br>on the hydrochloride | Prep 5 | |
| 4-(α-Methylbenzyl)semicarbazide | $C_9H_{13}N_3O$ | 179.22 | [80] | R-(+) | +31.2° (c 1.0, EtOH)<br>-31.2° (c 1.0, EtOH) | Prep 6 | |
| 5-(α-Methylbenzyl)semioxamazide | $C_{10}H_{13}N_3O_2$ | 207.23 | [83] | R-(+) | +102.0° (c 1.04, $CHCl_3$) (t = 25°)<br>-102.5° (c 0.625, $CHCl_3$) (t = 25°) | Prep 7 | +/-: KK, N |
| Tartramidic acid hydrazide[c] | $C_4H_9N_3O_4$ | 163.13 | [84] | 2R,3R-(+) | +124.0° ($H_2O$) (t = 21°) | Prep 8 | |

* References to Table 5 are on page 35.

a These products may be mixtures of meso and racemic diastereomers. Optically active product appears to be available from British Drug Houses and from Burdick and Jackson Laboratories, Inc., Muskegon, Michigan 49442.

b Commonly named menthydrazide; also called menthyl carbazate.

c Commonly named tartramazide.

## References to Table 5 - Resolving Agents for Aldehydes and Ketones

(1) (a) A. C. Neish, Can. J. Res., 23B, 10 (1945); (b) J. J. Plattner and H. Rapoport, J. Amer. Chem. Soc., 93, 1758 (1971).

(2) E. J. Corey and R. B. Mitra, J. Amer. Chem. Soc., 84, 2938 (1962).

(3) (a) C. Demetrescu, A. Popa and M. Vanghelovici, Acad. Rep. Populare Romine, Baza Cercetari Stiint. Timisoara, Studii Cercetari, Stiinte Chim., 10 (2) 189 (1963) [Chem. Abstr., 61, 609c (1964)]; (b) A. Gvozdjakova and J. Schnittova, Acta Fac. Rerum Natur. Univ. Comenianae, Chim., 9, 585 (1966) [Chem. Abstr., 66, 37602m (1967)].

(4) (a) R. B. Woodward, T. P. Kohman and G. C. Harris, J. Amer. Chem. Soc., 63, 120 (1941); (b) L. Velluz, Substances Naturelles de Synthèse, Vol. IX, Masson, Paris, 1954, p. 168.

(5) N. Kijner, J. Prakt. Chem., [2] 52, 424 (1895); see also J. A. Berson, C. J. Olsen and J. S. Walia, J. Amer. Chem. Soc., 84, 3337 (1962).

(6) (a) I. V. Hopper and F. J. Wilson, J. Chem. Soc., 1928, 2483; (b) W. Theilacker and W. Schmid, Justus Liebigs Ann. Chem., 570, 11 (1950); (c) W. Theilacker, in Methoden der Organischen Chemie (Houben-Weyl), 4th ed., Vol. 4, Part 2, E. Müller, ed., 1955, p. 529; (d) L. Velluz, Substances Naturelles de Synthèse, Vol. IX, Masson, Paris, 1954, p. 172.

(7) N. J. Leonard and J. H. Boyer, J. Org. Chem., 15, 42 (1950). The preparation is also given in L. Velluz, Substances Naturelles de Synthèse, Vol. IX, Masson, Paris, 1954, p. 173.

(8) F. Nerdel and E. Henkel, Chem. Ber., 85, 1138 (1952).

TABLE 6 RESOLVING AGENTS FOR AMINO ACIDS*

| NAME | Molecular Formula | MW | Structural Formula | Configuration | $[\alpha]_D$ | References | Suppliers |
|---|---|---|---|---|---|---|---|
| *threo*-2-Amino-1-(*p*-nitrophenyl)-1,3-propanediol | see Table 2 | | | | | | |
| Cholestenonesulfonic acid | $C_{27}H_{44}O_4S$ | 464.68 | | | | Prep 1 | |
| Ephedrine | see Table 2 | | | | | | |
| Menthyl dihydrogen phosphate | $C_{10}H_{21}O_4P$ | 236.25 | [96] | | | Prep 2 | |
| $\alpha$-Methylbenzylamine | see Table 2 | | | | | | |
| Pantolactone[a] | $C_6H_{10}O_3$ | 130.14 | [98] | L-(+) | $+50.1^o$ (c 2, $H_2O$) (t = $25^o$) | Res 3[b] | rac: KK, KL, PB, S<br>+/-: F, KK, PB, S |
| 2-Phenoxypropionic acid | see Table 3 | | | | | | |
| 2-Pyrrolidone-5-carboxylic acid | see Table 3 | | | | | | |
| Tyrosine hydrazide | $C_9H_{13}N_3O_2$ | 195.25 | [97] | L-(+) | $+70.2^o$ (c 2, 3N HCl) (t = $23^o$)<br>$-70.3^o$ (c 2, 3N HCl) (t = $23^o$) | Prep 4<br>Res 5 | (+): A |

* References to Table 6 are on p. 37.

a Pantoyl lactone; $\alpha$-hydroxy-$\beta$, $\beta$-dimethyl-$\gamma$-butyrolactone.

b see also Table 8b and ref. A115.

## References to Table 6 - Resolving Agents for Amino Acids

(1) A. Windaus and E. Kuhr, Ann., 532, 52 (1937).

(2) S. Watanabe and K. Suga, Isr. J. Chem., 7, 483 (1969).

(3) (a) M. Dunkel, I. Loter and H. C. Klein, U. S. Patent 3,185,710, May 25, 1965 (to NOPCO Chemical Co.) [Chem. Abstr., 63, 5563g (1965)]; (b) E. S. Zhdanovich, G. S. Kozlova, T. D. Marieva, T.V. Mel'nikova and N. A. Preobrazhenskii, Zh. Org. Khim., 4, 1359 (1968); Engl. transl., p. 1310.

(4) T. Curtius and W. Donselt, J. Prakt. Chem., [2] 95, 327 (1917) (see p. 353).

(5) K. Vogler and P. Lanz, Helv. Chim. Acta, 49, 1348 (1966).

TABLE 7 MISCELLANEOUS RESOLVING AGENTS AND CHIRAL ADSORBENTS*

| NAME | Molecular Formula | MW | Structural Formula | Configuration | $[\alpha]_D$ | References | Suppliers |
|---|---|---|---|---|---|---|---|
| Bisdemethylbrucine | $C_{21}H_{22}N_2O_4$ | 366.40 | [110] | | | Prep 1 | |
| N-sec-Butylpicramide | $C_{10}H_{12}N_4O_6$ | 284.22 | [12] | | $-2.6^{\circ}$ ($CHCl_3$) (t = $15^{\circ}$) | Res 2 | |
| Cellulose acetate | | | | | | 3 | a |
| α-Cyclodextrin[b] | $C_{36}H_{60}O_{30}$ | 972.85 | [18] | | $+150.5^{\circ}$ (c 1, $H_2O$) | Prep 4, 5 | c |
| β-Cyclodextrin[d] | $C_{42}H_{70}O_{35}$ | 1135.0 | | | $+162.5^{\circ}$ (c 1, $H_2O$) | Prep 5, 6 | (+): KK[c] |
| Deoxycholic acid[e] 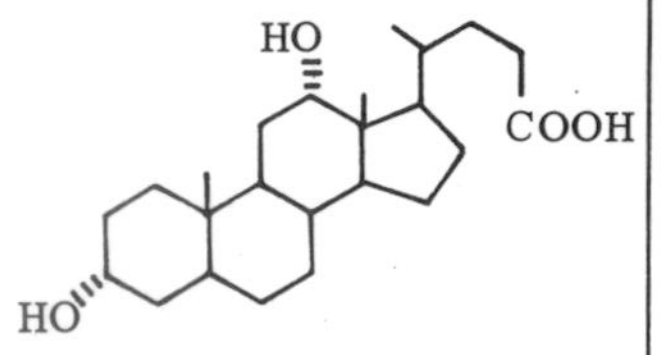 | $C_{24}H_{40}O_4$ | 392.58 | | | $+55^{\circ}$ (c 1, EtOH) (t = $20^{\circ}$)<br>$+60^{\circ}$ (c 1, EtOH) (t = $20^{\circ}$)<br>[latter value: Aldrich] | | (+): A, F, KK, KL, PB, S |
| trans-Dichloro(ethylene)(α-methylbenzylamine)platinum (II) | $C_{10}H_{15}NCl_2Pt$ | 415.2 | [14] | | $+13^{\circ}$ (c 1.4, $CH_2Cl_2$) (t = $31^{\circ}$)<br>$-16.7^{\circ}$ (c 0.9, $CH_2Cl_2$) (t = $28^{\circ}$) | Prep 7 | |
| Digitonin | $C_{56}H_{92}O_{29}$ | 1229.34 | [13] | | $-54^{\circ}$ (MeOH) (t = $20^{\circ}$) | | (-): F, KK, KL, PB, S |
| Diphenacyl malate | $C_{20}H_{18}O_7$ | 370.36 | [108] | S-(+) | $+10.0^{\circ}$ (acetone) | Prep 8 | |
| Lactose | see Table 3 | | | | | | |
| Mandelic acid | see Table 3 | | | | | | |

* References to Table 7 are on p. 40.

a Available from M. Woelm, Eschwege, M. Germany or its distributors in other countries; other firms sell cellulose acetate as well (e.g., F) but the extent of acetylation and hydration may be different.

b Cyclohexaamylose.

c Available from Pierce Chemical Co., Rockford, Illinois, 61105.

d Cycloheptaamylose.

e Desoxycholic acid.

TABLE 7 (continued)

| NAME | Molecular Formula | MW | Structural Formula | Configuration | $[\alpha]_D$ | References | Suppliers |
|---|---|---|---|---|---|---|---|
| Menthol | see Table 2 | | | | | | |
| Methyl camphor-3-sulfonate | $C_{11}H_{18}O_4S$ | 246.33 | [ 89 ] | | +98.6° ($CHCl_3$) | Prep 9 | |
| Silver camphor-10-sulfonate | $C_{10}H_{15}O_4SAg$ | 339.16 | | | +14.6° (c 4, $H_2O$) (t = 16°) | Prep 10 | |
| Silver 3-bromocamphor-9-sulfonate | $C_{10}H_{14}O_4SBrAg$ | 418.07 | | | +62.75° ($H_2O$) (t = 20°) on the monohydrate | Prep 11 | |
| Silver O,O'-dibenzoylhydrogen tartrate | $C_{18}H_{13}O_8Ag$ | 465.17 | | D-(-) | -98° (c 0.865, 1 N $NH_4OH$) (t = 25°) | Prep 12<br>Res: Table 12, ref. M75 (p.265) | |
| Silver menthoxyacetate | $C_{12}H_{21}O_3Ag$ | 321.17 | | | | Prep 13 | |
| Sodium mandelate | $C_8H_7O_3Ag$ | 259.02 | | | | Prep 14 | |
| 2,3,4,6-Tetra-O-acetylglucose [a] | $C_{14}H_{20}O_{10}$ | 348.31 | | D-(+) | +82.7° (EtOH) (t = 22°) [final value] | Prep 15 | |
| α-(2,4,5,7-Tetranitrofluorenylideneaminoöxy)-propionic acid [TAPA] | $C_{16}H_9N_5O_{11}$ | 447.27 | [ 7 ] | | +97° (t = 25°)[b]<br>-97° (t = 25°)[b] | Prep } 16<br>Res } 16 | |
| Tri-o-thymotide | $C_{33}H_{36}O_6$ | 528.63 | Chapt. II ref.1, p.128 | | +83° ($CHCl_3$) (t = 21.6°) | Prep 17<br>Res 18 | |

a Either the α/β equilibrium mixture or the crystalline β-isomer may be used; the latter has $[\alpha]_D^{20}$ +2.2° (EtOH) initially.

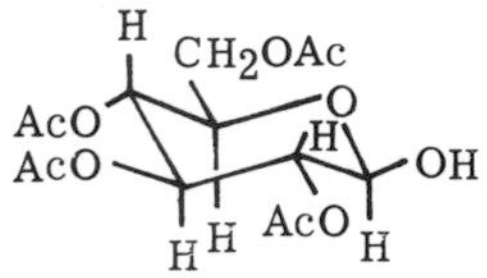


β-isomer

b No solvent reported.

## References to Table 7 - Miscellaneous Resolving Agents and Chiral Adsorbents


(1) H. Leuchs and R. Anderson, Ber., 44, 2136 (1911); H. Leuchs, H. Seeger and K. Jaegers, ibid., 71, 2023 (1938); A. G. Davies and B. P. Roberts, J. Chem. Soc. B, 1967, 17.

(2) R. Weiss and A. Abeles, Monatsh. Chem., 59, 238 (1932).

(3) A. Lüttringhaus and H.-J. Rosenbaum, Woelm Information Sheet 25; M. Woelm, 344 Eschwege, W. Germany.

(4) D. French, M. L. Levine, J. H. Pazur and E. Norberg, J. Amer. Chem. Soc., 71, 353 (1949).

(5) D. French, A. O. Pulley and W. J. Whelan, Stärke, 15, 280 (1963) [Chem. Abstr., 60, 1917f (1964)].

(6) N. Wiedenhof and J. N. J. J. Lammers, Carbohydrate Res., 7, 1 (1968).

(7) A. C. Cope, C. R. Ganellin, H. W. Johnson, Jr., T. V. Van Auken and H. J. S. Winkler, J. Amer. Chem. Soc., 85, 3276 (1963).

(8) J. B. Rather and E. E. Reid, J. Amer. Chem. Soc., 41, 75 (1919).

(9) M. Frèrejacque, Compt. rend., 183, 607 (1926); Ann. Chim. (Paris), [10] 14, 147 (1930) (see p. 164).

(10) W. J. Pope and C. S. Gibson, J. Chem. Soc., 97, 2211 (1910).

(11) H. Regler and F. Hein, J. Prakt. Chem., 148, 1 (1937).

(12) D. M. Coyne, W. E. McEwen and C. A. Van der Werf, J. Amer. Chem. Soc., 78, 3061 (1956).

(13) F. G. Mann and J. Watson, J. Chem. Soc., 1947, 511.

(14) D. G. Neilson and D. A. V. Peters, J. Chem. Soc., 1962, 1309.

(15) A. Georg, Helv. Chim. Acta, 15, 924 (1932); B. Helferich and W. Portz, Chem. Ber., 86, 604 (1953); C. M. McCloskey and G. H. Coleman, Organic Syntheses, Coll. Vol. III, 1955, p. 434.

(16) P. Block, Jr. and M. S. Newman, Org. Syn., 48, 120 (1968).

(17) W. Baker, B. Gilbert and W. D. Ollis, J. Chem. Soc., 1952, 1443.

(18) A. C. D. Newman and H. M. Powell, J. Chem. Soc., 1952, 3747.

# Chapter IV

## Tables of Resolutions

The tables in this chapter record resolutions of about 1200 organic compounds. The use of all common and many less well-known resolving agents is thereby illustrated. Almost all of the listed resolutions are from the literature of the past 20 years or so (ca. 1950 to mid-1971). Emphasis has been placed on practical resolutions; few resolutions by induced or spontaneous crystallization or by chromatographic methods are included. Enzymatic resolutions have been omitted from the survey (1) and are also excluded from these tables; also excluded are compounds optically activated by asymmetric synthesis.

The arrangement of the compounds in Tables 8 through 14 was intended to permit retrieval of information about resolutions and compounds resolved without the use of an index. To achieve this end, compounds were assigned to well recognized albeit arbitrary functional group classes. The tables are arranged roughly in the order in which resolving agents were reviewed in reference (1). The order is given in detail in the Table of Contents. Compounds are listed in each of the functional group tables according to their molecular formulas in order of increasing number of carbons and hydrogens, with heteroatoms following in the order N, O, and all others in alphabetical order.

Since the tables show structural formulas for almost all compounds included, the arrangement described makes possible two kinds of searches. One type of search is for illustrative resolutions which may help in planning new resolutions. Scanning of structural formulas should readily permit this. A second type of search is one for information on the resolution of a particular substance either in connection with the repetition

(1) S.H. Wilen, "Resolving Agents and Resolutions in Organic Chemistry," in Topics in Stereochemistry, Vol. 6, N. L. Allinger and E.L. Eliel, Eds., Wiley-Interscience, New York, 1971.

of the resolution, or to find, e.g., rotation data, on the chiral substance itself. Few names are included in the tables since they were felt to be largely superfluous for the purposes just described.

It is hoped that the reader will forgive any apparent inconsistencies in the structural formulas in the tables. In part, the overriding desire to conserve space in this book led to some abbreviations which, it is hoped, will not be annoying and which, in any event, should not reduce the clarity of presentation or make retrieval of information more difficult. Structures shown are not intended to designate absolute configurations. Relative configurations (*cis*/*trans*; *erythro*/*threo*; etc.) are specified, if known, in appropriate cases.

In cases where derivatives of substances were actually employed in the resolutions and this was made clear by the experimenters, the information is given in the tables. Particularly in Tables 10a-d (Alcohols) and Table 13 (Amino acids), a separate column shows the name of the derivative employed. Abbreviations utilized for derivatives are as follows:

| Alcohols | |
|---|---|
| Acid or hydrogen phthalate | H Phthal |
| Acid or hydrogen succinate | H Succ |
| Acid or hydrogen 3-nitrophthalate | H(3-$NO_2$)Phthal |
| **Amino Acids** | |
| N-Acetyl* | Ac |
| N-Benzoyl* | Bz |
| N-Benzyl* | Bzl |
| Benzyloxy | O-Benzyl |
| Benzylthio | S-Benzyl |
| N-Benzylthiomethyl* | Btm |
| Benzyloxy (benzyl ester) | O-Bzl |
| N-Benzyloxycarbonyl* | Z |
| O-Benzyloxycarbonyl | O-Cbzo |
| *tert*-Butoxy (*tert*-butyl ester)* | O-*tert*Bu |
| N-*t*-Butyloxycarbonyl* | Boc |

| | |
|---|---|
| Ethoxy (ethyl ester)* | OEt |
| Ethoxy (ethyl ether) | O-Ethyl |
| Methoxy (methyl ester)* | OMe |
| Methoxy (methyl ether) | O-Methyl |
| N-Phthaloyl* | Pht |
| N-Trifluoroacetyl* | Tfa |
| N-Tosyl (N-p-toluenesulfonyl)* | Tos |

* These abbreviations are those of the IUPAC-IUB Commission on Biochemical Nomenclature; see H.A. Sober, Ed., "Handbook of Biochemistry," Chemical Rubber Co., Cleveland, Ohio, 44128, 2nd Ed., 1970.

Full names of resolving agents are given together with the sign of rotation of the enantiomer employed (except for alkaloids). For certain resolving agents, e.g., tartaric acid, there remain ambiguities which could not be resolved by careful study of original papers. If the resolving agent is described only as D-tartaric acid, it is not clear whether the (+) or (-)-isomer was utilized since there is no complete agreement on the configurational symbols to be used for these acids (2,3). In a few cases, the D-symbol may also have been confused with the unrelated (+)-rotation. Similar problems exist for the derived O,O'-dibenzoyltartaric acids. Educated guesses were made as to the sign of the enantiomer employed in the resolutions mainly from the fact that (+)-tartaric acid, and hence its derivatives, is the common isomer.

Where two resolving agents were required to obtain both enantiomers of the compound resolved (or if both enantiomers of resolving agent were utilized), this is indicated. Note that usually the second of the listed resolving agents will have been employed not with racemic substrate but rather on the mother liquor or its residue after concentration (or pre-

(2) E.L. Eliel, Stereochemistry of Carbon Compounds, McGraw-Hill, New York, 1962, p. 90.

(3) J.N. Baxter, J. Chem. Educ., 41, 619 (1964).

ferably either of these after conversion from salt or covalent diastereomer back to the original substance). This second resolution thus takes place on a mixture enriched in the opposite enantiomer than that resolved with the first resolving agent.

Derivatives or salts of resolving agents are listed as such with the exception of hydrates. The state of hydration of specific samples of resolving agents such as amines is often not known with any accuracy and should be considered an unknown variable. For resolutions involving diastereomeric esters or amides note that an acyl halide rather than the carboxylic acid may actually have been employed (either as substrate or resolving agent) in the initial reaction.

The solvent(s) in which a given resolution was <u>begun</u> are reported under the name of the resolving agent employed. In a limited number of cases, solvents utilized in the recrystallization of diastereomeric derivatives have been given instead (shown as <u>Recr.</u>) especially for covalent diastereomers. Note that a distinction between reaction solvent and solvent used in the recrystallization of diastereomeric salts is not always made clear especially in the increasingly abbreviated experimental sections of papers.

While more specifics could not be given in tables of modest length, the information provided may be sufficient to attempt a resolution without further details and with some hope for success should the reference publication on the particular resolution not be directly accessible to the researcher.

Symbols used for solvents employed in the determination of specific rotations and for recrystallization of diastereomers are as follows:

| | | | |
|---|---|---|---|
| Acetic acid | HOAc | Ethyl acetate | EtOAc |
| Acetonitrile | $CH_3CN$ | Isopropyl alcohol | isoPrOH |
| Benzene | $C_6H_6$ | Methanol | MeOH |
| Bis-(2-methoxyethyl)ether | Diglyme | Methyl acetate | MeOAc |
| 1-Butanol | BuOH | Methyl cellosolve = 2-methoxy-ethanol | |
| Carbon disulfide | $CS_2$ | Pentane | $C_5H_{12}$ |
| Carbon tetrachloride | $CCl_4$ | Petroleum ether (ligroin or light petroleum) | PetE |
| Chloroform | $CHCl_3$ | 1-Propanol | PrOH |
| Dichloromethane | $CH_2Cl_2$ | Pyridine | py |
| Diethyl ether | $Et_2O$ | Tetrahydrofuran | THF |
| 1,2-Dimethoxyethane | glyme | Water | $H_2O$ |
| Dimethylacetamide | DMA | No solvent (in substantia) | neat |
| Dimethylformamide | DMF | | |
| Dimethylsulfoxide | DMSO | | |
| Ethanol* | EtOH | | |

* Unless specified, absolute ethanol should be assumed.

The term partial resolution has been occasionally used in describing tabulated resolutions where the original investigators themselves have stated that their resolutions are incomplete. Obviously, many other resolutions are also partial. Indeed, a statement by authors indicating that resolution was complete on the basis of rotation data alone (even when both enantiomers were obtained in identical specific rotation with different resolving agents) is clearly insufficient proof of such complete resolution.

Rotations given are those of the compounds whose structures are shown. Where derivatives have been employed in resolutions and the rotations of these are tabulated (in particular, for amino acids), the name of the derivative has been explicitly given in the rotation column. Abbreviations used in this column are as follows:

| | |
|---|---|
| $\alpha$ | Measured rotation |
| [$\alpha$] | Specific rotation, all given at the sodium D line (589 nm) unless explicitly indicated |
| l | Sample length in decimeters (dm) |
| c | Concentration in grams solute/100 ml. (% if so specified in original papers where grams solute/100 g. solvent may have been meant) |
| C.D. | Circular dichroism; [$\theta$] molecular ellipticity |
| [M] or [Ø] | Molecular rotation |
| nm | nanometers |

Since this book is principally a source of data on resolutions rather than a critical compilation of physical constants, rotations tabulated are those of highest magnitude reported for the corresponding resolutions in at least one of the articles cited. The rotations listed are thus not necessarily the highest reported for a given substance in the literature or even in the same reference as that cited. Optically purer materials may also have been obtained by non-resolution routes. Space limitations precluded citing rotation data for other optical activations or for older resolutions.

Unless specified (and this is rarely the case), no assumption should be made regarding the optical purity of the compound whose rotation is reported. Independent nmr or isotope dilution evidence (4) is required to demonstrate the extent of optical purity and while these methods are now increasingly utilized, few of the tabulated resolutions have been so analyzed.

The rotation of the enantiomer obtained from the less soluble

---

(4) For a recent review of these and other methods, see M. Raban and K. Mislow, "Modern Methods for the Determination of Optical Purity," in Topics in Stereochemistry, Vol. 2, N.L. Allinger and E.L. Eliel, Eds., Wiley-Interscience, New York, 1967.

diastereomeric compound or salt formed in a resolution is listed first in these tables (usually after the letter (a)). The rotation shown after the letter (b) is that of the enantiomer obtained usually by workup of the mother liquor with or without use of a second resolving agent.

More than one reference for the resolution of a given substance with the _same_ resolving agent may be given particularly if in the opinion of this author consultation of the several descriptions may be of assistance in the repetition of a resolution - even if the resolution is not fully described. Reports of resolutions subsequent to the original one are reinforcing and often the new rotation data may be useful as well. Citations of resolutions with more than one resolving agent may be of help to the experimenter both to provide choice, in the absence of a full assortment of resolving agents, or to permit assessment of the "best" procedure.

Procedures employing less accessible resolving agents or for which little information was provided in the references consulted have been listed last in entries for compounds often without rotation data. It will be evident which resolutions were considered the most practical although it must be emphasized that the order of the entries was arrived at subjectively. The choice of preferred resolutions is subject to change as information on and accessibility of resolving agents increases.

Where a _Chemical Abstracts_ reference is given in addition to the original reference, it should not be assumed that only the former has been consulted.

Table 12 on resolutions of organometallic and kindred compounds should be considered more representative than comprehensive. Data on metallocene resolutions tabulated by K. Schlögl in "Stereochemistry of Metallocenes" (_Topics in Stereochemistry_, Vol. 1, N. L. Allinger and E. L. Eliel, Eds., Wiley-Interscience, New York, 1967) are not repeated in Table 12.

The recent review "Stereochemical Aspects of Phosphorus Chemistry" by M. J. Gallagher and I. D. Jenkins (_Topics in Stereochemistry_, Vol. 3, E. L. Eliel and N. L. Allinger, Eds., Wiley-Interscience, New York, 1968) contains references and information on resolutions of chiral organophosphorus compounds that have not been duplicated in Table 12.

A substantial number of germanium, phosphorus, and silicon compounds have been obtained in optically active form by a combination of resolution and asymmetric synthesis. A number of these very useful optical activations have been included in Table 12. Optical activations of sulfur compounds with (-)-menthol have been excluded from the tables, however.

Many of the phosphorus acids and silanols represent cases where the compound whose formula is shown in the table is not actually that produced in an optically enriched state. The term formal or virtual resolution (5) may be applied to such optical activations. These compounds have been entered in Table 12 under the parent acid (or silanol) for convenience.

To avoid wasteful duplication as much as possible, resolutions of amino acids from the literature of and prior to 1958 tabulated or described in J.P. Greenstein and M. Winitz' Chemistry of the Amino Acids (Wiley, New York, 1961), have been omitted from Table 13.

The Greenstein and Winitz treatise should in any event be consulted by investigators interested in the resolution of amino acids or their derivatives.

---

(5) This term was suggested by Dr. Neil McKelvie to whom the author is also indebted for a stimulating discussion concerning the resolution of organometalloidal compounds.

## RESOLUTIONS OF ACIDS

### TABLE 8a Simple Carboxylic Acids*

| | Substance resolved | Resolving agent and solvent | Rotation $[\alpha]_D$ of resolved compound | Reference |
|---|---|---|---|---|
| $C_2H_2O_2BrCl$ | Br-C(Cl)(H)-COOH | Brucine<br>Cinchonidine | a) + 10.2° (t = 25°)<br>b) - 6.4° (t = 25°) | A 1 |
| $C_2H_2O_2ClF$ | Cl-C(F)(H)-COOH | Dehydroabietylamine<br>in EtOAc | a) + 24.7° (neat) (t = 27°)<br>b) - 29.9° (neat) (t = 28°) | A 2 |
| $C_4H_6O_2Cl_2$ | $CH_2$(Cl)-C($CH_3$)(Cl)-COOH | (+)-α-Methylbenzylamine<br>in hexane<br>(also with quinine or cinchonine) | - 15° (MeOH) (t = 27°) | A 3 |
| $C_4H_7O_2Br$ | $CH_3CH_2CH$(Br)-COOH | Dehydroabietylamine<br>in acetone<br>Partial resolution | - 14.70° (c 5.0, $Et_2O$) (t = 25°) | A 4 |
| | | Brucine<br>in acetone | $\alpha_D^{25}$ + 8.83° (l 0.5 dm) | A 5 |
| $C_4H_7O_2Br$ | $CH_3CH$(Br)$CH_2$-COOH | Morphine<br>in MeOH | $\alpha_D^{25}$ + 17.3° (l 0.5 dm) | A 5 |
| $C_5H_8O_2$ | 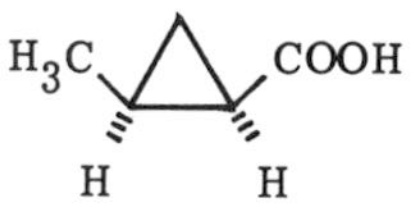 $H_3C$, COOH (cyclopropane), H, H | Quinine<br>in acetone<br>Partial resolution | + 17.9° (95% EtOH) (t = 25°) | A 6 |

* References to Tables 8a-e begin on p. 121

TABLE 8a (continued)

| | Substance resolved | Resolving agent and solvent | Rotation $[\alpha]_D$ of resolved compound | Reference |
|---|---|---|---|---|
| $C_5H_8O_2$ | trans-2-methylcyclopropanecarboxylic acid (H, COOH, $H_3C$, H) | Quinine<br>in acetone<br>Partial resolution | $- 46.4^o$ (95 % EtOH) (t = $25^o$) | A 6 |
| $C_5H_8O_3$ | tetrahydrofuran-3-carboxylic acid (COOH, O) | Quinine<br>in acetone | $+ 10.3^o$ (acetone)<br>$- 14.1^o$ (acetone) | A 7, 8 |
| $C_5H_9O_2D$ | $CH_3CH_2CD(CH_3)$-COOH | Brucine<br>Recr.: $H_2O$<br>Partial resolution | a) $\alpha_D^{25}$ $- 14.6^o$<br>b) $\alpha_D^{25}$ $+ 14.45^o$ | A 9 |
| $C_5H_9O_2Br$ | $CH_3$-C(H)($CH_3$) — CH(Br)-COOH | (-)-Deoxyephedrine<br>in EtOH<br>(+)-Deoxyephedrine<br>in EtOH | a) $- 22.2^o$ (c 10, $C_6H_6$) (t = $12^o$)<br>b) $+ 22.2^o$ (c 10, $C_6H_6$) (t = $12^o$) | A 10 |
| $C_5H_{10}O_2$ | $CH_3CH_2CH(CH_3)$-COOH | (-)- $\alpha$-Methylbenzylamine<br>in Pet E-$Et_2O$<br>(+)- $\alpha$-Methylbenzylamine<br>in Pet E-$Et_2O$ | a) $- 19.34^o$ (neat) (t = $21^o$)<br>b) $+ 19.30^o$ (neat) (t = $21.2^o$) | A 11 |
| $C_6H_9NO_2$ | $CH_3CH_2C(CH_3)(CN)$-COOH | Strychnine<br>in acetone | $+ 3.85^o$ (c 5.99, EtOH) | A 12 |
| $C_6H_{10}O_2$ | $CH_2$ =CHCH$_2$CH($CH_3$)-COOH | Quinine<br>in acetone<br>(+)- $\alpha$-Methylbenzylamine<br>in $Et_2O$ | a) $+ 8.25^o$ (neat) (t = $25^o$)<br>b) $- 8.25^o$ (neat) (t = $25^o$) | A 13<br><br>A 13 |

TABLE 8a (continued)

| | Substance resolved | Resolving agent and solvent | Rotation $[\alpha]_D$ of resolved compound | Reference |
|---|---|---|---|---|
| $C_6H_{10}O_2$ | H, $CH_3$, H, COOH, $CH_3$, H | Quinine in EtOH-$H_2O$ | a) - 10.0° (c 0.190, EtOH) (t = 25°)<br>b) - 4.56° (c 0.0473, 95% EtOH) (t = 24°) | A 14 |
| $C_6H_{11}O_2Cl$ | $CH_3$-C($CH_3$)($CH_3$)—CH(Cl)-COOH | Cinchonidine | - 9.9° (c 0.323, MeOH) (t = 27°) | A 15 |
| $C_6H_{14}O_2$ | $CH_2$-COOH | Brucine in acetone-$H_2O$<br>Cinchonidine in acetone-$H_2O$ | a) + 109.2° (c 5.9, $CHCl_3$) (t = 30°)<br>b) - 64.2° (c 4.6, $CHCl_3$) (t = 29°) | A 16 |
| $C_7H_{10}O_2$ | COOH | Brucine in acetone | a) + 94.5° (c 7, MeOH)<br>b) - 44.0° (c 4, MeOH) (578 nm) | A 17 |
| $C_7H_{12}O_2$ | O, $CH_2COOH$ | Quinine in $C_6H_6$ | - 5.67° (c 15, 95% EtOH) (t = 27°) | A 18 |
| $C_7H_{14}O_2$ | $CH_3$-C($CH_3$)($CH_3$)—CH($CH_3$)-COOH | Dehydroabietylamine in $Et_2O$ | a) - 41.0° (c 2%, EtOH) (t = 25°)<br>b) + 22° (t = 25°) | A 19 |
| $C_7H_{14}O_2$ | $CH_3CH(CH_3)CH_2CH(CH_3)$-COOH | Brucine in acetone-$H_2O$ | a) + 13.08° (neat) (t = 20°)<br>b) - 7.39° (neat) (t = 20°) | A 20 |
| $C_7H_{14}O_2$ | $CH_3CH_2CH_2CH(CH_3)CH_2$-COOH | Quinine in acetone<br>Cinchonidine in $H_2O$-EtOH | a) + 2.49° (neat) (t = 25°)<br>b) - 2.63° (neat) (t = 24°) | A 21<br>A 21, 22 |

TABLE 8a (continued)

| | Substance resolved | Resolving agent and solvent | Rotation $[\alpha]_D$ of resolved compound | Reference |
|---|---|---|---|---|
| $C_8H_{10}O_2$ | 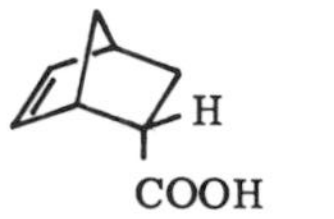 | Cinchonidine in acetone | a) - 70.4° (c 1.4, 95% EtOH) (t = 26°) ca. 49% optical purity<br>b) + 60.3° (c 1.4, 95% EtOH) (t = 29°) | A 23 |
| $C_8H_{10}O_2$ | 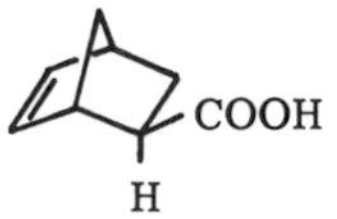 | Cinchonidine in EtOH | a) + 10.0° (EtOH)<br>b) - 11.7° (EtOH)<br>68% optical purity | A 24 |
| $C_8H_{10}O_2$ | 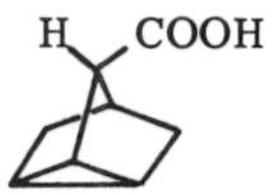 | Cinchonidine in acetone | + 15.3° (95% EtOH) (t = 24.1°) on the methyl ester | A 25 |
| $C_8H_{12}O_2$ |  | Cinchonidine in EtOH | a) - 30.6° (c 1.2, 95% EtOH) (t = 26°) optically pure<br>b) + 30.1° (c 1.3, 95% EtOH) (t = 26°) | A 23 |
| $C_8H_{12}O_2$ | 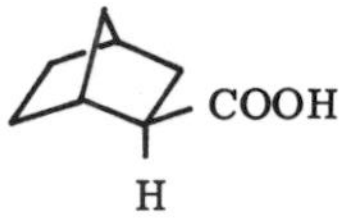 | Quinine · 3 $H_2O$ in acetone | - 8.06° (c 2.57, 95% EtOH) (t = 25°) 29% optical purity | A 26, 23 |
| $C_8H_{14}O_2$ | 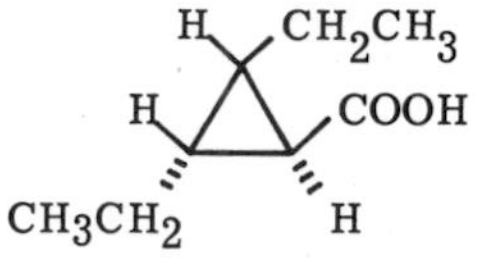 | Quinine in EtOH-$H_2O$ | a) - 4.55° (c 0.0314, 95% EtOH) (t = 26°)<br>b) + 1.74° (c 0.110, 95% EtOH) (t = 23°) | A 14 |

TABLE 8a (continued)

| | Substance resolved | Resolving agent and solvent | Rotation $[\alpha]_D$ of resolved compound | Reference |
|---|---|---|---|---|
| $C_8H_{14}O_2$ | 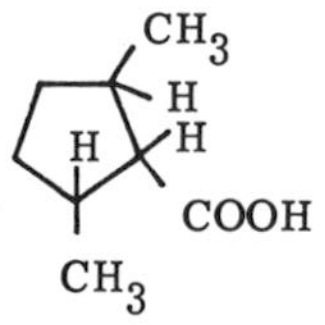 $CH_3$, H, H, H, COOH, $CH_3$ | Cinchonine methoxyhydroxide<br>Recr.: EtOH | - 9.0° (c 3.9, acetone) (t = 20.5°) | A 27 |
| $C_8H_{14}O_2$ | 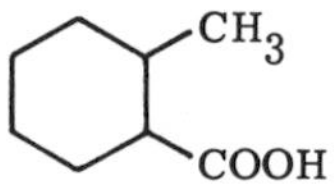 $CH_3$, COOH | Dehydroabietylamine<br>in EtOH | - 7.6° (c 5.95, EtOH) (t = 21.5°)<br>[contains 3% trans isomer] | A 28 |
| $C_8H_{16}O_3$ | $CH_3CH_2CH_2CH_2CH(CH_2CH_3)$-COOH | Quinine<br>in acetone-$H_2O$ | - 8.30° (t = 30°) | A 29 |
| $C_9H_7O_2F_3$ | 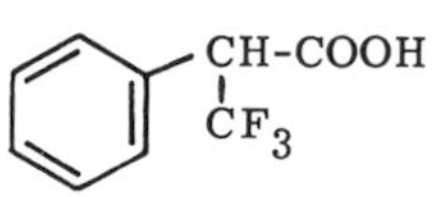 CH-COOH, $CF_3$ | (-)-α-Methylbenzylamine<br>(+)-α-Methylbenzylamine<br>both in EtOH-$H_2O$ (2:1) | a) - 71.4 ± 0.5° (c 2.24, $CHCl_3$) (t = 24°)<br>b) + 73.4 ± 0.5° (c 4, $CHCl_3$) (t = 24°) | A 30 |
| $C_9H_8O_3$ | 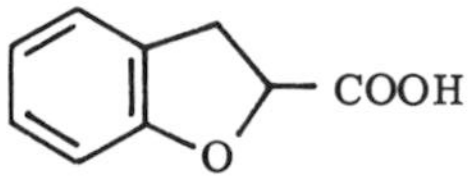 COOH, O | Brucine<br>in $H_2O$<br>(-)-Amphetamine<br>in $H_2O$ | a) + 55.2° (c 2.473, $C_6H_6$) (t = 25°)<br>b) - 55.0° (c 2.4, $C_6H_6$) (t = 25°) | A 31<br>A 31 |
| $C_9H_9NO_4$ | 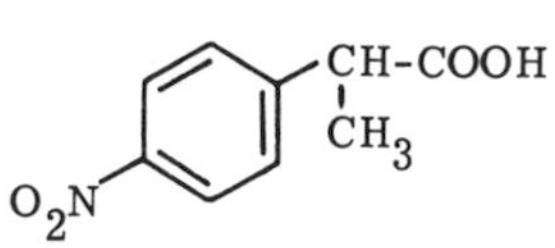 CH-COOH, $CH_3$, $O_2N$ | (-)-α-Methylbenzylamine<br>+ KOH in EtOH-$H_2O$<br>Cinchonine<br>in $H_2O$-EtOH | a) - 55.3° (c 4.015, EtOH) (t = 25°)<br>b) + 55.2° (c 3.914, EtOH) (t = 25°) | A 32a<br>A 32a |
| $C_9H_9NO_4$ | meta isomer | (-)-α-Methylbenzylamine<br>in MeOH | a) $[M]_D^{24}$ - 92° (MeOH)<br>b) $[M]_D^{24}$ + 91° (MeOH) | A 32b |

TABLE 8a (continued)

| Formula | Substance resolved | Resolving Agent and solvent | Rotation $[\alpha]_D$ of resolved compound | Reference |
|---|---|---|---|---|
| $C_9H_{10}O_2$ | $C_6H_5$-CH(CH$_3$)-COOH | Strychnine in EtOH-$H_2O$ (3:1) or $CHCl_3$ | a) + 98.8° (neat) (t = 21°)<br>+ 79.0° (c 1.674, EtOH) (t = 25°) | A 32a, 33 |
| | | Quinine in acetone | b) - 79.1° (c 1.567, EtOH) (t = 25°) | A 32a |
| | | (-)-Ephedrine in $H_2O$-EtOH | -109.6° (neat) (t = 19°) (546 nm) | A 34 |
| | Hydratropic acid | Resolution with (-)- and (+)-α-methylbenzylamine in $C_6H_6$-EtOH | | A 35 |
| $C_9H_{12}O_2$ | 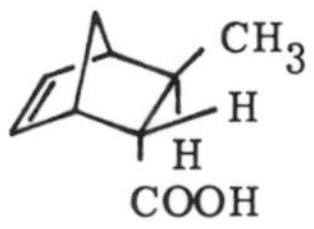 $CH_3$, H, H, COOH | Quinidine in EtOH | - 151° (95% EtOH) | A 36 |
| $C_9H_{12}O_2$ | 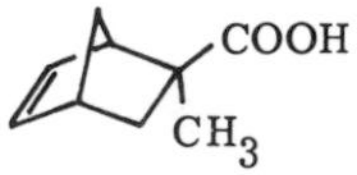 COOH, $CH_3$ | Quinine · 3 $H_2O$ in acetone | a) + 67.3° (95% EtOH)<br>b) - 45.4° (95% EtOH) | A 24 |
| $C_9H_{12}O_2$ | 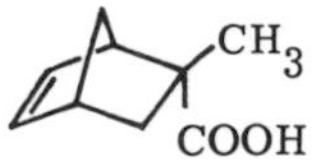 $CH_3$, COOH | Cinchonidine in 95% EtOH<br>Partial resolution | + 58.9° (95% EtOH) | A 24 |
| $C_9H_{14}O_2$ | (H)($CH_3$)C=C=C($CH_2CH_2CH_2CH_3$)(COOH) | Strychnine in $CHCl_3$-EtOAc | - 30.69° ($CHCl_3$) (t = 25°) | A 37 |
| $C_9H_{16}O_2$ | $CH_3CH(CH_3)CH_2C(CH_3)(CH=CH_2)$-COOH | (+)-α-Methylbenzylamine<br>Recr.: EtOAc<br>Partial resolution | + 13.79° (neat) (t = 25°) | A 38 |
| $C_9H_{16}O_2$ | 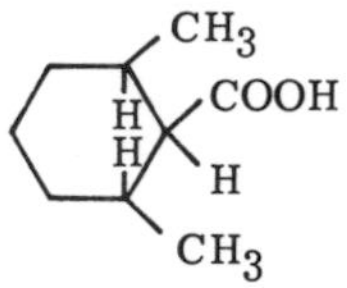 $CH_3$, COOH, H, H, H, $CH_3$ | Cinchonine methohydroxide<br>Recr.: acetone | a) - 16.18° (c 9.8, acetone) (t = 23.7°)<br>b) + 8.26° (c 9.8, acetone) (t = 24°) | A 39 |

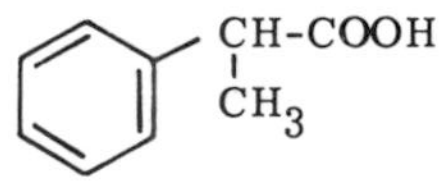

TABLE 8a (continued)

| | Substance resolved | Resolving agent and solvent | Rotation $[\alpha]_D$ of resolved compound | Reference |
|---|---|---|---|---|
| $C_9H_{16}O_2$ | 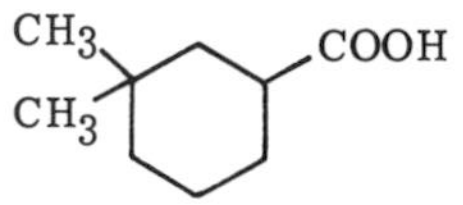 3,3-dimethylcyclohexanecarboxylic acid ($CH_3$, $CH_3$, COOH) | Quinine in acetone | $+ 1.14^{o}$ (neat) ($t = 26^{o}$) | A 40 |
| $C_9H_{18}O_2$ | $CH_3CH(CH_3)CH_2C(CH_3)(CH_2CH_3)$-COOH | Brucine<br>Recr.: EtOAc | a) $+ 21.55^{o}$ (neat) ($t = 24^{o}$)<br>b) $- 5.30^{o}$ (neat) ($t = 25^{o}$) | A 38 |
| $C_{10}H_{10}O_2$ |  2-phenylcyclopropanecarboxylic acid ($C_6H_5$, COOH) cis | Quinine<br>Recr.: $H_2O$-MeOH<br>Partial resolution | $- 20^{o}$ ($CHCl_3$) ($t = 23^{o}$) | A 41 |
| | trans | Brucine<br>Recr.: acetone | a) $- 368^{o}$ (c 0.931, $CHCl_3$) ($t = 24^{o}$) | A 41 |
| | | Quinine | b) $+ 381^{o}$ (c 0.960, $CHCl_3$) ($t = 14^{o}$) | A 42 |
| $C_{10}H_{10}O_2$ | 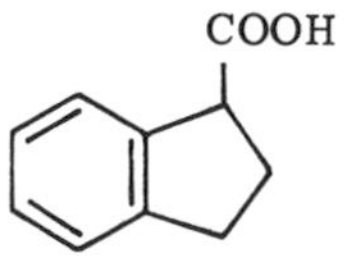 indan-1-carboxylic acid (COOH) | Brucine<br>in acetone-$H_2O$ (2:3)<br>Cinchonine<br>in acetone-$H_2O$ | a) $+ 43.3^{o}$ (c 2.59, $C_6H_6$) ($t = 25^{o}$)<br>b) $- 43.4^{o}$ (c 2.72, $C_6H_6$) ($t = 25^{o}$) | A 43 |
| $C_{10}H_{11}O_2D$ | 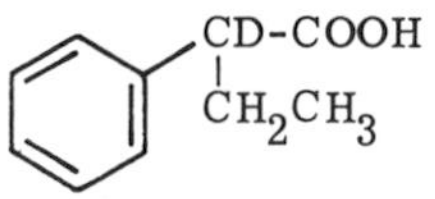 $C_6H_5$-CD(COOH)-$CH_2CH_3$ | Cinchonidine<br>in EtOH-$H_2O$ | a) $\alpha_D^{25}$ $+ 93.9^{o}$ (l 1 dm, neat)<br>b) $\alpha_D^{27}$ $- 78.0^{o}$ (l 1 dm, neat) | A 44 |
| $C_{10}H_{12}O_2$ | 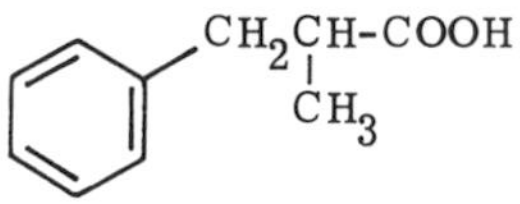 $C_6H_5$-$CH_2CH(CH_3)$-COOH | (-)-$\alpha$-Methylbenzylamine<br>in EtOAc<br>(+)-$\alpha$-Methylbenzylamine<br>in EtOAc | a) $+ 23.51^{o}$ (neat) ($t = 20^{o}$)<br>b) $- 24.56^{o}$ (neat) ($t = 22^{o}$) | A 45, 46<br>A 45 |

TABLE 8a (continued)

| Formula | Substance resolved | Resolving agent and solvent | Rotation $[\alpha]_D$ of resolved compound | Reference |
|---|---|---|---|---|
| $C_{10}H_{12}O_2$ | $C_6H_5$-CH(COOH)-$CH_2CH_3$ | Cinchonidine in EtOH-$H_2O$ | + 93.2° (neat) (t = 26°) | A 47, 48 |
| | | (+)-α-Methylbenzylamine in $C_6H_6$-EtOH | a) - 95.8° (neat) (t = 23°) | A 35 |
| | | (-)-α-Methylbenzylamine in $C_6H_6$-EtOH | b) + 95.4° (neat) (t = 23°) | A 35 |
| $C_{10}H_{12}O_2$ | $C_6H_5$-CH($CH_3$)-$CH_2$-COOH | (-)-α-Methylbenzylamine in EtOH-$H_2O$ | a) + 52.3° ($C_6H_6$) (t = 25°) | A 49, 50 |
| | | (+)-α-Methylbenzylamine in EtOH-$H_2O$ | b) - 51.1° ($C_6H_6$) (t = 25°) | A 49, 50 |
| | | Resolution (also of 3-deuterio analog) through ester formation with menthol | | A 51, 52 |
| $C_{10}H_{16}O_2$ | Chrysanthemic acid, cis | Quinine in EtOH | a) + 83.3° (c 1.597%, $CHCl_3$) (t = 22°) | A 53 |
| | | (+)-α-Methylbenzylamine in EtOH | b) - 83.3° (c 1.555%, $CHCl_3$) (t = 19°) | A 53 |
| | Chrysanthemic acid, trans | Quinine in EtOH | a) - 25.8° (c 2.930%, $CHCl_3$) (t = 20°) | A 53, A 54a |
| | | (+)-α-Methylbenzylamine in EtOH-$H_2O$ | b) + 25.8° (c 2.459%, $CHCl_3$) (t = 20°) | A 53 |
| | | Resolution of trans-isomer with (+)-threo-2-(N, N-dimethylamino)-1-(p-nitrophenyl)-1,3-propanediol (A 271) and of cis- and trans-isomers with (+)-lysine (A 54b) | | - |
| $C_{11}H_9NO_2$ | 2-$C_6H_5$-1-CN-cyclopropane-1-COOH, E-isomer | Brucine in MeOH | a) - 239° (c 0.5, EtOAc) (t = 25°) (546 nm)<br>b) + 235° (c 0.805, EtOAc) (t = 25°) (546 nm) | A 55 |
| | Z-isomer | Quinine in MeOH | a) + 199° (c 0.16, EtOAc) (t = 25°) (546 nm)<br>b) - 143° (c 0.28, EtOAc) (t = 25°) (546 nm) | A 55 |

TABLE 8a (continued)

| | Substance resolved | Resolving agent and solvent | Rotation $[\alpha]_D$ of resolved compound | Reference |
|---|---|---|---|---|
| $C_{11}H_{11}NO_2$ | 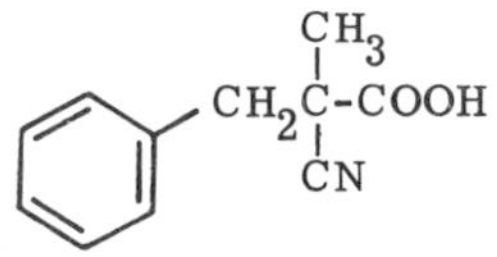 $CH_3$<br>$CH_2C$-COOH<br>CN | Brucine<br>in $H_2O$-acetone | a) + 25.1° (c 2.43, $CHCl_3$) (t = 19°)<br>b) - 25.7° (c 2.54, $CHCl_3$) (t = 18.5°) | A 12 |
| $C_{11}H_{11}NO_2$ | 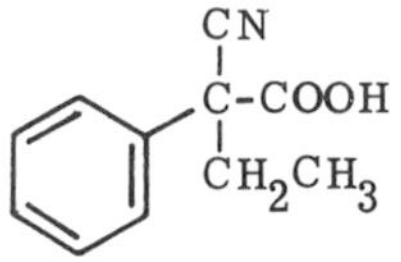 CN<br>C-COOH<br>$CH_2CH_3$ | Quinine<br>in MeOH | - 21.0° (c 10%, $CHCl_3$) (t = 25°) | A 56 |
| $C_{11}H_{12}O_2$ | 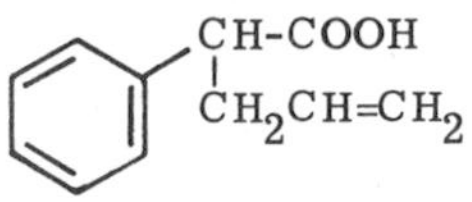 CH-COOH<br>$CH_2CH=CH_2$ | (+)-α-Methylbenzylamine<br>(-)-α-Methylbenzylamine<br>both in $H_2O$ | a) - 104.7° (t = 20°)<br>b) + 104.2° (t = 20°) | A 57, 58 |
| $C_{11}H_{12}O_2$ | 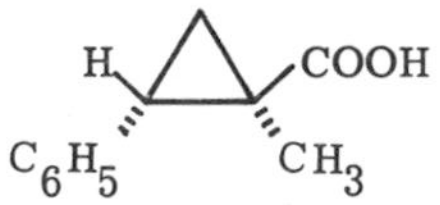 H, COOH<br>$C_6H_5$, $CH_3$ | Brucine<br>in acetone | a) - 155.1 ± 0.4° (c 5%, EtOH) (25°)<br>b) + 152.7 ± 0.4° (c 2%, EtOH) (25°) | A 59 |
| $C_{11}H_{12}O_2$ | COOH | Cinchonine<br>in EtOH-$H_2O$ (3 :2)<br>Quinine<br>in EtOH-$H_2O$<br>Resolution with (-)-Menthylamine | a) + 63.6° (c 1.534, $C_6H_6$) (t = 25°)<br>b) - 63.8° (c 1.614, $C_6H_6$) (t = 25°) | A 60<br>A 60<br>A 57 |
| $C_{11}H_{14}O_2$ | $CH_3$<br>C-COOH<br>$CH_2CH_3$ | Quinine<br>in EtOH-$H_2O$<br>Brucine<br>in acetone<br>Resolution with dehydroabietylamine | a) - 30.0° (c 4.8, $C_5H_6$) (t = 23°)<br>b) + 30.2° (c 4.5, $C_6H_6$) (t = 23°) | A 61, 48<br>A 61<br>A 62 |
| $C_{11}H_{14}O_2$ | 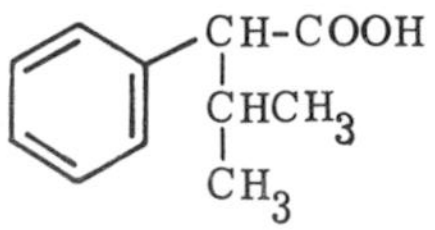 CH-COOH<br>$CHCH_3$<br>$CH_3$ | (+)-α-Methylbenzylamine<br>(-)-α-Methylbenzylamine<br>both in EtOH-$H_2O$ | a) - 62.4° (c 4.46, $CHCl_3$) (t = 24°)<br>b) + 62.5° (c 2, $CHCl_3$) (t = 25°) | A 30 |

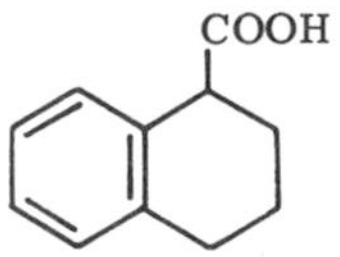

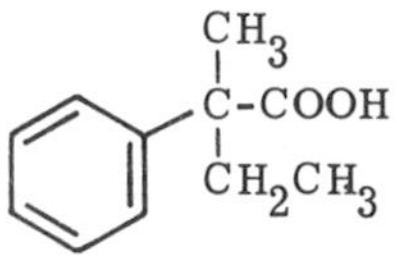

TABLE 8a (continued)

| Formula | Substance resolved | Resolving agent and solvent | Rotation $[\alpha]_D$ of resolved compound | Reference |
|---|---|---|---|---|
| $C_{11}H_{14}O_2$ | 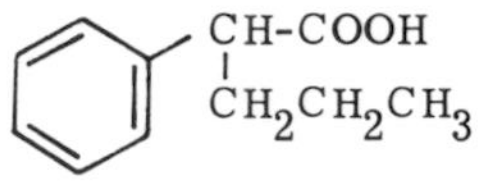 $CH\text{-}COOH$<br>$CH_2CH_2CH_3$ | (+)-α-Methylbenzylamine<br>(-)-α-Methylbenzylamine<br>both in EtOH | a) - 64.0° (c 1.534, EtOH) (t = 25°)<br>b) + 63.4° (c 0.905, EtOH) (t = 25°) | A 63<br>A 63 |
| $C_{11}H_{14}O_2$ | 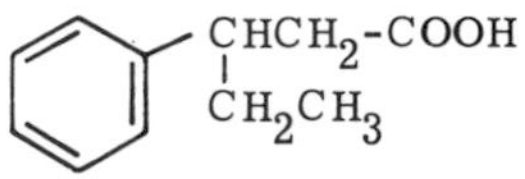 $CHCH_2\text{-}COOH$<br>$CH_2CH_3$ | (-)-α-Methylbenzylamine in EtOH-$H_2O$ (3:1)<br>(+)-α-Methylbenzylamine in EtOH-$H_2O$ (5:2) | a) + 46.3° ($C_6H_6$) (t = 25°)<br>b) - 45.3° ($C_6H_6$) (t = 25°) | A 64<br>A 64 |
| $C_{11}H_{20}O_2$ | 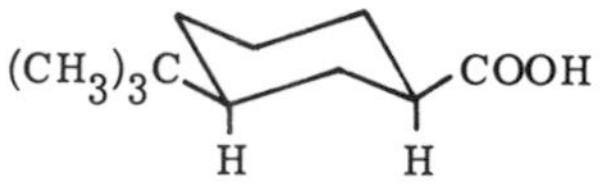 $(CH_3)_3C$ … COOH<br>H H | (-)-Menthylamine · HCl<br>Recr.: acetone | a) + 21.1° (c 1.6, $CHCl_3$) (t = 25°)<br>b) - 19.1° (c 1.6, $CHCl_3$) (t = 25°) | A 65 |
| $C_{12}H_{10}O_2$ | 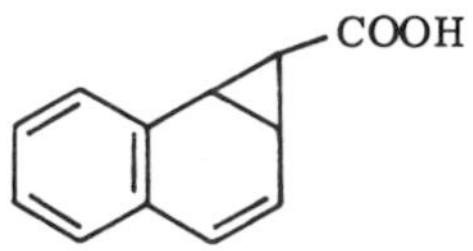 COOH | (-)-α-Methylbenzylamine in MeOH<br>(+)-α-Methylbenzylamine in MeOH-$H_2O$ | a) + 121.3° (c 1.206, acetone) (t = 25°)<br>b) - 121.6° (c 1.073, acetone) (t = 25°) | A 66<br>A 66 |
| $C_{12}H_{16}O_2$ | 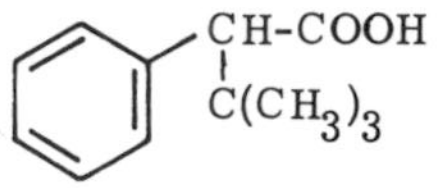 $CH\text{-}COOH$<br>$C(CH_3)_3$ | (-)-α-Methylbenzylamine<br>(+)-α-Methylbenzylamine<br>both in EtOH-$H_2O$ | a) + 62.9° (c 7, $CHCl_3$) (t = 27°)<br>b) - 62.9° (c 5, $CHCl_3$) (t = 27°) | A 30 |
| $C_{12}H_{16}O_2$ | 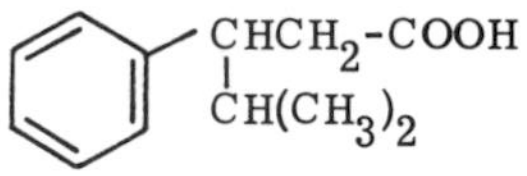 $CHCH_2\text{-}COOH$<br>$CH(CH_3)_2$ | Quinine in EtOH-$H_2O$<br>Cinchonidine in EtOH-$H_2O$ | a) + 40.6° ($C_6H_6$) (t = 25°)<br>b) - 40.5° ($C_6H_6$) (t = 25°) | A 64<br>A 64 |
| $C_{12}H_{16}O_2$ | 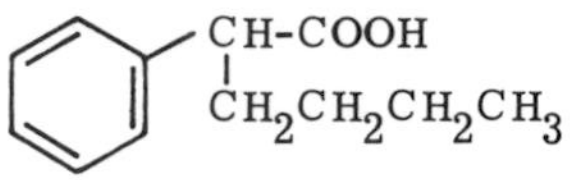 $CH\text{-}COOH$<br>$CH_2CH_2CH_2CH_3$ | (-)-α-Methylbenzylamine in EtOH<br>(+)-α-Methylbenzylamine in EtOH | a) + 62.4° (c 0.967, EtOH) (t = 25°)<br>b) - 61.6° (c 1.160, EtOH) (t = 25°) | A 63<br>A 63 |

TABLE 8a (continued)

| | Substance resolved | Resolving agent and solvent | Rotation $[\alpha]_D$ of resolved compound | Reference |
|---|---|---|---|---|
| $C_{12}H_{16}O_2$ | 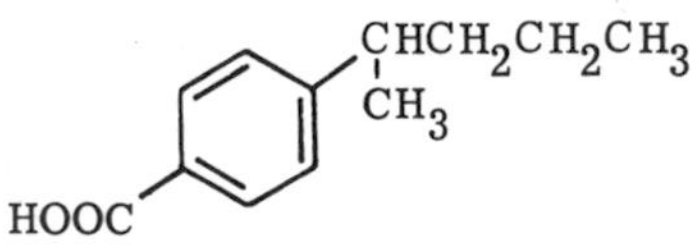 HOOC, CHCH$_2$CH$_2$CH$_3$, CH$_3$ | Quinine in 95% EtOH | a) $-23.3^o$ (c 3.3%, EtOH) (t = $23^o$)<br>b) $+12.3^o$ (t = $23^o$) | A 67 |
| $C_{12}H_{17}O_2Br$ | 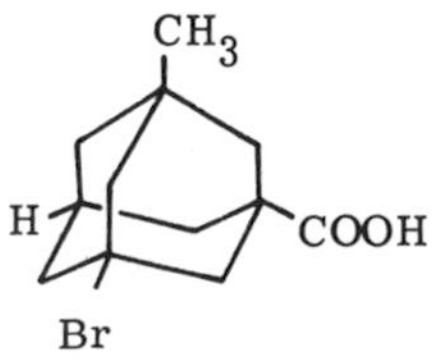 CH$_3$, H, COOH, Br | Quinine in $H_2O$-acetone<br>Resolution with dehydroabietylamine | a) $-0.36 \pm 0.03^o$ (c 6, $CHCl_3$) (t = $25^o$)<br>b) $+0.22^o$ | A 68<br>A 69 |
| $C_{12}H_{18}O_2$ | 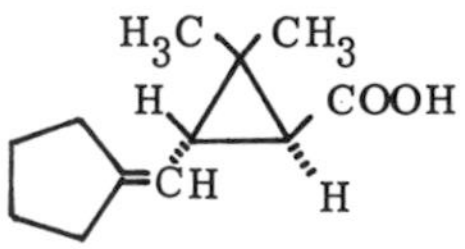 H$_3$C, CH$_3$, H, COOH, CH, H | (-)-Ephedrine in EtOAc | $+2^o$ (c 1, $CHCl_3$) | A 70 |
| $C_{12}H_{20}O_2$ | 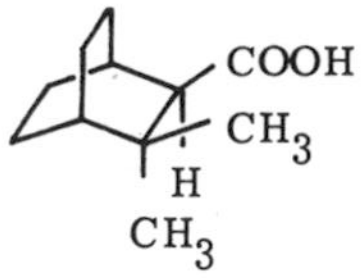 COOH, CH$_3$, H, CH$_3$ | Quinine in $C_6H_6$ | $-14.5^o \pm 0.2^o$ (c 5.20, EtOH) | A 71 |
| $C_{12}H_{22}O_2$ | 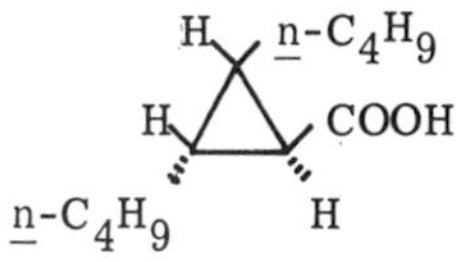 H, n-C$_4$H$_9$, H, COOH, n-C$_4$H$_9$, H | (-)-Ephedrine in $C_6H_6$ | a) $-2.78^o$ (c 0.4005, 95% EtOH) (t = $25^o$)<br>b) $+2.37^o$ (c 0.1023, 95% EtOH) (t = $25^o$) | A 14 |
| $C_{13}H_{10}O_2$ | 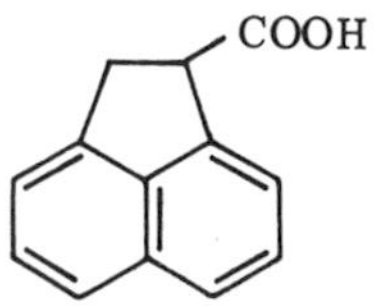 COOH | Cinchonine in MeOH<br>(+)-$\alpha$-(2-Naphthyl)-ethylamine in MeOH | a) $-120.1^o$ (c 2.013, EtOH) (t = $25^o$)<br>b) $+120.1^o$ (c 2.061, EtOH) (t = $25^o$) | A 72<br>A 72 |

TABLE 8a (continued)

| Formula | Substance resolved | Resolving agent and solvent | Rotation $[\alpha]_D$ of resolved compound | Reference |
|---|---|---|---|---|
| $C_{13}H_{11}O_2Br$ | $CH_3$-CH-COOH (naphthalene ring with substituent Y); Y=Br | Cinchonidine in $H_2O$-acetone | a) + 99.09° (c 0.333, acetone) (t = 20°) | A 73 |
| | | Brucine in $H_2O$-acetone | b) - 99.09° (c 0.333, acetone) (t = 20°) | A 73 |
| $C_{13}H_{11}NO_5$ | 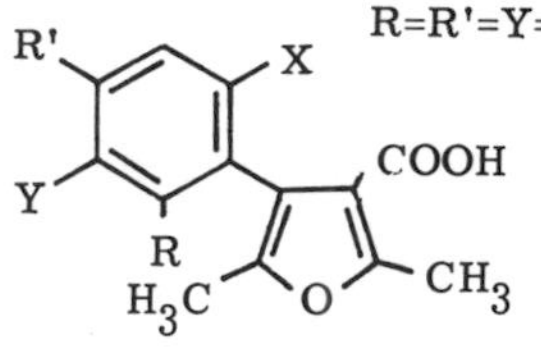 R', X, Y, R, COOH, $H_3C$, O, $CH_3$; R=R'=Y=H; X=$NO_2$ | Quinine in $H_2O$-EtOH | a) - 7.8° (c 1.533, $CHCl_3$) (t = 25°)<br>b) + 12.3° (c 1.216, $CHCl_3$) (t = 25°) | A 74 |
| | R'=X=H; R=$CH_3$; Y=$NO_2$ | Quinine in $H_2O$-EtOH | a) - 22.8° (c 1.290, $CHCl_3$) (t = 25°)<br>b) + 22.1° (c 0.760, $CHCl_3$) (t = 25°) | A 74 |
| | Y=H; R=R'=$CH_3$; X=$NO_2$ | Quinine in $H_2O$-EtOH | a) - 33.6° (c 1.540, $CHCl_3$) (t = 25°)<br>b) + 26.2° (c 1.378, $CHCl_3$) (t = 25°) | A 74 |
| $C_{13}H_{12}O_2$ | see $C_{13}H_{11}O_2Br$ above<br>Y = H | Brucine in MeOH | a) +120.3° (c 2.076, EtOH) (t = 25°) | A 75 |
| | | Cinchonidine in acetone-$H_2O$ | b) -120.1° (c 2.021, EtOH) (t = 25°) | A 75 |
| $C_{13}H_{12}O_2$ | 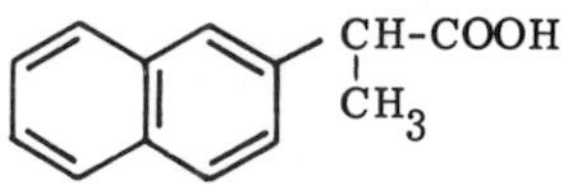 (2-naphthyl) CH-COOH, $CH_3$ | Cinchonidine in MeOH | a) + 68.8° (c 1.183, EtOH) (t = 25°)<br>b) - 68.6° (c 1.080, EtOH) (t = 25°) | A 76 |
| $C_{13}H_{18}O_2$ | (phenyl) $CHCH_2$-COOH, $C(CH_3)_3$ | Brucine in MeOH | a) + 22.2° (c 2, $CHCl_3$) (t = 25°) | A 52, 77 |
| | | Cinchonidine in MeOH | b) - 22.4° (c 2, $CHCl_3$) (t = 25°) | A 52 |

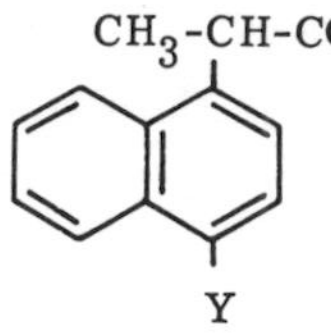

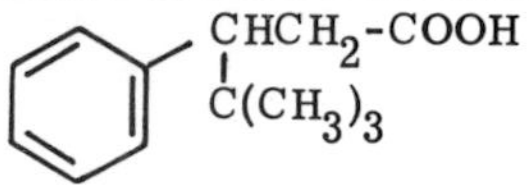

TABLE 8a (continued)

| | Substance resolved | Resolving agent and solvent | Rotation $[\alpha]_D$ of resolved compound | Reference |
|---|---|---|---|---|
| $C_{13}H_{18}O_2$ | $C_6H_5$-C($CH_3$)(COOH)-$CH_2CH_2CH_2CH_3$ | Quinine<br>in EtOH-$H_2O$ | + 11.6° (c 6.41, $Et_2O$) (t = 15°) | A 78 |
| $C_{13}H_{18}O_2$ | $C_6H_5$-CH(COOH)-$CH_2CH_2CH_2CH_2CH_3$ | (-)-α-Methylbenzylamine<br>(+)-α-Methylbenzylamine<br>both in EtOH | a) + 60.2° (c 0.967, EtOH) (t = 25°)<br>b) - 59.1° (c 1.031, EtOH) (t = 25°) | A 63 |
| $C_{14}H_{11}NO_4$ | $CH_3$; $O_2N$ COOH | Quinidine<br>in EtOH<br>Brucine<br>in MeOH | a) + 65° (c 1.5%, MeOH) (t = 23°)<br>b) - 69° (c 2%, MeOH) (t = 25°) | A 79<br>A 79 |
| $C_{14}H_{13}NO_5$ | see $C_{13}H_{11}NO_5$ above | R' = X = H ; R = $CH_3$ ; Y = $NO_2$ | | |
| $C_{14}H_{14}O_2$ | see $C_{13}H_{11}O_2Br$ above<br>Y = $CH_3$ | Quinidine<br>in EtOAc<br>Cinchonidine<br>in acetone-$H_2O$ | a) + 108.5° (c 0.643, EtOH) (t = 25°)<br>b) - 109.0° (c 0.751, EtOH) (t = 25°) | A 80<br>A 80 |
| $C_{14}H_{14}O_2$ | $C_{10}H_7$-$CH_2$CH($CH_3$)-COOH | Cinchonidine<br>in EtOH-$H_2O$ (1:1)<br>(+)-α-Methylbenzylamine<br>in EtOH-$H_2O$ | a) + 24.1° (c 0.966, EtOH) (t = 25°)<br>b) - 23.1° (c 0.980, EtOH) (t = 25°) | A 81<br>A 81 |
| $C_{14}H_{16}O_2$ | H, $C_6H_5$, H, COOH | Cinchonidine<br>Recr.: EtOH | - 19° (c 4.3, EtOH) (t = 25°) | A 82 |

TABLE 8a (continued)

| Formula | Substance resolved | Resolving agent and solvent | Rotation $[\alpha]_D$ of resolved compound | Reference |
|---|---|---|---|---|
| $C_{15}H_{12}O_2$ | H, $CH_3$; Y Y=COOH; Y' Y'=H | Quinine in acetone-$CHCl_3$ (5 :1) | a) - 40.1° (dioxane) (t = 25°) (546 nm) | A 83 |
| | | Cinchonidine in acetone-$CHCl_3$ (5 :1) | b) + 43.1° (c 6.65, dioxane) (t = 25°) (546 nm) | A 83 |
| | | Also resolution of 9-deuterio analog | | A 83 |
| | Y =H; Y'=COOH | Dehydroabietylamine in $H_2O$-MeOH | a) - 13.5° (c ca. 1, dioxane) (t =25°) (546 nm) | A 84 |
| | | | b) + 10.8° (c ca. 1, dioxane) (t =25°) (546 nm) | A 84 |
| | | Also resolution of 9-deuterio analog | | |
| $C_{15}H_{14}O_2$ | $CH_2CH$-COOH, $C_6H_5$ | (-)-Ephedrine · 2 $H_2O$ in acetone | a) - 134.3° (c 0.491, acetone) (t = 25°) | A 85 |
| | | (-)-α-Methylbenzylamine in EtOH-acetone-$H_2O$ | b) + 133.8° (c 0.55, acetone) (t = 25°) | A 85 |
| $C_{15}H_{15}NO_5$ | See $C_{13}H_{11}NO_5$ above | Y = H ; R = R' = $CH_3$ ; X = $NO_2$ | | A 74 |
| $C_{15}H_{17}O_2$ | 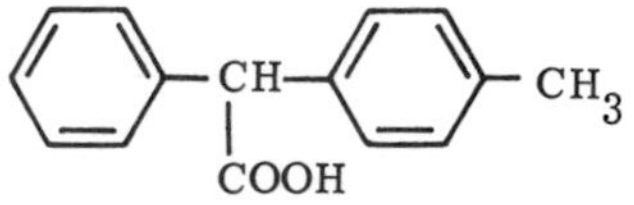 CH, COOH, $CH_3$ | Cinchonidine in EtOH | + 12.77° (c 1.68, acetone) (t = 24°) | A 86 |
| $C_{16}H_{13}O_2Br$ | $C_6H_5$, $C_6H_5$, X, COOH; X=Br | Brucine | a) + 112.2 ± 2.5° (c 1.061, $CHCl_3$) (t =23°) | A 87 |
| | | | b) - 109 ± 2.5° (c 1.104, $CHCl_3$) (t =22°) | |
| $C_{16}H_{13}O_2Cl$ | X=Cl | Quinine Recr.: DMF | - 77.6 ± 2.1° (c 1.074, $CHCl_3$) (t =21.5°) | A 88 |
| $C_{16}H_{14}O_2$ | X=H | Brucine | + 230 ± 5° (c 0.280, $CHCl_3$) (t = 24°) | A 87 |
| $C_{16}H_{14}O_2$ | H, $C_6H_5$, H, COOH, $C_6H_5$, H | Quinine in EtOH | a) - 23.87° (c 0.023, EtOH) (t = 20°) | A 14, 89 |
| | | (-)-Ephedrine in $CHCl_3$ | b) + 23.70° (c 0.021, EtOH) (t = 20°) | |

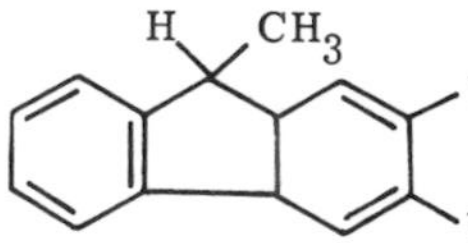

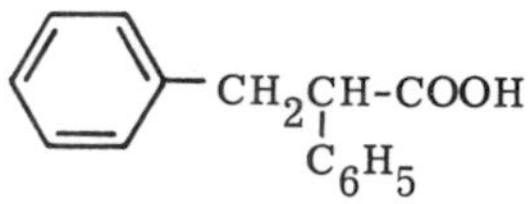

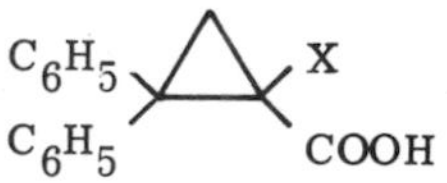

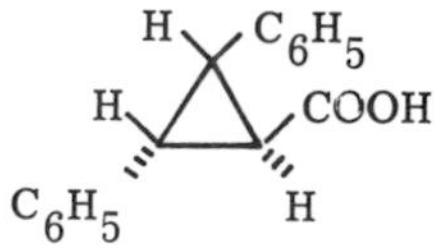

TABLE 8a (continued)

| | Substance resolved | Resolving agent and solvent | Rotation $[\alpha]_D$ of resolved compound | Reference |
|---|---|---|---|---|
| $C_{16}H_{14}O_2$ | 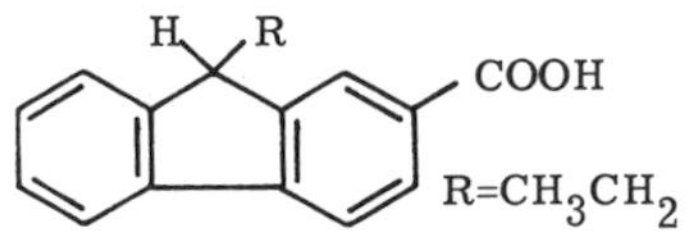 H R COOH R=$CH_3CH_2$ | Brucine<br>Cinchonidine | a) - 50.00° (MeOH) (t = 20°)<br>b) + 53.00° (MeOH) (t = 20°) | A 90<br>A 90 |
| $C_{16}H_{16}O_2$ | 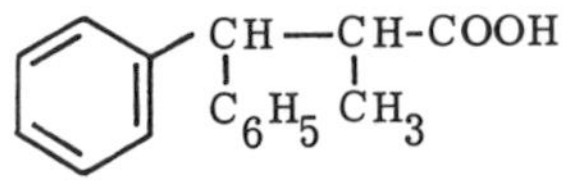 CH—CH-COOH; $C_6H_5$ $CH_3$ | Quinine in acetone | - 52.6 ± 1.7° (c 1.578, $CHCl_3$) (t = 26.5°) | A 91 |
| $C_{16}H_{16}O_2$ | 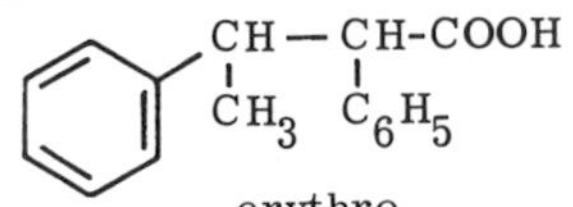 CH — CH-COOH; $CH_3$ $C_6H_5$; erythro | Brucine in EtOH-$H_2O$ | + 18° (EtOH) (t = 22°) | A 92 |
| $C_{17}H_{10}N_2O_6$ | 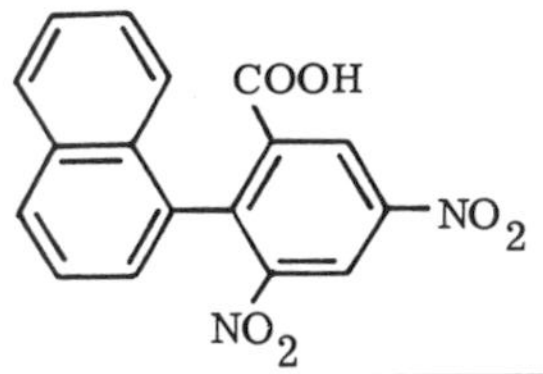 COOH $NO_2$ $NO_2$ | Quinidine in EtOH and by mechanical separation | a) + 50° (c 0.54, acetone) (579 nm)<br>b) - 60° (c 0.60, acetone) (579 nm) | A 93 |
| $C_{17}H_{13}NO_2$ | See $C_{16}H_{13}O_2Br$ above X = CN | Brucine Recr.: MeOH | a) + 128° (c 0.71, EtOAc) (t = 25°) (546 nm)<br>b) - 121° (c 0.53, EtOAc) (t = 25°) (546 nm) | A 94 |
| $C_{17}H_{13}NO_2$ | 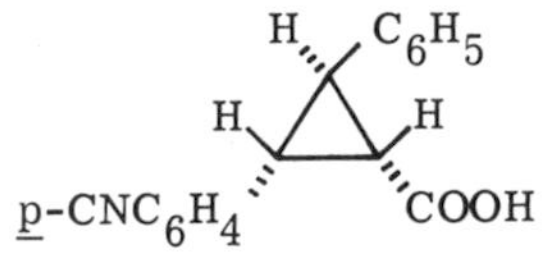 H $C_6H_5$ H H p-$CNC_6H_4$ COOH | Cinchonidine in EtOAc-$CHCl_3$ | a) + 165 ± 1° (c 0.00317, $CHCl_3$) (t = 25°) (436 nm)<br>b) - 164° (c 0.00532, $CHCl_3$) (t = 25°) (436 nm) | A 95 |
| $C_{17}H_{13}NO_2$ | 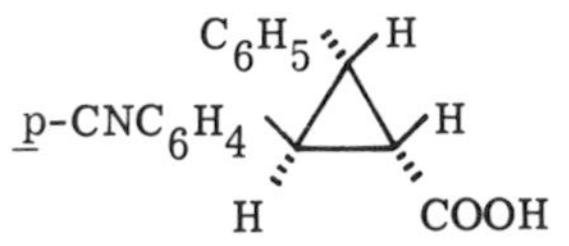 $C_6H_5$ H p-$CNC_6H_4$ H H COOH | Cinchonidine in EtOAc | a) - 85 ± 1° (c 0.00254, $CHCl_3$) (t = 25°) (436 nm)<br>b) + 85 ± 2° (c 0.00304, $CHCl_3$) (t = 25°) (436 nm) | A 95 |

TABLE 8a (continued)

| | Substance resolved | Resolving agent and solvent | Rotation $[\alpha]_D$ of resolved compound | Reference |
|---|---|---|---|---|
| $C_{17}H_{13}NO_2$ | 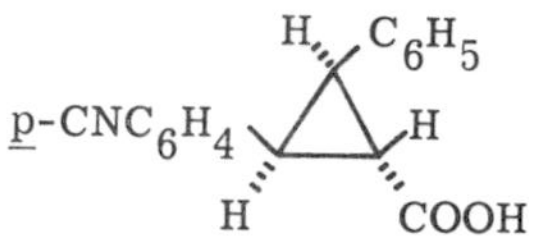 p-$CNC_6H_4$, H, $C_6H_5$, H, H, COOH | Cinchonidine in EtOAc | $+217 \pm 2^o$ (c 0.00300, $CHCl_3$) (t = $25^o$) (436 nm) | A 95 |
| $C_{17}H_{16}O_2$ | See $C_{16}H_{13}O_2Br$ above<br>X = $CH_3$ | Brucine | $+34 \pm 1^o$ (c 1.025, $CHCl_3$) (t = $23^o$) | A 87 |
| $C_{17}H_{16}O_2$ | 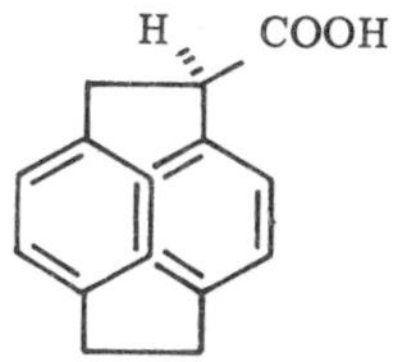 H, COOH | $\alpha$-Methylbenzylamine | a) $-44.1^o$ (c 0.54, $CHCl_3$) (t = $25^o$) (546 nm)<br>b) $+42.0^o$ (c 0.56, $CHCl_3$) (t = $25^o$) (546 nm) | A 96 |
| $C_{17}H_{16}O_2$ | 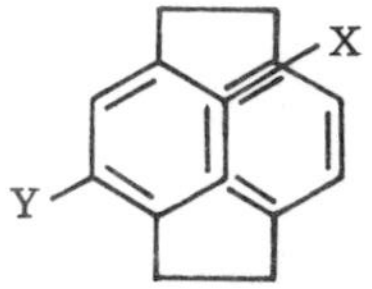 X=COOH<br>Y=H | Brucine in acetone | a) $-157^o$ (c 0.93, $CHCl_3$) (t = $24^o$)<br>b) $+161^o$ (c 0.943, $CHCl_3$) (t = $23^o$) | A 97 |
| | | (-)-$\alpha$-Methylbenzylamine in $CHCl_3$ | $+164^o$ (c 0.5, $CHCl_3$) (t = $20^o$) | A 98 |
| $C_{17}H_{16}O_2$ | 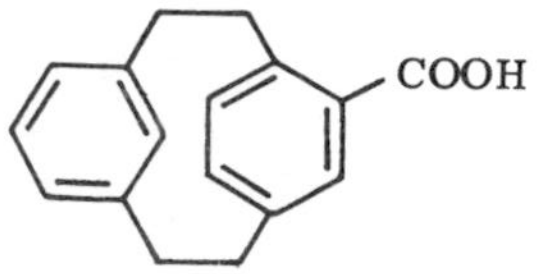 COOH | Brucine<br>Recr.: MeOH | $+34.2^o$ (c 1.07, $CHCl_3$) (t = $25^o$) (546 nm) | A 99 |
| $C_{17}H_{16}O_2$ | See $C_{16}H_{14}O_2$ above<br>R = $CH_3CH_2CH_2$ | Brucine<br>Cinchonidine | a) $-60.00^o$ (MeOH) (t = $20^o$)<br>b) $+57.70^o$ (MeOH) (t = $20^o$) | A 90 |

TABLE 8a (continued)

| | Substance resolved | Resolving agent and solvent | Rotation $[\alpha]_D$ of resolved compound | Reference |
|---|---|---|---|---|
| $C_{17}H_{22}O_2$ | 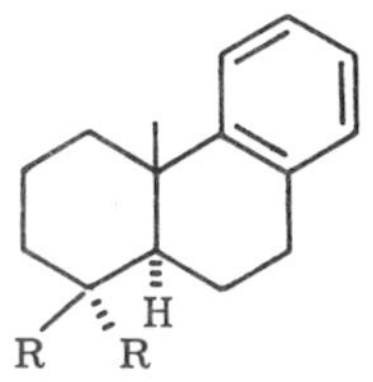 R=COOH R'=$CH_3$ | Cinchonine in MeOH-$H_2O$ | + 138.4° (EtOH) (t = 25°) | A 100 |
| | R=$CH_3$ R'=COOH | Cinchonine methohydroxide in EtOH | + 65° (c 1.58, EtOH) (t = 26°) | A 100 |
| $C_{17}H_{24}O_2$ | 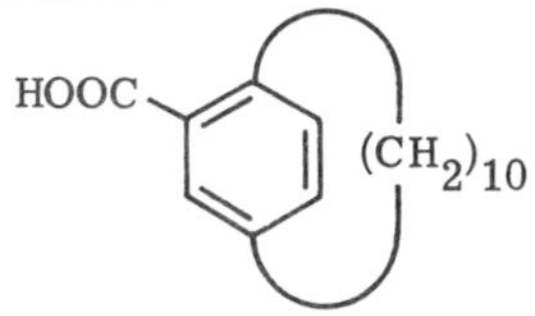 HOOC (CH$_2$)$_{10}$ | Cinchonidine in acetone | a) + 80.5 ± 2° (c 0.77, $CHCl_3$) (t = 24°)<br>b) - 82.3 ± 2° (c 0.96, $CHCl_3$) (t = 28°) | A 101 |
| $C_{18}H_{14}O_2$ | 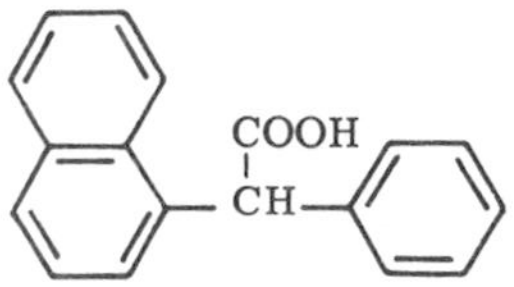 COOH CH | (-)-Menthol Esters Recr.: EtOH | a) - 35.8° (c 10, $CHCl_3$) (t = 23°)<br>b) + 35.2° (c 10, $CHCl_3$) (t = 23°)<br>both on the tosylate of the alcohol obtained on $LiAlH_4$ reduction | A 102 |
| $C_{18}H_{17}NO_2$ | 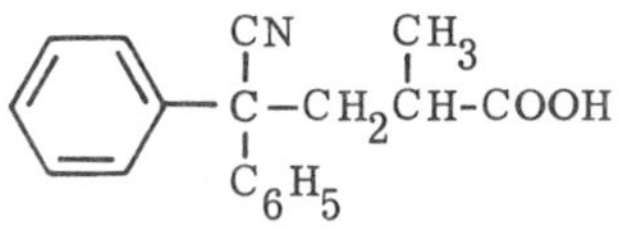 CN CH$_3$ C-CH$_2$CH-COOH C$_6$H$_5$ | (-)-α-Methylbenzyliso-thiouronium camphorate<br>(+)-α-Methylbenzyliso-thiouronium acetate<br>both in EtOH | a) - 58° (c 5, py) (t = 20°)<br>b) + 58° (c 5, py) (t = 20°) | A 103<br>A 103 |
| $C_{18}H_{18}O_2$ | See $C_{16}H_{14}O_2$ above<br>R = n-butyl | Brucine<br>Cinchonidine | a) - 71.60° (MeOH) (t = 20°)<br>b) + 69.00° (MeOH) (t = 20°) | A 90<br>A 90 |

TABLE 8a (continued)

| | Substance resolved | Resolving agent and solvent | Rotation $[\alpha]_D$ of resolved compound | Reference |
|---|---|---|---|---|
| $C_{18}H_{18}O_2$ | See (A) below | Quinidine in acetone | $-48.4^o$ (c 1.0, $C_6H_6$) (t = $28^o$) | A 104 |
| $C_{18}H_{18}O_2$ | See $C_{17}H_{16}O_2$ above<br>X = COOH<br>Y = $CH_3$ | Brucine | a) $-284^o$ (c 0.76, $CHCl_3$) (t = $25^o$) (546 nm)<br>b) $+280^o$ (c 0.78, $CHCl_3$) (t = $25^o$) (546 nm) | A 96 |
| $C_{20}H_{14}O_3$ | 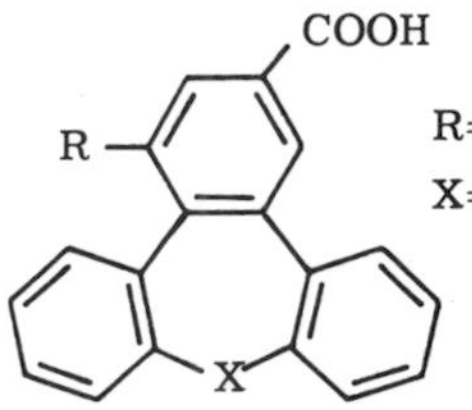 R=$CH_3$<br>X=O | Brucine in EtOH-MeOH | $+190^o$ (c 2, acetone) (t = $20.5^o$)<br>(436 nm) | A 105 |
| $C_{19}H_{16}O_2$ | See (B) below | Cinchonidine in acetone | a) $0^o$ (c 0.023, EtOH) (480 nm)<br>$+230^o$ (c 0.023, EtOH) (393 nm)<br>b) $0^o$ (c 0.023, EtOH) (480 nm)<br>$-182^o$ (c 0.023, EtOH) (393 nm) | A 106 |
| $C_{20}H_{20}O_3$ | See $C_{17}H_{16}O_2$ above<br>X = $CO(CH_2)_2COOH$<br>Y = H | Cinchonidine in $CHCl_3$<br>Partial resolution | $+37.4^o$ (c 0.5-1, $CHCl_3$) | A 98 |

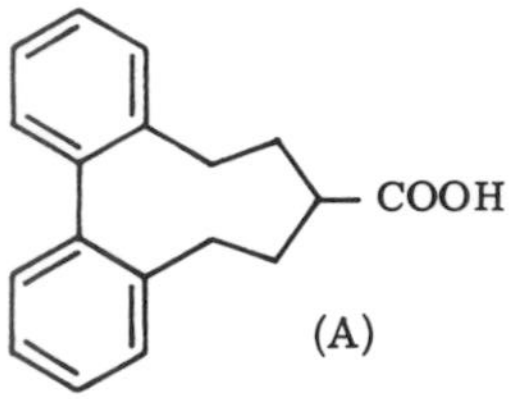

(A)

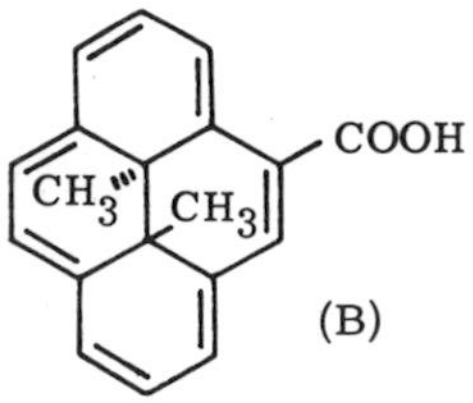

(B)

TABLE 8a (continued)

| | Substance resolved | Resolving agent and solvent | Rotation $[\alpha]_D$ of resolved compound | Reference |
|---|---|---|---|---|
| $C_{20}H_{22}O_2$ | See (A) below | Brucine in acetone | a) $-154^o$ (c 1.3, $CHCl_3$) (t = $27^o$) | A 107 |
| | | Strychnine in acetone-EtOH | b) $+148^o$ (c 1.6, $CHCl_3$) (t = $27^o$) | A 107 |
| $C_{21}H_{14}O_2$ | 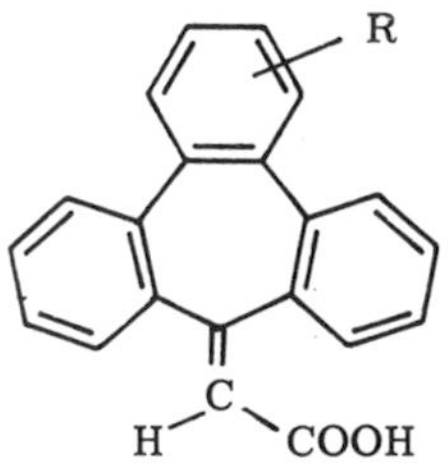 R, H, C, COOH; R=H | Brucine in acetone | a) $+31.7 \pm 0.15^o$ (c 2.6, acetone)<br>b) $-30.9 \pm 0.15^o$ (c 2.6, acetone)<br>both at t = $25^o$ (546 nm) | A 108 |
| $C_{22}H_{16}O_2$ | R=$CH_3$ | Brucine in acetone | $-62^o$ (c 2.6, $CHCl_3$) (t = $25^o$) (546 nm) | A 108 |
| $C_{22}H_{18}O_2$ | See (B) below | Cinchonidine in acetone | a) $-362.7 \pm 2.5^o$ (c 0.35, acetone) (t = $25^o$)<br>b) $+187^o$ (c 1.62, acetone) (t = $23^o$) | A 109 |
| $C_{26}H_{15}O_2I$ | See (C) below | Strychnine in EtOH | a) $+37.9^o$ (t = $28^o$)<br>b) $-16.8^o$ (t = $28^o$) | A 110 |

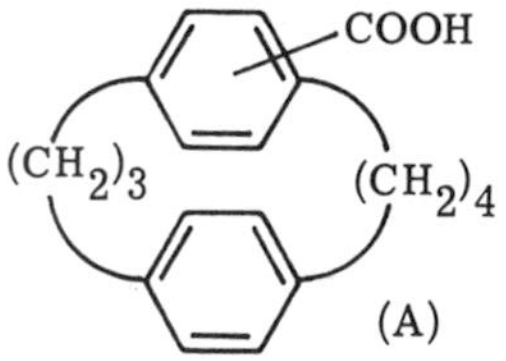

COOH, $(CH_2)_3$, $(CH_2)_4$ (A)

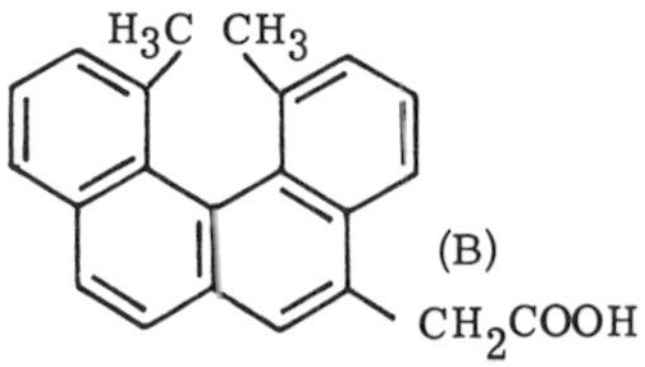

$H_3C$ $CH_3$, $CH_2COOH$ (B)

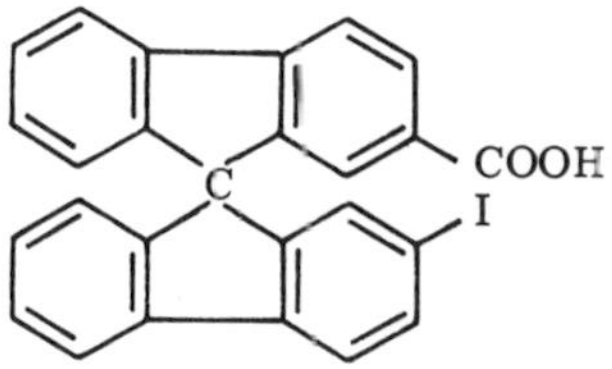

C, COOH, I (C)

TABLE 8b Hydroxy and Keto Monocarboxylic Acids and Lactones

| | Substance resolved | Resolving agent and solvent | Rotation $[\alpha]_D$ of resolved compound | Reference |
|---|---|---|---|---|
| $C_4H_8O_2$ | $CH_3CH(OH)-CH(OH)-COOH$ erythro | Quinine in $H_2O$ | - 9.5° (c 1.0, $H_2C$) (t = 25°) | A 111 |
| | threo | Quinidine in $H_2O$ | - 17.75° (c 1.0, $H_2O$) (t = 25°) | A 111 |
| $C_4H_8O_3$ | See $C_6H_{10}O_4$ below | | | |
| $C_5H_{10}O_3$ | $CH_3CH(OH)CH_2CH_2-COOH$ | Cinchonidine in $H_2O$<br>Partial resolution | a) - 14.7° (neat) (t = 24°)<br>b) + 7.91° (neat) (t = 16°)<br>both measured on the lactone | A 112 |
| $C_5H_{10}O_4$ | $CH_3CH(OH)-C(CH_3)(OH)-COOH$<br>m.p. 88 isomer | Brucine in EtOH | a) + 4° (c 5, $H_2O$) (t = 25°)<br>b) - 3° (c 6, $H_2O$) (t = 25°) | A 113 |
| $C_5H_{10}O_4$ | $CH_3C(CH_3)(OH)-CH(OH)-COOH$ | Quinine in EtOH | - 12.5° (c 2, 0.1 N HCl) (t = 23°) | A 114 |
| $C_6H_{10}O_3$ | [structure: 3,3-dimethyl-2-hydroxy-γ-butyrolactone]<br>Pantolactone<br>See also $C_6H_{12}O_4$ (Pantoic acid) | (+)-threo-2-Amino-1-(p-nitrophenyl)-1,3-propanediol in $H_2O$<br>Also, resolutions with $\alpha$-methylbenzylamine (A 116), ephedrine (A 117), and dehydroabietylamine (A 118) | a) - 50.25° (c 2, $H_2O$) (t = 20°)<br>b) + 50.3° (c 2, $H_2O$) (t = 20°) | A 115<br>A 115<br>See also Table 6 |
| $C_6H_{10}O_3$ | Mevalonolactone | See $C_8H_{16}O_4$ below | | |
| $C_6H_{10}O_4$ | Mevaldic acid | See $C_8H_{16}O_4$ below | | |

TABLE 8b (continued)

| | Substance resolved | Resolving agent and solvent | Rotation $[\alpha]_D$ of resolved compound | Reference |
|---|---|---|---|---|
| $C_6H_{10}O_4$ | $CH_3CH(OAc)CH_2$-COOH | Quinine<br>in EtOAc-Pet E | $+3.6 \pm 0.3^o$ (c 10.175, EtOH) (t = $24^o$) | A 119 |
| $C_6H_{11}NO_3$ | $CH_3CH(O\text{-}N=C(CH_3)_2)$-COOH | (-)-Ephedrine<br>in EtOAc-EtOH | a) $+32^o$ (c 1.6, $H_2O$) (t = $20^o$)<br>b) $-29^o$ (c 1.44, $H_2O$) (t = $20^o$) | A 120, A 121 |
| $C_6H_{12}O_3$ | $CH_3C(CH_3)_2$—CH(OH)-COOH | Brucine<br>in $H_2O$<br>Cinchonidine<br>in EtOH | a) $+4.5^o$ (c 4, $H_2O$) (t = $20^o$)<br>b) $-4.3^o$ (c 4, $H_2O$) (t = $20^o$) | A 122<br>A 122 |
| $C_6H_{12}O_4$ | $CH_2(OH)C(CH_3)_2$—CH(OH)-COOH<br>Pantoic acid; see also $C_6H_{10}O_3$ (Pantolactone) above | (+)-α-(1-Naphthyl)-ethylamine in $H_2O$<br>(-)-α-(1-Naphthyl)-ethylamine in $H_2O$ | a) $+36^o$<br>b) $-36^o$<br>both on the γ-lactone | A 123 |
| $C_6H_{12}O_4$ | $CH_3CH_2CH(OH)$-$C(CH_3)(OH)$-COOH<br><u>erythro</u> | (+)-threo-2-Amino-1-(<u>p</u>-nitrophenyl)-1,3-propanediol<br>(-)-enantiomer of resolving agent<br>both in EtOH | a) $+13.8^o$ (c 1, $H_2O$) (t = $20^o$)<br>b) $-13.3^o$ (c 1, $H_2O$) (t = $20^o$) | A 124 |
| | <u>threo</u> | Strychnine<br>Morphine<br>both in EtOAc | a) $+28^o$ (c 3, $H_2O$) (t = $20^o$)<br>b) $-27.1^o$ (c 5, $H_2O$) (t = $21^o$) | A 124 |

TABLE 8b (continued)

| | Substance resolved | Resolving agent and solvent | Rotation $[\alpha]_D$ of resolved compound | Reference |
|---|---|---|---|---|
| $C_6H_{12}O_4$ | $CH_3CH_2C(CH_3)(OH)-CH(OH)-COOH$ | Quinine in EtOH | $-15^o$ (c 2.3, dil. HCl, pH 1) (t = $23^o$) | A 114 |
| $C_6H_{12}O_4$ | $CH_2(OH)CH_2C(CH_3)(OH)CH_2-COOH$<br>Mevalonic acid | (+)-$\alpha$-Phenyl-1-naphthalene-methylamine on the acid lactone; via amide formation | $+29^o$ (c 1.1, $CHCl_3$) (t = $25^o$) on the amide | A 125 |
| $C_7H_{10}O_3$ | 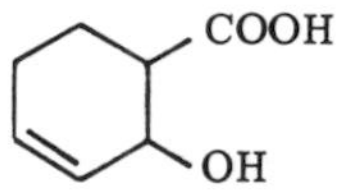 2-hydroxycyclohex-3-ene-1-carboxylic acid (structure) cis | Brucine in acetone or EtOAc | a) $-305^o$ (c 0.40, $H_2O$) (t = $23^o$)<br>b) $+312^o$ (c 0.40, $H_2O$) (t = $23^o$) | A 126 |
| | trans | Brucine in acetone or EtOAc | a) $-70^o$ (c 0.40, $H_2O$) (t = $23^o$)<br>b) $+77^o$ (c 0.40, $H_2O$) (t = $23^o$) | A 126 |
| $C_7H_{12}O_3$ | 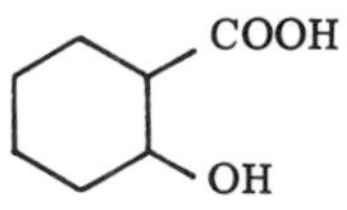 2-hydroxycyclohexanecarboxylic acid (structure) cis | Brucine in acetone | a) $-30.30^o$ (c 31.0, $CHCl_3$) (t = $13.5^o$)<br>$-34.7^o$ (c 1.74, $Et_2O$) (t = $25^o$)<br>b) $+27.16^o$ (c 28.9, $CHCl_3$) (t = $15^o$)<br>$+33.0^o$ (c 1.12, $Et_2O$) (t = $24^o$) | A 127, A 128<br>A 129<br>A 127, A 128<br>A 129 |
| | trans | Brucine in acetone | a) $+47.26^o$ (c 8.15, $CHCl_3$) (t = $16.5^o$)<br>b) $-47.13^o$ (c 7.35, $CHCl_3$) (t = $16.5^o$) | A 127 |
| $C_7H_{10}O_5$ | 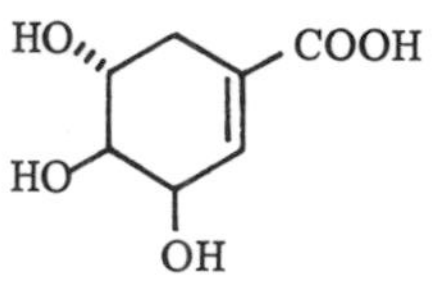 (structure: HO, HO, OH, COOH)<br>Shikimic acid | (-)-$\alpha$-Methylbenzylamine<br>(+)-$\alpha$-Methylbenzylamine<br>both in acetone on the 4,5-O-cyclohexylidene derivative | a) $-159^o$ (c 1, MeOH) (t = $20^o$)<br>b) $+159^o$ (c 1, MeOH) (t = $20^o$) | A 130 |
| $C_7H_{12}O_2$ | $\gamma$-Methyl-$\gamma$-caprolactone | See $C_7H_{14}O_3$ below | | |

TABLE 8b (continued)

| Formula | Substance resolved | Resolving agent and solvent | Rotation $[\alpha]_D$ of resolved compound | Reference |
|---|---|---|---|---|
| $C_7H_{12}O_3$ | HO, H / COOH, H (cyclohexane) | Quinine · 3 $H_2O$ in MeOH | a) + 9.82° (c 4, MeOH) (t = 23.6°) | A 131 |
| | | Cinchonidine in 95% EtOH | b) - 9.93° (c 4, MeOH) (t = 19.6°) | A 131 |
| $C_7H_{14}O_3$ | $CH_3CH_2C(CH_3)(OH)CH_2CH_2$-COOH | (+)-$\alpha$-(2-Naphthyl)-ethylamine sulfate in $H_2O$ | a) $\alpha^{25}_D$ + 17.99° (l 2 dm) | A 132 |
| | | $\alpha$-Methylbenzylisothiouronium chloride in $H_2O$ | b) $\alpha^{25}_D$ - 11.78° (l 2 dm) both on the lactone | A 132 |
| | | Resolution with brucine and with (-)-$\alpha$-methylbenzylamine | | A 133 |
| $C_7H_{14}O_4$ | $CH_3CH(OH)$-$C(OH)(CH(CH_3)_2)$-COOH erythro isomer m.p. 150° | (-)-$\alpha$-Methylbenzylamine | a) + 1.97° (c 1, $H_2O$) (t = 21°) | A 134 a |
| | | (+)-$\alpha$-Methylbenzylamine both in EtOH | b) - 2.0° (c 1, $H_2O$) (t = 21°) | A 134 a |
| | | Resolution with brucine | | A 134 b |
| | threo isomer m.p. 119° | (-)-$\alpha$-Methylbenzylamine | a) + 3.8° (c 1, $H_2O$) (t = 24°) | A 134 c |
| | | (+)-$\alpha$-Methylbenzylamine both in EtOH | b) - 2.4° (c 1, $H_2O$) (t = 22°) | A 134 c |
| | | Resolution with brucine | | |
| $C_8H_7O_3Br$ | 2-X-$C_6H_4$-CH(OH)-COOH X = Br | (-)-Ephedrine in $C_6H_6$ | a) - 166.2° (c 0.5-1.5, EtOH) (t =18-23°) (546 nm)<br>b) + 114.2° (c 0.5-1.5, EtOH) (t =18-23°) (546 nm) | A 135 |
| $C_8H_7O_3Cl$ | X = Cl | (-)-Ephedrine in EtOH | a) - 187.7° (c 1-1.5, acetone) (t =18-23°) (546 nm)<br>b) + 173.1° (c 1-1.5, acetone) (t =18-23°) (546 nm) | A 136 |

TABLE 8b (continued)

| | Substance resolved | Resolving agent and solvent | Rotation $[\alpha]_D$ of resolved compound | Reference |
|---|---|---|---|---|
| $C_8H_7NO_5$ | CH-COOH, OH, $NO_2$ | Brucine<br>in $H_2O$ | a) - 320° (EtOH) (t = 20°)<br>b) + 346° (EtOH) (t = 20°) | A 137 |
| $C_8H_8O_3$ | CH-COOH, OH<br>Mandelic acid | Cinchonine<br>in acetone-$CHCl_3$ (2:3) | a) - 158° (EtOH)<br>b) + 156° (EtOH) | A 138, A 139<br>A 138, A 139 |
| | | (-)-α-Methylbenzylamine<br>(+)-α-Methylbenzylamine<br>both in $H_2O$ | a) + 157.0° (c 1.070, $H_2O$) (t = 26°)<br>b) - 154.0° (c 2.060, $H_2O$) (t = 25°) | A 140<br>A 140 |
| | | Also, resolutions with (-)-ephedrine (A 141, A 142, A 143), strychnine (A 143), cinchonidine (A 144), and morphine (A 145) | | |
| $C_8H_8O_4$ | HO, CH-COOH, OH | Brucine<br>Recr.: $H_2O$ | a) - 128.5° (c 3.454, acetone) (t = 22°)<br>b) + 127.5° (c 3.454, acetone) (t = 23°) | A 146 |
| $C_8H_{10}O_3$ | $CH_3$, O, COOH | Quinine | - 385° ($CHCl_3$) (t = 25°) | A 147 |
| $C_8H_{12}O_4$ | OH, $CH_3$, O, COOH | Brucine<br>in acetone | a) - 48 ± 1° (c 1.3, $H_2O$) (t = 25°)<br>b) + 48 ± 1° (c 1.3, $H_2O$) (t = 25°) | A 148 |
| $C_8H_{16}O_4$ | $CH_3OCH(OCH_3)CH_2C(CH_3)(OH)CH_2$-COOH | Quinine sulfate<br>on the barium salt<br>of the acid<br>Recr.: EtOAc-hexane | - 10.0° ($CHCl_3$) (t = 20°)<br>on mevalonolactone<br>prepared from the acid | A 149 |
| $C_9H_7O_3F_3$ | $CF_3$, C-COOH, OH | (+)-α-(1-Naphthyl)-ethyl-amine in $C_6H_6$-EtOH (6:1) | a) - 22.5 ± 0.7° (c 2.70, $CHCl_3$) (t = 20°)<br>b) + 9.3 ± 0.3° (c 3.56, $CHCl_3$) (t = 19°) | A 150 |

TABLE 8b (continued)

| | Substance resolved | Resolving agent and solvent | Rotation $[\alpha]_D$ of resolved compound | Reference |
|---|---|---|---|---|
| $C_9H_8O_5$ | O, $CH_2$, O; CH-COOH, OH | (-)-Ephedrine in EtOH-$H_2O$ (9:1) | - 128.5° (c 0.5-1.5, EtOH) (t = 18-23°) (546 nm) | A 135 |
| $C_9H_{10}O_3$ | $CHCH_2$-COOH, OH | Brucine in EtOAc | a) + 21.1° (c 2.2%, EtOH) (t = 22°)<br>b) - 18.9° (c 2.3%, EtOH) (t = 22°) | A 151, A 152 |
| | | Resolution of 2-deuterio analog | | A 153 |
| $C_9H_{10}O_3$ | $CH_3$, C-COOH, OH<br>Atrolactic acid | (-)- α-Methylbenzylamine in $H_2O$ | a) + 37.7° (c 3.5, EtOH) (t = 25°) | A 154 |
| | | (+)- α-Methylbenzylamine in $H_2O$ or morphine | b) - 37.8° (c 3.23, EtOH) (t = 27°) | A 139, A 154 |
| $C_9H_{10}O_4$ | $CH_3O$; CH-COOH, OH | (-)-Ephedrine in $CHCl_3$-$CCl_4$ (2:3) | a) - 117.3° (c 0.5-1.5, EtOH) (t = 18-23°) (546 nm)<br>b) + 69.5° (c 0.5-1.5, EtOH) (t = 18-23°) (546 nm) | A 135 |
| | 4-$CH_3O$ isomer | (-)-Ephedrine | - 141.4° (c 0.5-1.5, $H_2O$) (t = 18-23°) | A 135 |
| $C_9H_{14}O_4$ | $CH_2$-COOH; $CH_2OCH_3$; OH | (-)-Ephedrine in $C_6H_6$ | - 11.7° (c 1.35, MeOH) (t = 29°) | A 155 |
| $C_{10}H_{12}O_3$ | $CH_3$, $CH_2$C-COOH, OH | Brucine in acetone | a) + 17.0° (c 5.595, dioxane)<br>b) - 16.9° (c 4.645, dioxane) | A 156 |
| $C_{10}H_{12}O_3$ | $CH_3$, $CCH_2$-COOH, OH | Morphine · $H_2O$ in $H_2O$ | a) + 7.5° (c 1.660, EtOH) (t = 20°) (546 nm)<br>b) - 8.9° (c 1.7544, EtOH) (t = 23°) (546 nm) | A 157 |

TABLE 8b (continued)

| | Substance resolved | Resolving agent and solvent | Rotation $[\alpha]_D$ of resolved compound | Reference |
|---|---|---|---|---|
| $C_{10}H_{12}O_3$ | 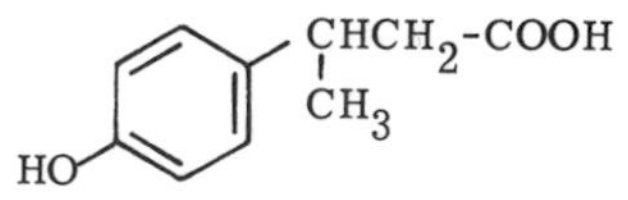 HO–C6H4–CH(CH3)CH2-COOH | Brucine in $H_2O$ | a) - 34.4° (c 4.12, EtOH) (t = 25°)<br>b) + 33.1° (c 3.37, EtOH) (t = 25°) | A 158 |
| $C_{11}H_{12}O_4$ | 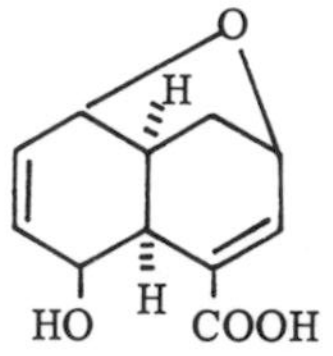 HO, H, H, O, COOH | (-)-Ephedrine in MeOH-acetone | a) + 22 ± 2° (c 1, MeOH) (t = 20°) | A 159 a |
| | | Brucine in acetone-MeOH | b) - 22 ± 3° (c 1, MeOH) (t = 20°) | A 159a, A 159b |
| $C_{11}H_{14}O_3$ | 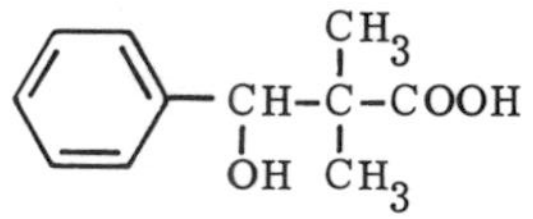 C6H5–CH(OH)–C(CH3)2-COOH | Morphine in $H_2O$-EtOH | a) + 9.8° (c 1.038, HOAc) (t = 25°) | A 160 |
| | | (+)-α-Methylbenzylamine in $H_2O$ | b) - 9.0° (c 1.086, HOAc) (t = 25°) | A 160 |
| $C_{11}H_{14}O_3$ | 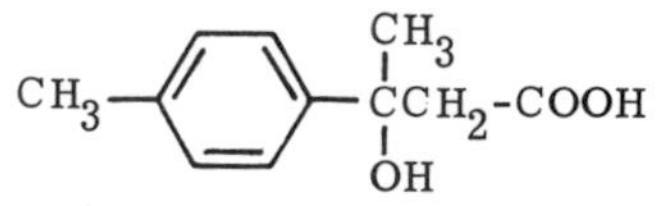 CH3–C6H4–C(CH3)(OH)CH2-COOH | Brucine in EtOAc | + 11.9° (c 0.10, $CHCl_3$) (t = 23°) | A 161 |
| $C_{11}H_{14}O_4$ | 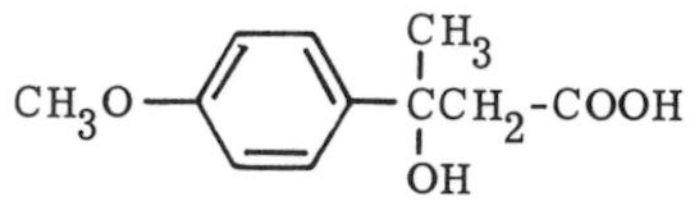 CH3O–C6H4–C(CH3)(OH)CH2-COOH | Brucine in EtOAc | + 7.40° (c 0.10, $CHCl_3$) (t = 25°) | A 161 |
| $C_{12}H_{14}O_4$ | 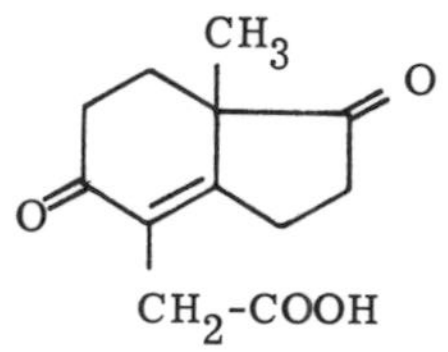 CH3, O, O, CH2-COOH | (-)-threo-(N, N-Dimethylamino)-1-(p-nitrophenyl)-1, 3-propanediol | a) + 225 ± 1° (c 0.5, $H_2O$) | A 162 |
| | | (+)-enantiomer of resolving agent<br>both in EtOAc-EtOH-$H_2O$ | b) - 225 ± 1 (c 0.5, $H_2O$) | |

TABLE 8b (continued)

| | Substance resolved | Resolving agent and solvent | Rotation $[\alpha]_D$ of resolved compound | Reference |
|---|---|---|---|---|
| $C_{14}H_{10}O_3$ | 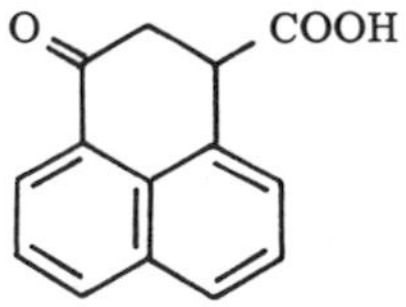 O, COOH | Cinchonidine in $H_2O$-MeOH<br>Dehydroabietylamine in $H_2O$-MeOH | a) + 278.8° (c 2.043, EtOH) (t = 25°)<br>b) - 279.0° (c 2.043, EtOH) (t = 25°) | A 163<br>A 163 |
| $C_{14}H_{18}O_3$ | 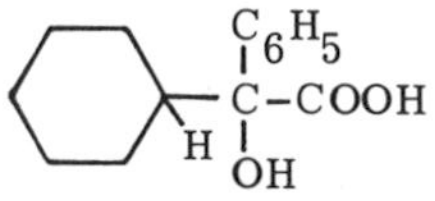 $C_6H_5$, C-COOH, H, OH | Quinine<br>Recr.: EtOH | a) + 25.8 ± 1° (c 4.483, EtOH)<br>b) - 25.2 ± 1° | A 164 |
| $C_{15}H_{14}O_3$ | 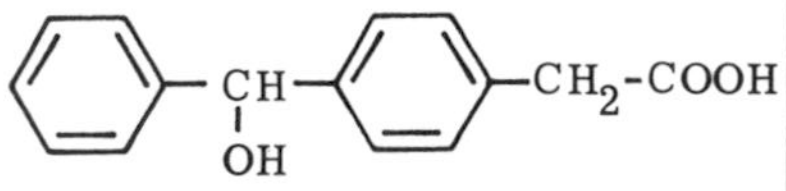 CH, OH, $CH_2$-COOH | Strychnine in dioxane-$H_2O$ | + 2.8° (c 5.9, dioxane) (t = 25°) | A 165 |
| $C_{15}H_{14}O_3$ | See (A) below | Cinchonine in EtOH-$H_2O$ | + 11.4° (c 6.87, EtOH) (t = 26°) | A 166 |
| $C_{15}H_{14}O_4$ | See (B) below | Cinchonine<br>Brucine<br>both in acetone | a) + 1.2° (c 5.76, dioxane) (t = 19°)<br>b) - 1.2° (c 2.52, dioxane) (t = 18°) | A 156<br>A 156 |
| $C_{15}H_{20}O_3$ | See (C) below | Cinchonine in acetone | + 61° ($CHCl_3$) (t = 20.5°) | A 167 |

$CH_3$, C, COOH, OH

(A)

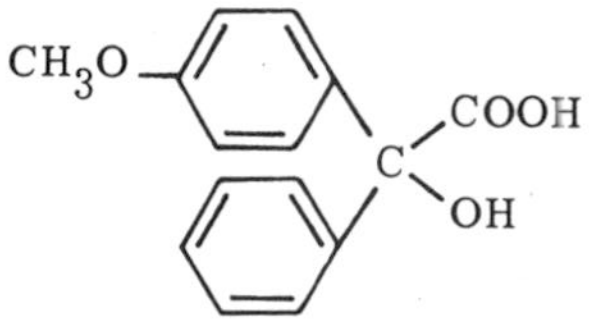

$CH_3O$, C, COOH, OH

(B)

$(CH_3)_3C$-C(=O), $CH_3$, $CH_3$, $CH_3$, COOH

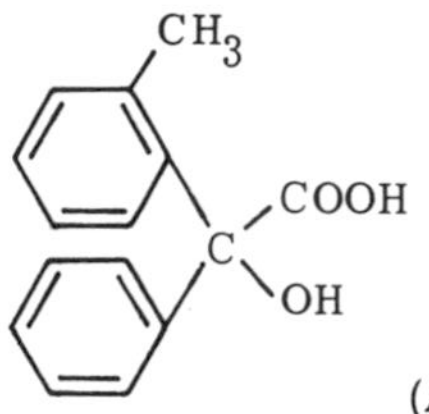

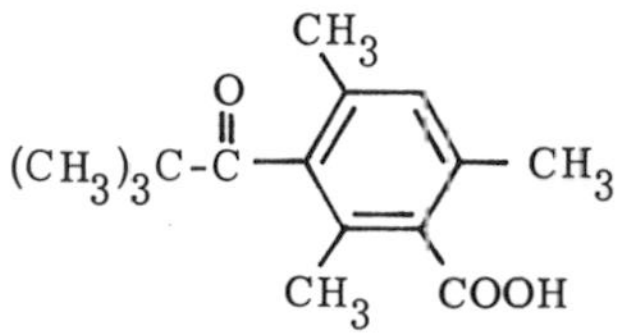

(C)

TABLE 8b (continued)

| | Substance resolved | Resolving agent and solvent | Rotation $[\alpha]_D$ of resolved compound | Reference |
|---|---|---|---|---|
| $C_{15}H_{24}O_4$ | Formula (A) below | See Table 10b $C_{16}H_{26}O_4$ | | |
| $C_{16}H_{14}O_3$ | 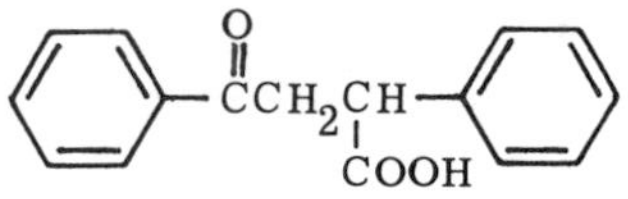 $C_6H_5$-CO-$CH_2$-CH(COOH)-$C_6H_5$ | Quinine in EtOAc | a) $+136^{\circ}$ (c 1.03, EtOAc) (t = $27^{\circ}$)<br>b) $-111^{\circ}$ (c 1.40, EtOAc) (t = $27^{\circ}$) | A 168 |
| $C_{16}H_{16}O_3$ | 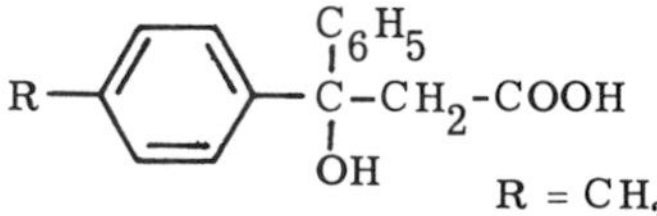 R-$C_6H_4$-C($C_6H_5$)(OH)-$CH_2$-COOH, R = $CH_3$ | Brucine · 4 $H_2O$ in EtOH | a) $-168^{\circ}$ (c 4.63, acetone) (t = $13^{\circ}$)<br>b) $+151^{\circ}$ (c 2.44, acetone) (t = $13^{\circ}$) | A 169 |
| $C_{16}H_{16}O_4$ | R = $OCH_3$ | Cinchonine in EtOH | a) $+14.2^{\circ}$ (c 1.46, dioxane) (t = $14^{\circ}$)<br>b) $-11.4^{\circ}$ (c 5.04, dioxane) (t = $12^{\circ}$) | A 169 |
| $C_{16}H_{18}O_4$ | See (B) below | (+)-threo-2-Amino-1-(p-nitrophenyl)-1,3-propanediol in MeOH | a) $-120^{\circ}$ (c 1, MeOH) (t = $20^{\circ}$)<br>b) $+118^{\circ}$ (c 1, MeOH) (t = $20^{\circ}$) | A 170 |
| | | Resolution with (-)-ephedrine in $H_2O$ | | A 171 |
| $C_{16}H_{20}O_4$ | See (C) below | Cinchonidine Recr.: EtOAc | a) $+8.3^{\circ}$ ($CHCl_3$) (t = $24^{\circ}$)<br>b) $-5.3^{\circ}$ ($CHCl_3$) (t = $24^{\circ}$) | A 172 |
| $C_{16}H_{24}O_3$ | See (D) below | Dehydroabietylamine in MeOH | a) $-49^{\circ}$ (c 0.10, MeOH) | A 173 |
| | trans isomer | (+)-$\alpha$-(1-Naphthyl)-ethylamine in MeOH | b) $+49^{\circ}$ (c 0.10, MeOH) | A 173 |

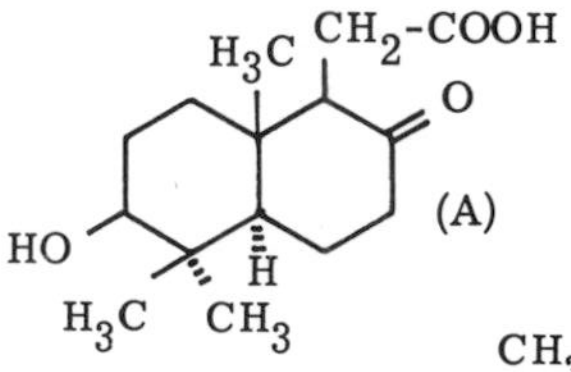

(A): HO, $H_3C$, $CH_3$, H, $H_3C$, $CH_2$-COOH, O

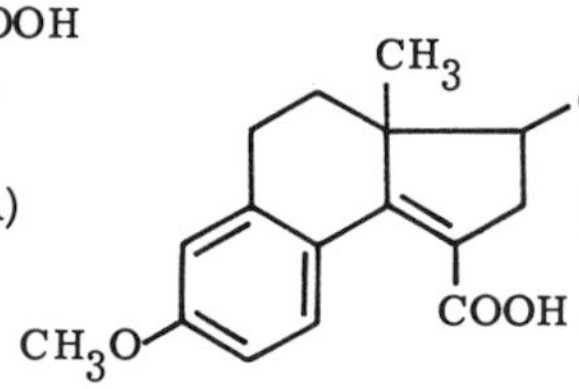

(B): $CH_3O$, $CH_3$, OH, COOH

(C): $CH_3O$, $CH_3CH_2$, $CH_2CH_2$-COOH, O

(D): HO, CH-COOH

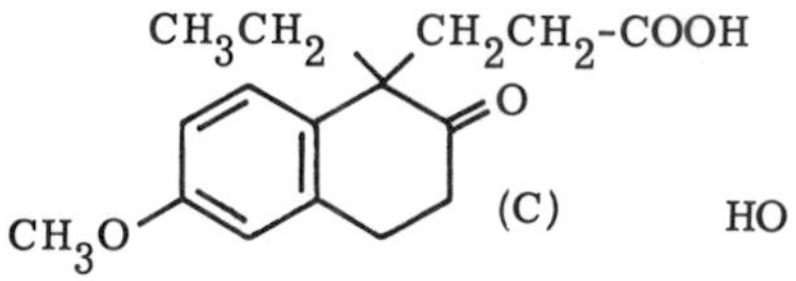

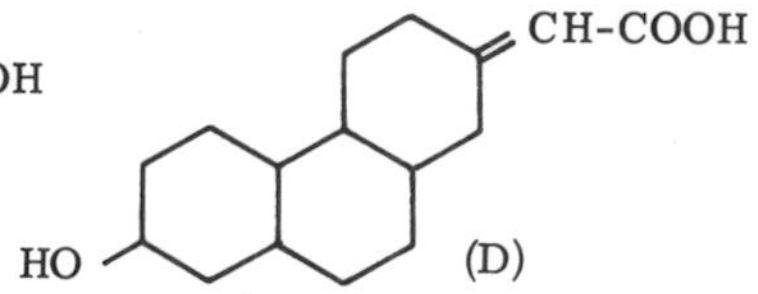

TABLE 8b (continued)

| | Substance resolved | Resolving agent and solvent | Rotation $[\alpha]_D$ of resolved compound | Reference |
|---|---|---|---|---|
| $C_{17}H_{16}O_3$ | 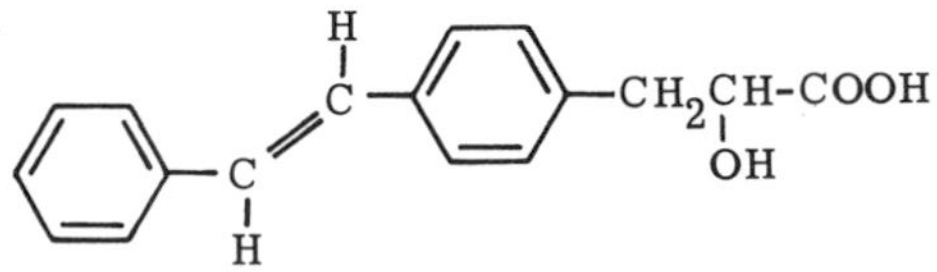 | Brucine<br>in dioxane-$H_2O$ (1:1) | a) $-21^o$ (dioxane) ($t = 20^o$)<br>b) $+20.5^o$ (dioxane) ($t = 20^o$) | A 174 |
| $C_{18}H_{16}O_3$ | See (A) below | Brucine<br>Recr.: acetone-$H_2O$ | $-27.7^o$ (c 1.3, $CHCl_3$) ($t = 22^o$) (578 nm) | A 175 |
| $C_{18}H_{16}O_7$ | Usnolic acid<br>See (B) below<br>For Usnic acid see Table 10d | Brucine<br>in acetone | a) $-19^o$ (c 1.0, MeOH) ($t = 20^o$)<br>b) $+31^o$ (c 1.0, MeOH) ($t = 20^o$) | A 176 |
| $C_{19}H_{16}O_3$ | See (C) below | Brucine · 4 $H_2O$<br>in EtOAc | a) $-205^o$ (c 1.56, acetone) ($t = 16^o$)<br>b) $+207^o$ (c 1.69, acetone) ($t = 15^o$) | A 169 |
| $C_{19}H_{16}O_4$<br>and<br>$C_{19}H_{18}O_3$ | Warfarin<br>See Table 10c | | | |
| $C_{19}H_{22}O_5$ | Zearalenone<br>See Table 10d | | | |
| $C_{20}H_{14}O_3$ | See Table 8a $C_{20}H_{14}O_3$<br>X = CH(OH) R = H | Brucine<br>in acetone | a) $+375^o$ (c 0.67, acetone) ($t = 24^o$) (436 nm)<br>b) $-370^o$ (c 0.67, acetone) ($t = 24^o$) (436 nm) | A 177 |

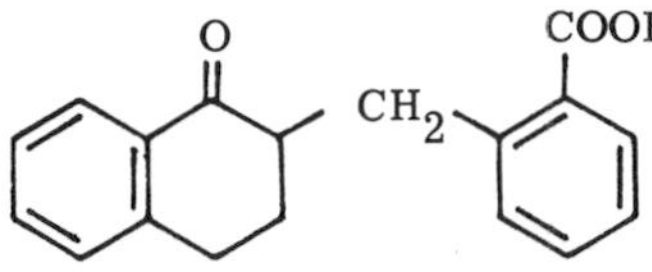

(A)

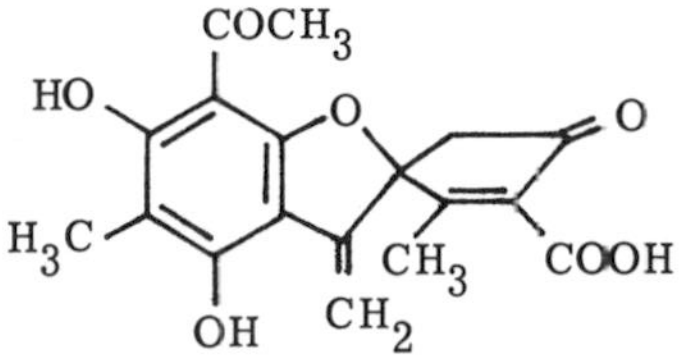

(B)

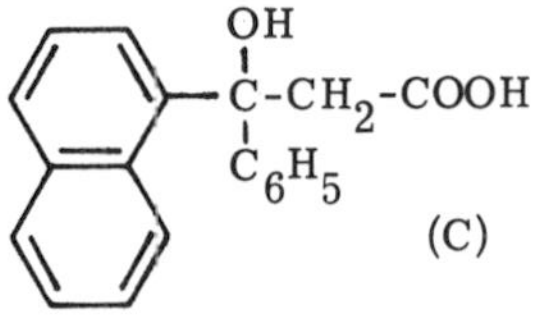

(C)

TABLE 8b (continued)

| | Substance resolved | Resolving agent and solvent | Rotation $[\alpha]_D$ of resolved compound | Reference |
|---|---|---|---|---|
| $C_{21}H_{23}NO_5$ | See (A) below | Brucine in EtOAc | $- 2.2^o$ (c 0.86, $CHCl_3$) | A 178 |
| $C_{22}H_{22}O_8$ | $\alpha$-Apopodophyllic acid See (B) below | Quinine in acetone | $- 158^o$ (c 0.59, $CHCl_3$) (t = $26^o$) | A 179 |
| $C_{24}H_{48}O_2$ | $CH_3-(CH_2)_{21}-CH(OH)-COOH$ | Brucine · 4 $H_2O$ in $CHCl_3$ | $+ 3.9^o$ (c 1.34, py) (t = $25^o$) | A 180 |

(A)

(B)

TABLE 8c Alkoxy and Aryloxy Monccarboxylic Acids

| Formula | Substance resolved | Resolving agent and solvent | Rotation $[\alpha]_D$ of resolved compound | Reference |
|---|---|---|---|---|
| $C_4H_6O_3$ | (epoxide: H, O, COOH; $CH_3$, H) | Brucine<br>in MeOH-acetone | - 82.5° (c 0.59, $C_6H_6$) (t = 25°) | A 181 |
| $C_6H_9O_3F_3$ | See Table 10a under $C_3H_5OF_3$ | | | |
| $C_6H_{11}NO_3$ | (Isopropylideneaminooxy)-propionic acid | See Table 8b | | |
| $C_7H_{14}O_3$ | $CH_3CH(OC(CH_3)_3)$-COOH | Brucine · 2 $H_2O$<br>in acetone | - 20.1° (c 3, EtOH) (t = 20°) | A 182 |
| $C_8H_{14}O_3$ | (cyclohexane: COOH, $OCH_3$) | Brucine<br>in acetone<br>Partial resolution | - 43.3° (or -43.8° - unclear)<br>(c 0.99, $Et_2O$) | A 183 |
| $C_9H_6O_3Cl_4$<br>$C_9H_7O_3Cl_3$ | See under $C_9H_{10}O_3$<br>See under $C_9H_{10}O_3$ | | | |
| $C_9H_8O_3$ | (epoxide: H, O, COOH; $C_6H_5$, H) | (-)- α-Methylbenzylamine<br>(+)- α-Methylbenzylamine<br>both in $Et_2O$ | a) - 154.0° (c 1.03, $H_2O$) (t = 25°)<br>b) + 153.3° (c 1.10, $H_2O$) (t = 25°)<br>both on the $NH_4^+$ salt | A 184 |
| $C_9H_8O_3$ | Hydrocoumarilic acid | See Table 8a | | |
| $C_9H_8O_3Cl_2$<br>$C_9H_8NO_5Cl$<br>$C_9H_9O_3Br$<br>$C_9H_9O_3Cl$<br>$C_9H_9O_3I$<br>$C_9H_9NO_5$ | See under $C_9H_{10}O_3$ | | | |

TABLE 8c (continued)

| | Substance resolved | Resolving agent and solvent | Rotation $[\alpha]_D$ of resolved compound | Reference |
|---|---|---|---|---|
| $C_9H_{10}O_3$ | $C_6H_5$-CH(OCH$_3$)-COOH<br>O-Methylmandelic acid | (-)-Ephedrine in EtOH | a) - 164° (c 1.5, $H_2O$) (t = 25°)<br>- 150.7° (c 0.754, EtOH) (t = 20°)<br>b) + 150.0° (c 0.494, EtOH) (t = 20°) | A 185<br>A 186<br>A 186 |
| $C_9H_{10}O_3$ | $C_6H_5$-OCH(CH$_3$)-COOH | Dehydroabietylamine in EtOH-$H_2O$ or MeOH-$H_2O$ | + 40.0° (c 1, EtOH) (t = 23°) | A 187, A 188 |
| | | Yohimbine in EtOH-$H_2O$ | a) + 39.8° (c 1, EtOH) (t = 25°)<br>b) - 39.5° (c 1, EtOH) (t = 23.5°) | A 189, A 190 |
| | | Also resolved with L-lysine | | A 191 |
| | Substituted $\alpha$-Phenoxypropionic Acids | | | |
| $C_9H_6O_3Cl_4$ | 2,3,4,6-Tetrachloro | Cinchonidine in $C_6H_6$ | a) + 26.4° (c 1.478, acetone) (t = 25°)<br>b) - 26.5° (c 1.602, acetone) (t = 25°) | A 192 |
| $C_9H_7O_3Cl_3$ | 2,4,6-Trichloro | Strychnine in EtOH | a) + 18.1° (c 1.19, EtOH) (t = 25°) | A 193 |
| | | (-)-Amphetamine in acetone-$H_2O$ | b) - 17.7° (c 0.96, EtOH) (t = 25°) | A 193 |
| $C_9H_7O_3Cl_3$ | 3,4,5-Trichloro | Cinchonidine in EtOH-$H_2O$ | a) - 39.9° (c 2.053, EtOH) (t = 25°) | A 194 |
| | | Dehydroabietylamine in MeOH-$H_2O$ | b) + 39.7° (c 2.285, EtOH) (t = 25°) | A 194 |

TABLE 8c (continued)

| Formula | Substance resolved | Resolving agent and solvent | Rotation $[\alpha]_D$ of resolved compound | Reference |
|---|---|---|---|---|
| | Substituted α-Phenoxypropionic Acids (continued) | | | |
| $C_9H_8O_3Cl_2$ | 2,3-Dichloro | Strychnine in EtOAc | a) + 5.1° (acetone) (t = 25°)<br>- 49.9° (c 1.834, $H_2O$, neutralized) (t = 25°)<br>b) + 49.7° (c 1.821, $H_2O$, neutralized) (t = 25°) | A 195 |
| $C_9H_8O_3Cl_2$ | 2,4-Dichloro | Quinine in acetone-$H_2O$ | a) + 28.1° (c 1.004, EtOH) (t = 25°)<br>b) - 28.5° (c 0.978, EtOH) (t = 25°) | A 196 |
| $C_9H_8O_3Cl_2$ | 2,5-Dichloro | Strychnine in acetone-$H_2O$ | a) - 61.3° (c 2.080, acetone) (t = 25°)<br>b) + 61.6° (c 1.298, acetone) (t = 25°) | A 195 |
| $C_9H_8O_3Cl_2$ | 2,6-Dichloro | Strychnine in acetone-$H_2O$ | a) + 23.4° (c 2.518, acetone) (t = 25°)<br>b) - 23.5° (c 2.572, acetone) (t = 25°) | A 195 |
| $C_9H_8O_3Cl_2$ | 3,4-Dichloro | Brucine in EtOH-$H_2O$<br>Strychnine in acetone-$H_2O$ | a) - 39.3° (c 0.996, EtOH) (t = 25°)<br>b) + 39.1° (c 1.063, EtOH) (t = 25°) | A 196<br>A 196 |
| $C_9H_8O_3Cl_2$ | 3,5-Dichloro | Quinine in EtOAc<br>Cinchonidine in acetone-$H_2O$ | a) + 55.3° (c 1.800, acetone) (t = 25°)<br>b) - 55.4° (c 1.809, acetone) (t = 25°) | A 195<br>A 195 |
| $C_9H_8NO_5Cl$ | 2-Chloro-4-nitro | Cinchonine in EtOH-$H_2O$ (3:2) | a) - 49.7° (c 1.520, acetone) (t = 25°)<br>b) + 49.5° (c 1.623, acetone) (t = 25°) | A 197 |
| $C_9H_8NO_5Cl$ | 4-Chloro-2-nitro | Strychnine<br>Cinchonine<br>both in EtOH-$H_2O$ (3:1) | a) - 100.7° (c 1.236, acetone) (t = 25°)<br>b) + 100.5° (c 1.052, acetone) (t = 25°) | A 197<br>A 197 |

TABLE 8c (continued)

| Formula | Substance resolved | Resolving agent and solvent | Rotation $[\alpha]_D$ of resolved compound | Reference |
|---|---|---|---|---|
| | Substituted $\alpha$-Phenoxypropionic Acids (continued) | | | |
| $C_9H_9O_3Br$ | 2-Bromo | Strychnine in EtOH-$H_2O$ (1:1) | a) - 16.9° (c 2.520, acetone) (t = 25°) | A 198 |
| | | Dehydroabietylamine acetate in EtOH | b) + 17.0° (c 2.507, acetone) (t = 25°) | A 198 |
| $C_9H_9O_3Br$ | 3-Bromo | Strychnine<br>(-)-Amphetamine<br>both in EtOH-$H_2O$ | a) - 46.4° (c 2.003, acetone) (t = 25°)<br>b) + 46.4° (c 1.895, acetone) (t = 25°) | A 199<br>A 199 |
| $C_9H_9O_3Br$ | 4-Bromo | Brucine in MeOH-$H_2O$ | a) - 50.5° (c 2.144, acetone) (t = 25°)<br>b) + 50.4° (c 2.006, acetone) (t = 25°) | A 200 |
| $C_9H_9O_3Cl$ | 2-Chloro | Strychnine in EtOH-$H_2O$ | a) - 30.1° (c 2.036, acetone) (t = 25°)<br>b) + 30.2° (c 2.036, acetone) (t = 25°) | A 201 |
| $C_9H_9O_3Cl$ | 3-Chloro | Strychnine<br>(-)-Amphetamine<br>both in EtOH-$H_2O$ | a) - 54.4° (c 2.050, acetone) (t = 25°)<br>b) + 54.5° (c 2.061, acetone) (t = 25°) | A 201<br>A 201 |
| $C_9H_9O_3Cl$ | 4-Chloro | Brucine<br>(-)-Amphetamine<br>both in EtOH-$H_2O$ | a) - 40.1° (c 1.0, EtOH) (t = 25°)<br>b) + 41.0° (c 0.88, EtOH) (t = 25°) | A 202<br>A 202 |
| $C_9H_9O_3I$ | 2-Iodo | Strychnine in EtOH-$H_2O$ (1:1) | a) + 40.8° (c 2.280, dil. $NH_3$) (t = 25°)<br>b) - 41.0° (c 2.175, $H_2O$) (t = 25°) | A 203 |
| $C_9H_9O_3I$ | 3-Iodo | Cinchonidine in MeOH-$H_2O$ (4:1) | a) - 39.8° (c 2.069, acetone) (t = 25°)<br>b) + 39.9° (c 2.001, acetone) (t = 25°) | A 203 |

TABLE 8c (continued)

| Formula | Substance resolved | Resolving agent and solvent | Rotation $[\alpha]_D$ of resolved compound | Reference |
|---|---|---|---|---|
| | Substituted $\alpha$-Phenoxypropionic Acids (continued) | | | |
| $C_9H_9O_3I$ | 4-Iodo | Brucine · 4 $H_2O$ in acetone-$H_2O$ | a) - 46.4° (c 1.921, acetone) (t = 25°) | A 203 |
| | | (-)-Amphetamine in EtOH-$H_2O$ | b) + 46.7° (c 1.842, acetone) (t = 25°) | A 203 |
| $C_9H_9NO_5$ | 3-Nitro | Strychnine | a) - 51.68° (c 2.202, EtOH) (t = 25°) | A 204 |
| | | (+)-Amphetamine both in EtOH-$H_2O$ | b) + 51.65° (c 2.210, EtOH) (t = 25°) | A 204 |
| $C_{10}H_{10}O_4Cl_2$ | 4,5-Dichloro-2-methoxy | Cinchonidine in EtOH-$H_2O$ | a) - 60.5° (c 1.243, acetone) (t = 25°)<br>b) + 60.4° (c 1.015, acetone) (t = 25°) | A 205 |
| $C_{10}H_{10}O_4Cl_2$ | 2,4-Dichloro-5-methoxy | (-)-$\alpha$-Methylbenzylamine in 96% EtOH | a) - 52.8° (c 1.125, acetone) (t = 25°) | A 206 |
| | | (+)-$\alpha$-Methylbenzylamine in EtOH | b) + 52.9° (c 1.065, acetone) (t = 25°) | A 206 |
| $C_{10}H_{10}O_4Cl_2$ | 2,5-Dichloro-4-methoxy | Strychnine in EtOH-$H_2O$ (3:1) | a) - 56.5° (c 1.118, acetone) (t = 25°) | A 206 |
| | | (-)-$\alpha$-Methylbenzylamine in 96% EtOH | b) + 56.6° (c 1.037, acetone) (t = 25°) | A 206 |
| $C_{10}H_{11}O_3Cl$ | 2-Chloro-4-methyl | Cinchonidine in EtOH-$H_2O$ | a) - 37.6° (c 2.077, acetone) (t = 25°) | A 207 |
| | | Dehydroabietylamine in EtOH | b) + 37.5° (c 2.053, acetone) (t = 25°) | A 207 |

TABLE 8c (continued)

| Formula | Substance resolved | Resolving agent and solvent | Rotation $[\alpha]_D$ of resolved compound | Reference |
|---|---|---|---|---|
| | Substituted $\alpha$-Phenoxypropionic Acids (continued) | | | |
| $C_{10}H_{11}O_3Cl$ | 2-Chloro-5-methyl | Cinchonidine in EtOAc<br>Strychnine in MeOH-$H_2O$ | a) $-43.6^o$ (c 0.932, acetone) (t = $25^o$)<br>b) $+43.4^o$ (c 0.869, acetone) (t = $25^o$) | A 208<br>A 208 |
| $C_{10}H_{11}O_3Cl$ | 3-Chloro-4-methyl | (-)-Amphetamine in EtOH-$H_2O$ (1:1)<br>Quinidine in EtOH-$H_2O$ | a) $+53.4^o$ (c 2.084, acetone) (t = $25^o$)<br>b) $-53.1^o$ (c 2.061, acetone) (t = $25^o$) | A 209<br>A 209 |
| $C_{10}H_{11}O_3Cl$ | 4-Chloro-2-methyl | (+)-$\alpha$-Methylbenzylamine<br>(-)-$\alpha$-Methylbenzylamine<br>both in EtOH-$H_2O$ | a) $+19.0^o$ (c 0.92, EtOH) (t = $25^o$)<br>b) $-18.8^o$ (c 0.87, EtOH) (t = $25^o$) | A 210<br>A 210 |
| $C_{10}H_{11}O_3Cl$ | 4-Chloro-3-methyl | Brucine · 4 $H_2O$ in EtOH-$H_2O$<br>Morphine in acetone-$H_2O$ | a) $-50.0^o$ (c 1.128, acetone) (t = $25^o$)<br>b) $+50.4^o$ (c 1.143, acetone) (t = $25^o$) | A 209<br>A 209 |
| $C_{10}H_{11}O_3Cl$ | 5-Chloro-2-methyl | (-)-Ephedrine in EtOH-$H_2O$ (2:3)<br>Strychnine in acetone-$H_2O$ | a) $+62.1^o$ (c 1.899, acetone) (t = $25^o$)<br>b) $-62.2^o$ (c 1.096, acetone) (t = $25^o$) | A 208<br>A 208 |
| $C_{10}H_{12}O_3$ | 2-Methyl | Strychnine in EtOH-$H_2O$ (1:1) | a) $-23.0^o$ (c 2.613, acetone) (t = $25^o$)<br>b) $+22.7^o$ (c 2.544, acetone) (t = $25^o$) | A 198 |

TABLE 8c (continued)

| Formula | Substance resolved | Resolving agent and solvent | Rotation $[\alpha]_D$ of resolved compound | Reference |
|---|---|---|---|---|
| | Substituted $\alpha$-Phenoxypropionic Acids (continued) | | | |
| $C_{10}H_{12}O_3$ | 3-Methyl | Cinchonine in EtOH-$H_2O$ | a) $- 47.4^o$ (c 1.99, acetone) (t = $25^o$) | A 199 |
| | | (-)-Amphetamine in $H_2O$ | b) $+ 47.3^o$ (c 1.039, acetone) (t = $25^o$) | A 199 |
| $C_{10}H_{12}O_3$ | 4-Methyl | Brucine · 4 $H_2O$ in EtOH | a) $- 55.3^o$ (c 2.034, acetone) (t = $25^o$) | A 211 |
| | | Quinine in EtOH-$H_2O$ | b) $+ 55.2^o$ (c 2.122, acetone) (t = $25^o$) | A 211 |
| $C_{10}H_{12}O_4$ | 2-Methoxy | Strychnine in EtOH-$H_2O$ (1:1) | a) $- 52.8^o$ (c 2.226, acetone) (t = $25^o$)<br>b) $+ 52.4^o$ (c 2.067, acetone) (t = $25^o$) | A 212 |
| $C_{10}H_{12}O_4$ | 3-Methoxy | (-)-Amphetamine in EtOH-$H_2O$ | a) $+ 34.6^o$ (c 2.063, acetone) (t = $25^o$)<br>b) $- 34.4^o$ (c 2.144, acetone) (t = $25^o$) | A 212 |
| $C_{10}H_{12}O_4$ | 4-Methoxy | Brucine<br>Cinchonidine<br>both in EtOH-$H_2O$ | a) $- 58.9^o$ (c 2.052, acetone) (t = $25^o$)<br>b) $+ 59.0^o$ (c 2.032, acetone) (t = $25^o$) | A 213<br>A 213 |
| $C_{11}H_{12}O_5$ | 4-Formyl-2-methoxy | Cinchonine in $H_2O$ | a) $- 32.5^o$ (c 1.008, EtOH) (t = $15^o$)<br>b) $+ 33.0^o$ (c 0.912, EtOH) (t = $15^o$) | A 214 |
| $C_{11}H_{14}O_3$ | 2,3-Dimethyl | Cinchonidine<br>Quinidine<br>both in EtOH-$H_2O$ | a) $+ 5.2^o$ (c 2.607, acetone) (t = $25^o$)<br>b) $- 5.4^o$ (c 2.030, acetone) (t = $25^o$) | A 215<br>A 215 |

TABLE 8c (continued)

| Formula | Substance resolved | Resolving agent and solvent | Rotation $[\alpha]_D$ of resolved compound | Reference |
|---|---|---|---|---|
| | Substituted $\alpha$-Phenoxypropionic Acids (continued) | | | |
| $C_{11}H_{14}O_3$ | 2,4-Dimethyl | Dehydroabietylamine<br>Quinidine<br>both in EtOH-$H_2O$ | a) + 28.5° (c 2.032, acetone) (t = 25°)<br>b) - 28.7° (c 2.162, acetone) (t = 25°) | A 207<br>A 207 |
| $C_{11}H_{14}O_3$ | 2,5-Dimethyl | (-)-Amphetamine<br>in EtOH-$H_2O$ (1 : 1)<br>Strychnine<br>in acetone-$H_2O$ | a) - 43.4° (c 2.112, acetone) (t = 25°)<br>b) + 43.5° (c 2.072, acetone) (t = 25°) | A 216<br>A 216 |
| $C_{11}H_{14}O_3$ | 2,6-Dimethyl | Strychnine<br>in EtOH-$H_2O$<br>(-)-Amphetamine<br>in 96 % EtOH | a) + 51.5° (c 2.527, acetone) (t = 25°)<br>b) - 51.9° (c 1.716, acetone) (t = 25°) | A 215<br>A 215 |
| $C_{11}H_{14}O_3$ | 3,4-Dimethyl | Dehydroabietylamine<br>in 96 % EtOH<br>Strychnine<br>in EtOH-$H_2O$ | a) + 47.3° (c 1.480, acetone) (t = 25°)<br>b) - 47.2° (c 1.574, acetone) (t = 25°) | A 217<br>A 217 |
| $C_{11}H_{14}O_3$ | 3,5-Dimethyl | Cinchonine<br>in EtOH-$H_2O$ (1:1)<br>Brucine<br>in $H_2O$ | a) - 44.3° (c 2.427, acetone) (t = 25°)<br>b) + 44.1° (c 2.243, acetone) (t = 25°) | A 218<br>A 218 |
| $C_{13}H_{17}O_3Cl$ | 4-Chloro-2-isopropyl-5-methyl | Cinchonidine<br>in acetone-$H_2O$<br>Morphine<br>in EtOH-$H_2O$ | a) - 22.9° (c 1.09, EtOH) (t = 25°)<br>b) + 23.3° (c 1.05, EtOH) (t = 25°) | A 193<br>A 193 |

TABLE 8c (continued)

| | Substance resolved | Resolving agent and solvent | Rotation $[\alpha]_D$ of resolved compound | Reference |
|---|---|---|---|---|
| | Substituted α-Phenoxypropionic Acids (continued) | | | |
| $C_{13}H_{18}O_3$ | 3-tert-butyl | Brucine<br>Cinchonidine<br>both in EtOH-$H_2O$ | a) - 35.3° (c 1.9541, acetone) (t = 25°)<br>b) + 35.3° (c 1.9845, acetone) (t = 25°) | A 219<br>A 219 |
| $C_{13}H_{18}O_3$ | 4-tert-butyl | Strychnine<br>Dehydroabietylamine<br>both in EtOH-$H_2O$ | a) - 49.9° (c 2.4461, acetone) (t = 25°)<br>b) + 50.0° (c 2.4132, acetone) (t = 25°) | A 219<br>A 219 |
| $C_{10}H_9O_3F_3$ | See Table 10c under $C_8H_7OF_3$ | | | |
| $C_{10}H_9O_3F_3$ | $C_6H_5$-C($CF_3$)($OCH_3$)-COOH | (+)-α-Methylbenzylamine<br>(-)-α-Methylbenzylamine | a) + 68.5 ± 1.3° (c 1.49, MeOH) (t = 25°)<br>b) - 71.8 ± 0.6° (c 3.28, MeOH) (t = 24°) | A 150 |
| $C_{10}H_{10}O_3Cl_2$ | 2,4-$Cl_2C_6H_3$-OCH($CH_2CH_3$)-COOH | Strychnine<br>(-)-α-Methylbenzylamine<br>both in EtOH-$H_2O$ | a) + 41.8° (c 2.052, EtOH) (t = 25°)<br>b) - 41.4° (c 0.14, EtOH) (t = 25°) | A 210<br>A 210 |
| $C_{10}H_{12}O_3$ | $C_6H_5$-C($CH_3$)($OCH_3$)-COOH | Brucine<br>in acetone<br><br>Resolution with quinine | a) - 32.5° (neat)<br>b) + 31.5° (neat) | A 220<br><br><br>A 221 |
| $C_{10}H_{12}O_3$ | R-$C_6H_4$-OCH($CH_2CH_3$)-COOH<br>R = H | Strychnine<br>Morphine<br>both in EtOH-$H_2O$<br><br>Resolution with L-lysine | a) - 51.4° (c 0.810, EtOH) (t = 25°)<br>b) + 51.2° (c 1.077, EtOH) (t = 25°) | A 222<br>A 222<br><br><br>A 191 |

TABLE 8c (continued)

| | Substance resolved | Resolving agent and solvent | Rotation $[\alpha]_D$ of resolved compound | Reference |
|---|---|---|---|---|
| $C_{11}H_{14}O_3$ | $C_{10}H_{12}O_3$ above, $R = OCH_3$ | Cinchonidine<br>Recr.: $EtOH-H_2O$ | $+64.76^o$ (c 4, EtOH) | A 30 |
| | | Also resolved with (+)-$\alpha$-methylbenzylamine | | A 30 |
| $C_{11}H_{14}O_3$ | $CH_3O-C_6H_4-CH(CH_3)CH_2-COOH$ | Quinine<br>in $CHCl_3$-Pet E | a) $+34^o$ (c 1.4 %, EtOH) (t = $32^o$)<br>b) $-12^o$ (c 1.2 %, EtOH) (t = $25^o$) | A 223 |
| $C_{11}H_{14}O_3$ | $C_6H_5-C(CH_3)(OCH_3)CH_2-COOH$ | Cinchonidine<br>in acetone | $+35.2^o$ (c 0.9, EtOH) | A 224 |
| $C_{11}H_{14}O_3$ | $C_6H_5-OCH(CH_2CH_2CH_3)-COOH$ | (-)-Amphetamine<br>(+)-Amphetamine<br>both in $EtOH-H_2O$ | a) $+45.5^o$ (c 1.05, EtOH) (t = $25^o$)<br>b) $-46.0^o$ (c 0.82, EtOH) (t = $25^o$) | A 225<br>A 225 |
| $C_{12}H_{16}O_3$ | $C_6H_5-OCH(C_4H_9\text{-}\underline{n})-COOH$ | (-)-Amphetamine<br>(+)-Amphetamine<br>both in $EtOH-H_2O$ | a) $+37.6^o$ (c 0.95, EtOH) (t = $25^o$)<br>b) $-37.5^o$ (c 0.87, EtOH) (t = $25^o$) | A 226<br>A 226 |
| $C_{12}H_{16}O_4$ | $(CH_3O)_2C_6H_3-CH_2CH(CH_3)-COOH$ | Quinine<br>in EtOH | a) $+27.5^o$ (c 4.01, $CHCl_3$) (t = $21^o$)<br>b) $-28.1^o$ (c 4.16, $CHCl_3$) (t = $21^o$) | A 45 |
| $C_{13}H_{11}O_3Cl$ | 1-X-$C_{10}H_5$-2-$OCH(R)-COOH$<br>$R=CH_3$<br>X=Cl | Cinchonidine<br>Morphine<br>in $EtOH-H_2O$ | a) $+19.0^o$ (c 0.79, EtOH) (t = $25^o$)<br>b) $-18.7^o$ (c 0.83, EtOH) (t = $25^o$) | A 227<br>A 227 |

TABLE 8c (continued)

| | Substance resolved | Resolving agent and solvent | Rotation $[\alpha]_D$ of resolved compound | Reference |
|---|---|---|---|---|
| $C_{13}H_{12}O_3$ | R' <br> OCH-COOH (1-naphthyl) <br> R' = $CH_3$ | Cinchonine <br> in EtOH-$H_2O$ | a) - 49.4° (c 2.265, EtOH) (t = 20°) <br> b) + 49.6° (c 2.380, EtOH) (t = 20°) | A 228, A 229 |
| $C_{13}H_{12}O_3$ | See $C_{13}H_{11}O_3Cl$ above <br> R = $CH_3$, X = H | Cinchonine <br> in EtOH-$H_2O$ | a) - 93.3° (c 2.122, EtOH) (t = 20°) <br> b) + 93.4° (c 2.102, EtOH) (t = 20°) | A 228 |
| $C_{14}H_{14}O_3$ | See $C_{13}H_{12}O_3$ above <br> R' = $CH_3CH_2$ | Cinchonine <br> in EtOH-$H_2O$ | + 6.6° (c 0.9135, EtOH) (t = 25°) | A 222 |
| $C_{14}H_{14}O_3$ | See $C_{13}H_{11}O_3Cl$ above <br> R = $CH_3CH_2$, X = H | Quinine <br> (+)-α-Methylbenzylamine <br> both in EtOH-$H_2O$ | a) + 91.8° (c 1.012, EtOH) (t = 25°) <br> b) - 90.8° (c 1.045, EtOH) (t = 25°) | A 222 <br> A 222 |
| $C_{15}H_{16}O_3$ | See $C_{13}H_{11}O_3Cl$ above <br> R = $CH_3CH_2CH_2$, X = H | (-)-Amphetamine <br> (+)-Amphetamine <br> both in EtOH-$H_2O$ | a) + 73.6° (c 0.999, EtOH) (t = 25°) <br> b) - 73.3° (c 1.06, EtOH) (t = 25°) | A 225 |
| $C_{16}H_{16}O_3$ | CH- <br> O <br> $CH_3$ $CH_2$ <br> COOH | (-)-Ephedrine <br> (+)-Ephedrine <br> both in xylene-Pet E-$Et_2O$ | a) + 16.1° (EtOH) <br> b) - 14.6° (EtOH) | A 230 |
| $C_{16}H_{18}O_3$ | See $C_{13}H_{11}O_3Cl$ above <br> R = n-$C_4H_9$ X = H | (-)-α-Methylbenzylamine <br> (+)-α-Methylbenzylamine <br> both in EtOH-$H_2O$ | a) + 58.6° (c 0.94, EtOH) (t = 25°) <br> b) - 58.5° (c 0.97, EtOH) (t = 25°) | A 226 |
| $C_{17}H_{16}O_3$ | $C_6H_5$ COOH <br> p-$CH_3O$-$C_6H_4$ H | Brucine <br> Recr.: MeOH | - 176° ($CHCl_3$) (t = 22°) | A 41 |

TABLE 8c (continued)

| | Substance resolved | Resolving agent and solvent | Rotation $[\alpha]_D$ of resolved compound | Reference |
|---|---|---|---|---|
| | $CH_3O-C_6H_4-CH(COOH)-CH(R)-C_6H_4-OCH_3$ | | | |
| $C_{18}H_{20}O_4$ | erythro R = $CH_3$ | Quinine<br>(-)-Ephedrine<br>both in EtOH | a) + 29.9° (c 1.13, dioxane)<br>b) - 28.1° (c 1.21, dioxane) | A 231<br>A 231 |
| $C_{19}H_{22}O_4$ | erythro R = $CH_3CH_2$ | Quinine<br>(-)-Ephedrine<br>both in EtOH | a) + 22.7° (c 1.27, dioxane)<br>b) - 22.4° (c 1.31, dioxane) | A 231<br>A 231 |
| $C_{21}H_{29}O_4$ | [structure: $CH_3O$, $CH_3O$, $C_5H_{11}$, H, $CH_3$, COOH] | (+)-α-(1-Naphthyl)-ethylamine<br>(-)-α-(1-Naphthyl)-ethylamine<br>Recr.: hexane | a) - 63° (c 1.00, $CHCl_3$) (t = 24°)<br>b) + 63° (c 1.07, $CHCl_3$) (t = 24°) | A 232 |
| $C_{27}H_{24}O_3$ | See Table 10c under $C_{24}H_{20}O$ | | | |

**TABLE 8d** Di- and Polycarboxylic Acids and their Derivatives

| | Substance resolved | Resolving agent and solvent | Rotation $[\alpha]_D$ of resolved compound | Reference |
|---|---|---|---|---|
| $C_4H_4O_5$ | 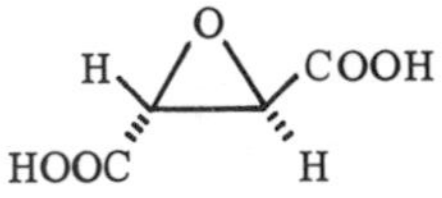 (epoxide structure: H, HOOC / O / COOH, H) | (-)-Ephedrine<br>in MeOH-acetone | + 117.8° (c 2.6, EtOH) (t = 25°) | A 233 |
| $C_4H_6O_6$ | HOOC-CH(OH)-CH(OH)-COOH | 2-(D-Glycero-D-gulo-hexahydrohexyl)-benzimidazole<br>in EtOH-$H_2O$ (1:3) | - 14° (c 4, $H_2O$) (t = 20°) | A 234a |
| | | Resolution with cinchonine | | A 234b |
| | | See also Table 3, ref. 35 for resolution with deoxyephedrine | | |
| $C_5H_4O_4$ | 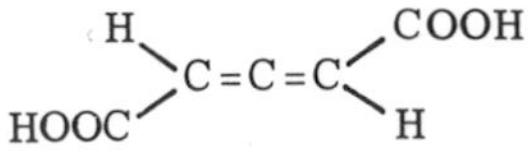 H(HOOC)C=C=C(COOH)H<br>Glutinic acid | Quinine<br>in $Et_2O$<br>Partial resolution | - 104° (MeOH) | A 235 |
| $C_5H_6O_4$ | 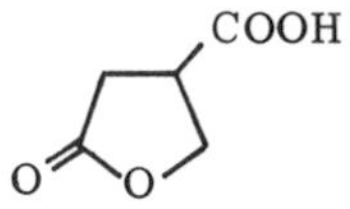 (lactone ring structure with COOH) | (-)-α-Methylbenzylamine<br>(+)-α-Methylbenzylamine<br>both in MeOH-$Et_2O$ (1:2) | a) + 49 ± 0.5° (c 1.1, MeOH) (t = 20°)<br>b) - 49° (MeOH) | A 236 |
| $C_5H_7O_5Cl$ | HOOC-C(OH)(CH$_2$Cl)CH$_2$-COOH | Brucine<br>in MeOH-$H_2O$ | - 44.9° ($H_2O$) (t = 25°) | A 237 |
| $C_5H_8O_4$ | HOOC-CH(CH$_3$)CH$_2$-COOH | Quinine<br>in EtOH-$H_2O$ | a) + 12° (c 2.4, EtOH) | A 236 |
| | | Strychnine<br>Recr.: $H_2O$ | + 9.28° (c 6.0%, $H_2O$) (t = 25°)<br>b) - 9.35° (c 6.0%, $H_2O$) (t = 25°) | A 238<br>A 238 |

TABLE 8d (continued)

| | Substance resolved | Resolving agent and solvent | Rotation $[\alpha]_D$ of resolved compound | Reference |
|---|---|---|---|---|
| $C_5H_8O_6$ | HOOC-C(OH)(CH₃)—CH(OH)-COOH | | | |
| | Isomer, m.p. 146° | Brucine in $H_2O$ | a) + 8.90° (c 3.7, $H_2O$) (t = 20°)<br>b) - 8.94° (c 3.7, $H_2O$) (t = 20°) | A 239 |
| | Isomer, m.p. 161° | Brucine in $H_2O$ | a) + 5.88° (c 3.2, $H_2O$) (t = 20°)<br>b) - 5.60° (c 3:2, $H_2O$) (t = 20°) | A 239 |
| $C_5H_9NO_3$ | $H_2NC(=O)-CH_2CH(CH_3)-COOH$ | Brucine in MeOH | - 21.4 ± 0.5° (c 2.088, EtOH) (t = 25°) | A 240 |
| $C_6H_8O_4$ | O= (ring) O, $CH_3$, COOH | Quinine in EtOH | + 15.39° (c 2.5, $H_2O$) (t = 28°) | A 241 |
| $C_6H_{10}O_4$ | $CH_3OC(=O)-CH(CH_3)CH_2-COOH$ | Quinine<br>Cinchonidine<br>both in acetone-$H_2O$ | a) $\alpha_D^{25}$ - 9.98° (l 1 dm, neat)<br>b) $\alpha_D^{25}$ + 9.98° (l 1 dm, neat)<br>+ 9.7 ± 0.3° (c 6.98, $CHCl_3$) (t = 25°) | A 242<br>A 242 |
| $C_6H_{10}O_6$ | HOOC-C(OH)(CH₃)—C(CH₃)(OH)-COOH | Brucine · 2 $H_2O$ in $H_2O$ | a) + 13.4° (c 4.0, $H_2O$) (t = 20°)<br>b) - 13.4° (c 4.0, $H_2O$) (t = 20°) | A 239 |
| $C_7H_{10}O_4$ | $H_3C$, $CH_3$, H, COOH, HOOC, H<br>trans-Caronic acid | (-)-Ephedrine in acetone-$H_2O$<br>Quinine in 96 % EtOH | a) - 37.5° (c 4.592, EtOH) (t = 25°) (546 nm)<br>b) + 37.6° (c 3.520, EtOH) (t = 25°) (546 nm) | A 243<br>A 243 |

TABLE 8d (continued)

| | Substance resolved | Resolving agent and solvent | Rotation $[\alpha]_D$ of resolved compound | Reference |
|---|---|---|---|---|
| $C_7H_{10}O_4Br_2$ | HOOC-CH(Br)(CH$_2$)$_3$CH(Br)-COOH | Cinchonidine<br>Brucine<br>both in acetone-$H_2O$ | a) - 56° (c 0.385, EtOH) (t = 25°)<br>b) + 57° (c 0.662, EtOH) (t = 25°) | A 244<br>A 244 |
| $C_7H_{12}O_4$ | $CH_3OC(=O)$-$CH_2CH(CH_3)CH_2$-COOH | Cinchonidine<br>Quinine<br>both in MeOH-$H_2O$ | a) $\alpha_D^{17.7}$ + 0.65° (l 1 dm, neat)<br>b) $\alpha_D^{18.8}$ - 0.63° (l 1 dm, neat) | A 245<br>A 245 |
| $C_7H_{12}O_5$ | HOOC-$CH_2CH_2CH(OCH_3)CH_2$-COOH | Strychnine<br>in $H_2O$ | a) - 14.5° ± 1.5° (c 1.792, $CHCl_3$) (t = 18°)<br>b) + 13.1° ± 1° (c 2.203, $CHCl_3$) (t = 17°) | A 246 |
| $C_8H_8O_5Br_2$ | 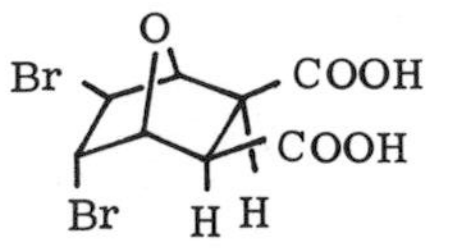 Br, O, COOH, COOH, Br, H H | Quinine<br>Partial resolution | - 77.5° | A 247 |
| $C_8H_{12}O_4$ | 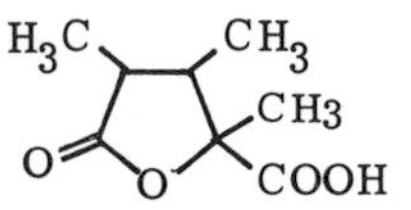 $H_3C$, $CH_3$, $CH_3$, COOH, O, O | Brucine<br>in $C_6H_6$-EtOH | + 60.00° (c 1.0, EtOH) (t = 17°) | A 241 |
| $C_8H_{12}O_4$ | 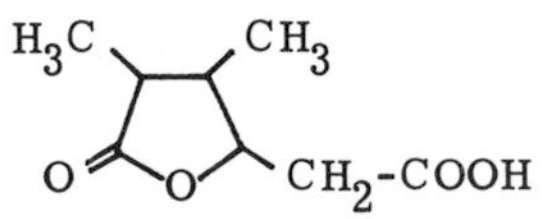 $H_3C$, $CH_3$, $CH_2$-COOH, O, O | Strychnine<br>in $H_2O$ | a) - 33.5° (c 1.2, EtOH) (t = 30°)<br>b) + 22.2° (c 3.0, EtOH) (t = 30°) | A 248 |
| $C_8H_{12}O_4$ | 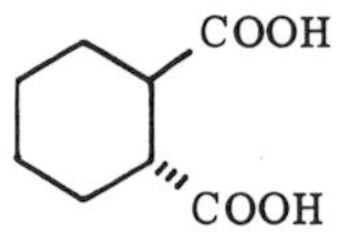 COOH, COOH | Quinine<br>in 95 % EtOH | + 22.3° (c 5.3, acetone) (t = 30°) | A 249 |
| $C_8H_{12}O_5$ | 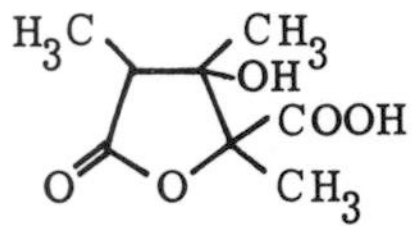 $H_3C$, $CH_3$, OH, COOH, $CH_3$, O, O — Isomer I | Brucine | - 5.04 ± 0.5° (c 1.25, EtOH) (t = 28°) | A 250 |
| | Isomer II | Brucine<br>both in EtOH | - 60.8 ± 1° (c 0.24, EtOH) (t = 25°) | A 250 |

TABLE 8d (continued)

| | Substance resolved | Resolving agent and solvent | Rotation $[\alpha]_D$ of resolved compound | Reference |
|---|---|---|---|---|
| $C_8H_{12}O_6$ | $CH_3OOC\text{-}CH_2CH(OCOCH_3)CH_2\text{-}COOH$ | Cinchonidine in EtOAc-$Et_2O$<br>Strychnine in $CHCl_3$ - $Et_2O$ | a) + 5.28 ± 0.03° (neat) (t = 25°)<br>b) - 5.27 ± 0.05° (neat) (t = 21°) | A 251<br>A 251 |
| $C_8H_{12}O_6$ | cyclohexane-1,2-diol-1,2-dicarboxylic acid (ring bearing OH and COOH on each of two adjacent carbons) | Brucine in $H_2O$ | a) + 31.0° (c 2, $H_2O$) (t = 20°)<br>b) - 30.5° (c 2, $H_2O$) (t = 20°) | A 252 |
| $C_8H_{14}O_4$ | $CH_3CH_2OOC\text{-}C(CH_3)(CH_2CH_3)\text{-}COOH$ | Quinine in EtOH - $H_2O$<br>Cinchonidine in acetone | a) + 3.38° (c 15.0, $CHCl_3$) (t = 18°)<br>b) - 3.47° (c 5.03, $CHCl_3$) (t = 25°) | A 253<br>A 253 |
| $C_8H_{14}O_4$ | $CH_3OOC\text{-}C(CH_3)(CH_2CH_3)CH_2\text{-}COOH$ | Quinine in acetone-$H_2O$ | - 11.50° (neat) (t = 23°) | A 254 |
| $C_8H_{14}O_4$ | $HOOC\text{-}CH(\underline{n}\text{-}C_4H_9)CH_2\text{-}COOH$ | Strychnine<br>Quinine<br>both in EtOH-$H_2O$ | a) + 22.6° (c 5.14, $H_2O$) (t = 25°)<br>b) - 22.5° (c 5.15, $H_2O$) (t = 25°) | A 255<br>A 255 |
| $C_8H_{14}O_4$ | $HOOC\text{-}CH(CH_2CH(CH_3)_2)CH_2\text{-}COOH$ | Strychnine<br>Quinidine<br>both in EtOH-$H_2O$ | a) + 26.8° (c 5.008, EtOH) (t = 25°)<br>b) - 27.0° (c 5.009, EtOH) (t = 25°) | A 255<br>A 255 |
| $C_8H_{14}O_4$ | $HOOC\text{-}CH(CH(CH_3)_2)CH_2CH_2\text{-}COOH$ | Brucine in $H_2O$ | a) + 19.1° (c 4.936, EtOH) (t = 25°)<br>b) - 19.2° (c 4.272, EtOH) (t = 25°) | A 256 |

TABLE 8d (continued)

| Formula | Substance resolved | Resolving agent and solvent | Rotation $[\alpha]_D$ of resolved compound | Reference |
|---|---|---|---|---|
| $C_9H_{10}O_4$ | COOH, H, H, COOH | Brucine<br>Recr.: acetone-$H_2O$ | + 147° (acetone) (t = 26°) | A 257 |
| $C_9H_{12}O_4$ | H, HOOC, H, COOH<br>Fecht acid | Brucine<br>in $H_2O$ | $[M]_{546}$ + 9.9° (c 5.3, acetone) | A 258 |
| $C_9H_{14}O_4$ | $CH_3$ H $CH_2$-COOH, $CH_3$, H, COOH<br>trans-Caryophyllenic acid | Quinine methohydroxide<br>in $H_2O$ | a) - 28° (c 2.125, $C_6H_6$) (t = 20°)<br>b) + 28.5° (c 1.452, $C_6H_6$) (t = 20°) | A 259 |
| $C_9H_{14}O_4$ | $CH_3$ H COOH, $CH_3$, H, $CH_2$-COOH | Quinine<br>in EtOH-$H_2O$ | a) + 24.2° (c 1.115, acetone) (t = 21°)<br>b) - 24.6° (c 1.084, acetone) (t = 21°) | A 260 |
| $C_9H_{14}O_4$ | $CH_3$ H H, $CH_3$, COOH, $CH_2$-COOH | Quinine<br>in EtOH-$H_2O$ | a) - 11.1° (c 3.650, acetone) (t = 19°)<br>b) + 10.9° (c 3.960, acetone) (t = 20°) | A 260 |
| $C_9H_{14}O_4$ | $CH_2$-COOH, $CH_3$, $CH_3$, O, O<br>Terpenylic acid | (+)-α-(2-Naphthyl)-ethyl-amine in $H_2O$<br>(-)-α-(2-Naphthyl)-ethyl-amine in $Et_2O$ | a) + 56.3° (c 0.9, $H_2O$) (t = 25°)<br>b) - 56.5° (c 1.1, $H_2O$) (t = 25°) | A 261<br>A 261 |

TABLE 8d (continued)

| | Substance resolved | Resolving agent and solvent | Rotation $[\alpha]_D$ of resolved compound | Reference |
|---|---|---|---|---|
| $C_9H_{16}O_4$ | HOOC-CHCH$_2$-COOH (n-$C_5H_{11}$) | Strychnine in EtOH-$H_2O$ | a) + 26.7° (c 4.773, EtOH) (t = 25°) | A 262 |
| | | Brucine in $H_2O$ | b) - 26.3° (c 4.768, EtOH) (t = 25°) | A 262 |
| $C_{10}H_{10}O_4$ | HOOC-CHCH$_2$-COOH ($C_6H_5$) | Brucine in EtOH | a) + 171.1° (c 2.010, acetone) (t = 25°)<br>b) - 171.0° (c 2.047, acetone) (t = 25°) | A 263 |
| $C_{10}H_{10}O_4$ | 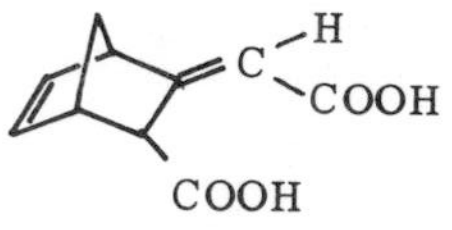 [structure: C(H)COOH, COOH] endo | Quinine in $CH_3OH$-EtOAc | a) - 122° (MeOH)<br>b) + 39.4 ± 0.2° (MeOH) | A 235 |
| | exo | Quinine in $CH_3OH$-EtOAc | - 81 ± 0.2° (MeOH) | A 235 |
| $C_{10}H_{14}O_4$ | 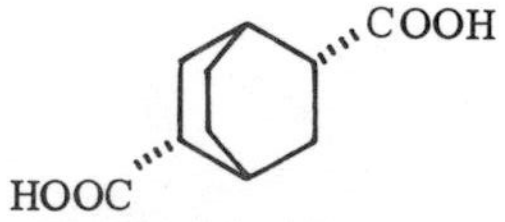 [structure: HOOC, COOH] | Brucine | - 125° (c 0.500, MeOH) (t = 25°) | A 264 |
| $C_{10}H_{14}O_4$ | 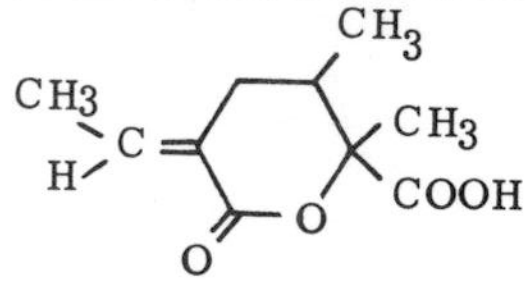 [structure: $CH_3$, C, H, $CH_3$, $CH_3$, COOH, O, O] | Brucine in EtOH | a) + 40° (c 0.94, EtOH) (t = 25°)<br>b) - 40° (c 1.20, EtOH) (t = 25°) | A 265 |
| $C_{10}H_{16}O_4$ | 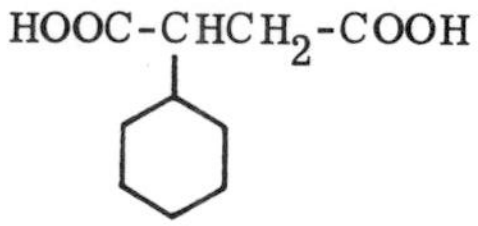 HOOC-CHCH$_2$-COOH (cyclohexyl) | Strychnine in EtOH-$H_2O$ | a) + 38.5° (c 2.597, acetone) (t = 25°) | A 263 |
| | | Quinine in MeOH-$H_2O$ | b) - 38.2° (c 2.576, acetone) (t = 25°) | A 263 |

TABLE 8d (continued)

| Formula | Substance resolved | Resolving agent and solvent | Rotation $[\alpha]_D$ of resolved compound | Reference |
|---|---|---|---|---|
| $C_{10}H_{18}O_4$ | $CH_3OOC-CH_2CH(CH_3)CH_2CH(CH_3)CH_2-COOH$ erythro | (+)-α-Methylbenzylamine<br>(-)-α-Methylbenzylamine<br>both in $Et_2O$-Pet E | a) - 1.96° (neat) (t = 22°)<br>b) + 1.96° (neat) (t = 22°) | A 266<br>A 266 |
| | threo | Cinchonidine in acetone-$H_2O$ | - 25.0° (neat) (t = 25°) | A 266 |
| $C_{11}H_9O_5Cl$ | 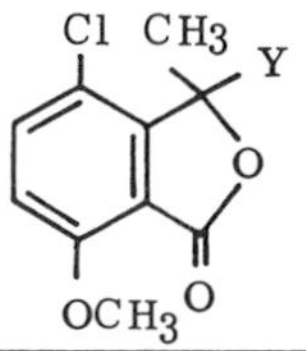 [structure: Cl, $CH_3$, Y, O, $OCH_3$, O] Y = COOH | Brucine in $H_2O$ | + 25° (c 1.2 %, EtOH) (t = 28°) | A 267a |
| $C_{11}H_{12}O_4$ | $HOOC-CH(CH_2C_6H_5)CH_2-COOH$ | (+)-α-Methylbenzylamine in $H_2O$<br>Strychnine in EtOH-$H_2O$ | a) - 29.0° (c 3.205, acetone) (t = 25°)<br>b) + 28.9° (c 2.321, acetone) (t = 25°) | A 268<br>A 268 |
| $C_{11}H_{12}O_4$ | $HOOC-CH(C_6H_5)CH_2CH_2-COOH$ | Quinine in 96 % EtOH<br>Brucine in MeOH | a) + 86.1° (c 1.063, EtOH) (t = 25°)<br>b) - 85.5° (c 0.889, EtOH) (t = 25°) | A 269<br>A 269 |
| $C_{11}H_{12}O_5$ | 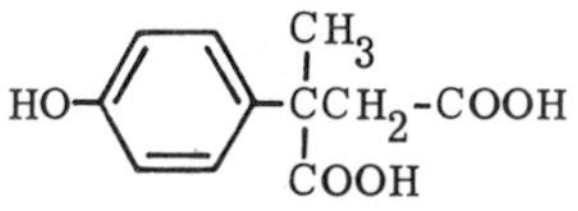 HO-$C_6H_4$-$C(CH_3)(COOH)CH_2$-COOH | Brucine in EtOH-$H_2O$ | a) $[M]^{20}_D$ + 70.5° (c 0.800, acetone)<br>b) $[M]^{20}_D$ - 70.0° (c 0.800, acetone) | A 270 |
| $C_{11}H_{12}O_5$ | 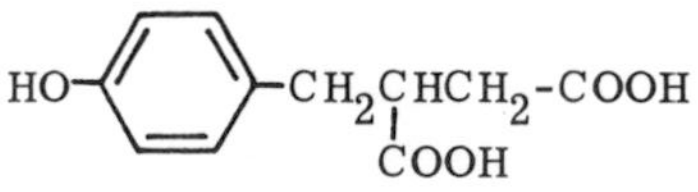 HO-$C_6H_4$-$CH_2CH(COOH)CH_2$-COOH | Brucine in EtOH-$H_2O$ | a) $[M]^{20}_D$ + 61.6° (c 0.800, acetone)<br>b) $[M]^{20}_D$ - 59.0° (c 0.800, acetone) | A 270 |

TABLE 8d (continued)

| Formula | Substance resolved | Resolving agent and solvent | Rotation $[\alpha]_D$ of resolved compound | Reference |
|---|---|---|---|---|
| $C_{11}H_{16}O_4$ | $CH_3OOC(CH_3)C{=}CH$–cyclopropane($CH_3$, $CH_3$)–COOH | (+)-$\alpha$-Methylbenzylamine in $EtOH-H_2O$ | - 12.5° ($CCl_4$) (t = 20°) | A 271 |
| $C_{11}H_{18}O_4$ | cyclohexane with $CH_3$, $COOCH_3$; H, $CH_2$-COOH | (-)-Menthol<br>Esters separated by column chromatography | a) - 9° (c 1.86, acetone)<br>b) + 9° (c 1.0, acetone) | A 272 |
| $C_{12}H_{14}O_4$ | $HOOC-CH(CH_2CH_2C_6H_5)CH_2-COOH$ | Brucine<br>Cinchonidine | a) - 38.4° (c 2.171, EtOH) (t = 25°)<br>b) + 38.1° (c 2.264, EtOH) (t = 25°) | A 273<br>A 273 |
| $C_{12}H_{14}O_4$ | $HOOC-CH(C_6H_5)(CH_2)_3-COOH$ | Strychnine in $EtOH-H_2O$ | a) + 63.8° (c 1.513, EtOH) (t = 25°)<br>b) - 63.6° (c 1.549, EtOH) (t = 25°) | A 274 |
| $C_{12}H_{14}O_5$ | adamantane skeleton with COOH, H, H, O, COOH | Dehydroabietylamine<br>Recr.: $MeOH-H_2O$ and acetone<br>Quinine in acetone | a) C.D. $[\theta]_{293}$ + 260 ± 26° (c 0.0375, 2 % aq. NaOH)<br>b) C.D. $[\theta]_{293}$ - 260 ± 26° (c 0.159, 2 % aq. NaOH) | A 275<br>A 275 |
| $C_{12}H_{16}N_4O_4$ | $HOOC-CH_2CH_2C(CH_3)(CN)-N{=}N-C(CH_3)(CN)CH_2CH_2-COOH$ | Quinine in acetone | a) + 45.3° (c 3.037, MeOH) (t = 25°)<br>b) - 44.8° (c 2.463, MeOH) (t = 25°) | A 276 |
| $C_{12}H_{20}O_4$ | cyclooctane(H)–CH(COOH)–$CH_2$-COOH | Brucine in $H_2O$ | a) + 29.3° (c 0.7515, EtOH) (t = 27°)<br>b) - 28.2° (c 0.9921, EtOH) (t = 27°) | A 277 |

TABLE 8d (continued)

| | Substance resolved | Resolving agent and solvent | Rotation $[\alpha]_D$ of resolved compound | Reference |
|---|---|---|---|---|
| $C_{13}H_{14}O_4$ | HOOC-CHCH$_2$-COOH, CH$_2$CH$_2$CH$_2$C$_6$H$_5$ | Dehydroabietylamine in $Et_2O$ | a) - 28.2$^o$ (c 2.054, EtOH) (t = 25$^o$)<br>b) + 28.3$^o$ (c 2.086, EtOH) (t = 25$^o$) | A 188 |
| $C_{13}H_{16}O_4$ | 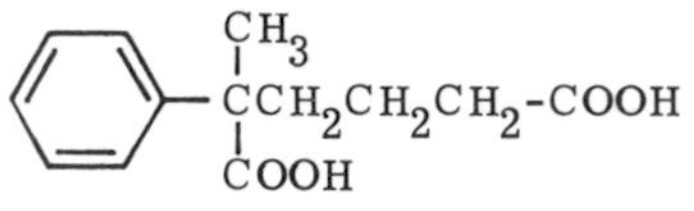 C$_6$H$_5$-C(CH$_3$)(COOH)CH$_2$CH$_2$CH$_2$-COOH | Brucine<br>Strychnine<br>both in acetone-$H_2O$ | a) - 25.65$^o$ (c 5.27, EtOH) (t = 25$^o$) (546 nm)<br>b) + 25.55$^o$ (c 5.22, EtOH) (t = 25$^o$) (546 nm) | A 278<br>A 278 |
| $C_{13}H_{16}O_5$ | 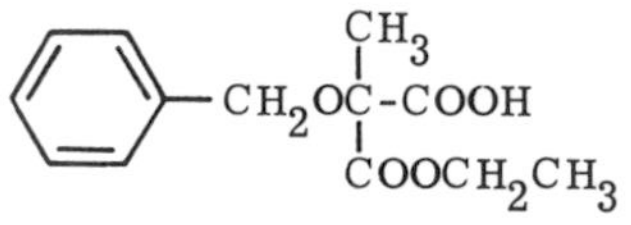 C$_6$H$_5$-CH$_2$OC(CH$_3$)(COOH)COOCH$_2$CH$_3$ | Anhydropilosine in isoPrOH | + 9.5$^o$ (c 2, EtOH) (t = 20$^o$) | A 279 |
| $C_{13}H_{20}O_4$ | 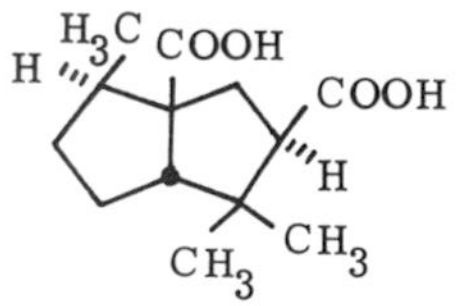 H, H$_3$C, COOH, COOH, H, CH$_3$, CH$_3$ | Quinine in acetone | - 38.9 ± 1.5$^o$ (c 1.08, acetone) (t = 27$^o$) | A 280 |
| $C_{14}H_6O_4I_4$ | 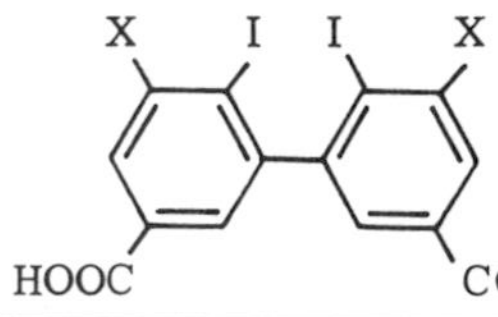 X, I, I, X, HOOC, COOH; X = I | Brucine in EtOH-$H_2O$ | a) - 112$^o$ (NaOH-Na$_2$CO$_3$, H$_2$O) (t = 25$^o$)<br>b) + 104$^o$ (NaOH-Na$_2$CO$_3$, H$_2$O) (t = 25$^o$) | A 281 |
| $C_{14}H_8N_2O_8$ | 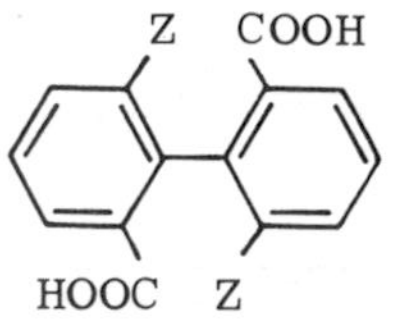 Z, COOH, HOOC, Z; Z = NO$_2$ | $\alpha$-Methylbenzylamine in acetone<br>Compare with Table 3, ref. 16 | a) + 125$^o$ (c 1.690, MeOH) (t = 28$^o$)<br>b) - 81.5$^o$ (c 1.853, MeOH) (t = 31$^o$) | A 282 |
| $C_{14}H_8O_4I_2$ | See $C_{14}H_6O_4I_4$ above, X = H | Brucine in EtOH | - 24.3$^o$ (NaOH, EtOH) (t = 25$^o$) | A 281 |

TABLE 8d (continued)

| | Substance resolved | Resolving agent and solvent | Rotation $[\alpha]_D$ of resolved compound | Reference |
|---|---|---|---|---|
| $C_{14}H_{11}O_5Br$ | 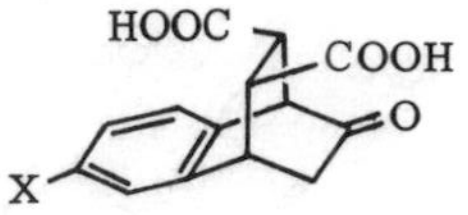 HOOC, COOH, O, X; X = Br | Cinchonidine in EtOH | a) + 192.3° (MeOH)<br>b) - 193.3° (MeOH) | A 283 |
| $C_{14}H_{12}O_6$ | X = OH | Cinchonidine in EtOH | + 249.8° (MeOH) | A 283 |
| | 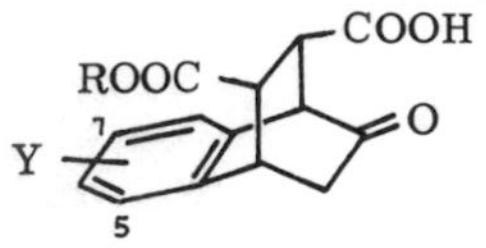 ROOC, COOH, O, Y, 7, 5 | | | |
| $C_{14}H_{12}O_6$ | Y = 5-OH<br>R = H | Cinchonidine in EtOH | + 209.5° (MeOH) | A 283 |
| $C_{14}H_{12}O_6$ | Y = 7-OH<br>R = H | Cinchonidine in EtOH | + 152.7° (MeOH) on the sodium salt | A 283 |
| $C_{15}H_{14}O_5$ | Y = H<br>R = $CH_3$ | Cinchonidine in MeOH | + 208.1 ± 4.6° (c 0.571, MeOH) on the dimethyl ester | A 283 |
| $C_{14}H_{13}O_7Cl$ | See $C_{11}H_9O_5Cl$ Y = -CH-COOH with $CH_2$-COOH on the CH | Brucine in EtOH | - 20.4° (c 5 %, EtOH) (t = 25°) | A 267b |
| $C_{14}H_{14}O_8Cl_2$ | 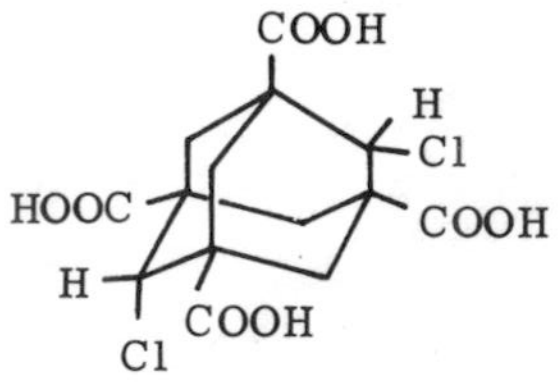 COOH, H, Cl, COOH, HOOC, H, COOH, Cl | Cinchonine · HCl · 2 $H_2O$ in 0.5 N aqueous NaOH | a) + 7.67° (c 1.50, $H_2O$) (t = 20°)<br>b) - 2.65° (c 2.0, $H_2O$) (t = 20°) | A 284 |

TABLE 8d (continued)

| | Substance resolved | Resolving agent and solvent | Rotation $[\alpha]_D$ of resolved compound | Reference |
|---|---|---|---|---|
| $C_{15}H_{14}O_5$ | See above under $C_{14}H_{12}O_6$ | | | |
| $C_{16}H_8O_4D_6$ | See $C_{14}H_8N_2O_8$ above<br>Z = $CD_3$ | Morphine<br>in EtOH | a) + 22$^o$ (c 1.3, MeOH) (t = 25$^o$)<br>b) - 20$^o$ (c 1.2, MeOH) (t = 25$^o$) | A 285 |
| $C_{16}H_{16}O_6$ | See (A) below<br>m.p. 216-7$^o$ isomer | Brucine<br>in dioxane | a) + 6.8$^o$ (c 1, 96 % EtOH)<br>b) - 6.7$^o$ (c 1, 96 % EtOH) | A 286 |
| | m.p. 144-5$^o$ isomer | Cinchonidine<br>Cinchonine<br>both in dioxane | a) - 87.8$^o$ (c 1, EtOH)<br>b) + 90.6$^o$ (c 1, EtOH) | A 286<br>A 286 |
| $C_{16}H_{16}O_6$ | See (B) below<br>m.p. 188$^o$ isomer | Cinchonine<br>Cinchonidine<br>both in dioxane | a) - 32$^o$ (c 1, EtOH)<br>b) + 30.5$^o$ (c 1, EtOH) | A 287<br>A 287 |
| | m.p. 166-8$^o$ isomer | Brucine<br>in dioxane<br>Cinchonine | a) + 128.5$^o$ (c 1, EtOH)<br>b) - 128$^o$ (c 1, EtOH) | A 287<br>A 287 |
| $C_{16}H_{18}O_5$ | See (C) below<br>11-Carboxysantonin | Brucine<br>in MeOH | a) - 75.1$^o$ (c 1.01, EtOH) (t = 18$^o$)<br>b) + 75.4$^o$ (c 1.03, EtOH) (t = 18$^o$) | A 288 |
| $C_{16}H_{20}O_5$ | See (D) below | Brucine<br>in MeOH | + 72.5$^o$ (c 0.40, EtOH) (t = 15$^o$) | A 288 |

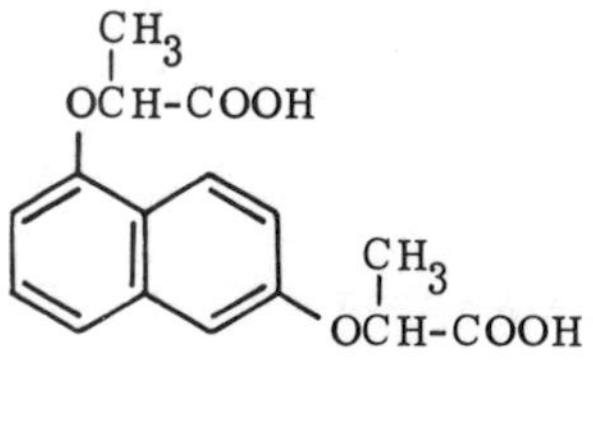

(A) 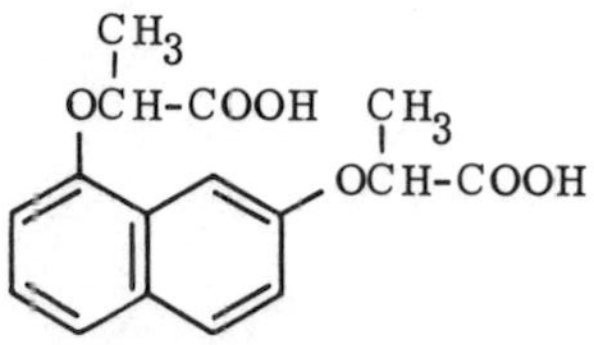 (B) 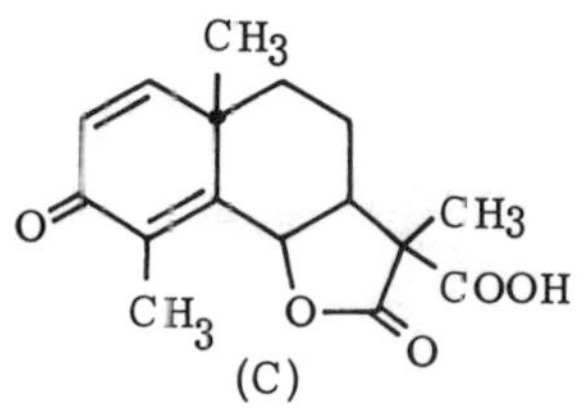 (C) 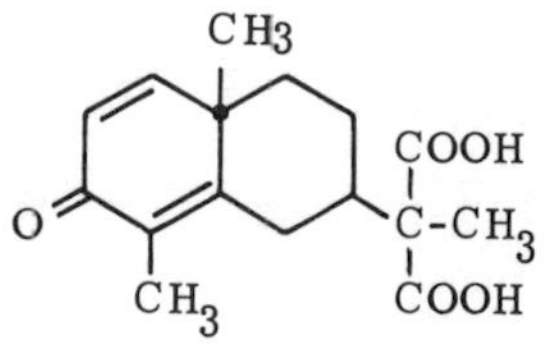 (D)

TABLE 8d (continued)

| | Substance resolved | Resolving agent and solvent | Rotation $[\alpha]_D$ of resolved compound | Reference |
|---|---|---|---|---|
| $C_{18}H_{11}O_4Br$ | 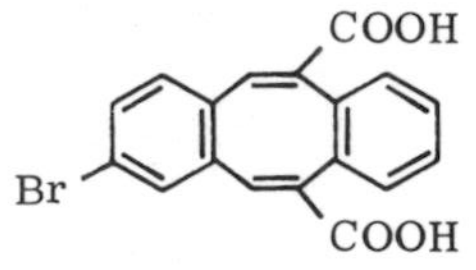 (structure: Br, COOH, COOH) | Brucine | a) + 71.2° (c 0.9, EtOH) (t = 21°)<br>b) - 61.6° (c 0.9, EtOH) (t = 26°) | A 289 |
| $C_{18}H_{14}O_8$ | O,O'-Dibenzoyltartaric acid | See Table 3, ref. 14 | | |
| $C_{18}H_{16}O_4$ | 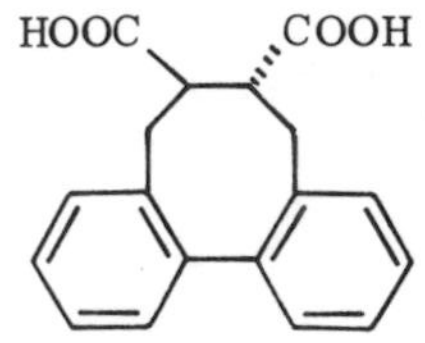 (structure: HOOC, COOH) m.p. 219-220° isomer | Brucine in acetone | a) + 159° (c 5.0, EtOH) (t = 31.5°)<br>b) - 158° (c 5.9, EtOH) (t = 31.5°) | A 290 |
| $C_{18}H_{16}O_4$ | See (A) below | Quinine in EtOH | + 5° (c 1.8, EtOH) (t = 22°) | A 291 |
| $C_{18}H_{18}O_4$ | (structure: $H_3C$, COOH, $CH_3$, $H_3C$, HOOC, $CH_3$) | Quinine · 3 $H_2O$ in EtOH-$H_2O$ | a) - 67° (c 0.521, $CHCl_3$-1% EtOH) (t = 20°) | A 292 |
| | | Cinchonine in 96% EtOH | b) + 67° (c 0.982, $CHCl_3$-1% EtOH) (t = 20°) | A 292 |
| $C_{18}H_{34}O_6$ | HOOC-$(CH_2)_7$-CH(OH)-CH(OH)-$(CH_2)_7$-COOH<br>Phloionic acid | Brucine in EtOH | + 4.6 ± 0.4° (c 2.24, EtOH) (t = 26°) | A 293 |

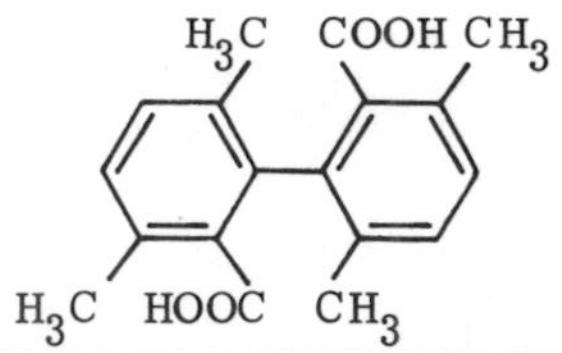

(structure: HOOC, COOH)

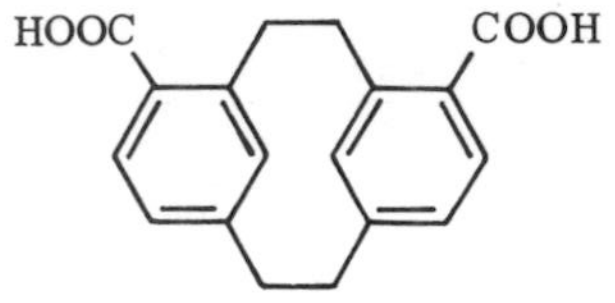

(A)

TABLE 8d (continued)

| | Substance resolved | Resolving agent and solvent | Rotation $[\alpha]_D$ of resolved compound | Reference |
|---|---|---|---|---|
| $C_{19}H_{20}O_4$ | See (A) below | (-)-Menthol<br>Ester Recr.: Pet E | - 173° ($C_6H_6$) (t = 29°) | A 294 |
| $C_{20}H_{18}O_3$ | See (B) below | Brucine<br>Asymmetric destruction<br>(partial hydrolysis in $C_6H_6$) | a) + 0.67° (c 8, $CH_2Cl_2$) (t = 25°) (365 nm)<br>b) - 2.68° (c 8, $CH_2Cl_2$) (t = 25°) (365 nm) | A 295 |
| $C_{20}H_{18}O_8$ | O,O'-Di-p-toluoyltartaric acid | See Table 3, ref. 18 | | |
| $C_{20}H_{20}O_6$ | 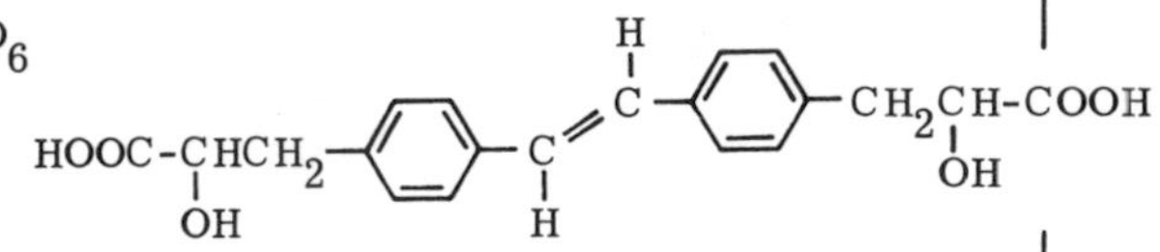 HOOC-CH(OH)CH$_2$-C$_6$H$_4$-CH=CH-C$_6$H$_4$-CH$_2$CH(OH)-COOH | Brucine<br>in EtOH-$H_2O$ (1:1) | a) + 37.7° (dioxane) (t = 20°)<br>b) - 37° (dioxane) (t = 20°) | A 174 |
| $C_{20}H_{22}O_6$ | See (C) below | Brucine<br>in $H_2O$ | a) + 47° (c 1.0, 95% EtOH) (t = 25°)<br>b) - 42° (c 1.0, 95% EtOH) (t = 25°) | A 296 |
| $C_{21}H_{24}O_8$ | 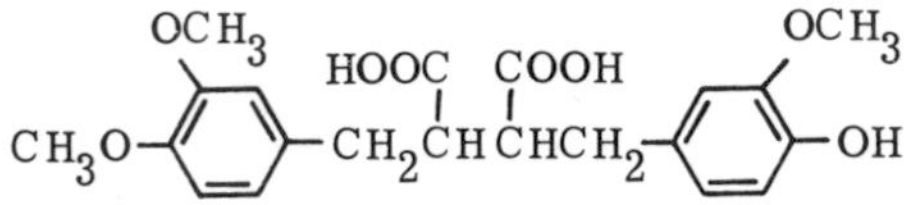 CH$_3$O(OCH$_3$)C$_6$H$_3$-CH$_2$CH(COOH)CH(COOH)CH$_2$-C$_6$H$_3$(OCH$_3$)OH | Strychnine<br>in MeOH-$H_2O$ | a) - 16.8°<br>b) + 17.4° | A 297 |

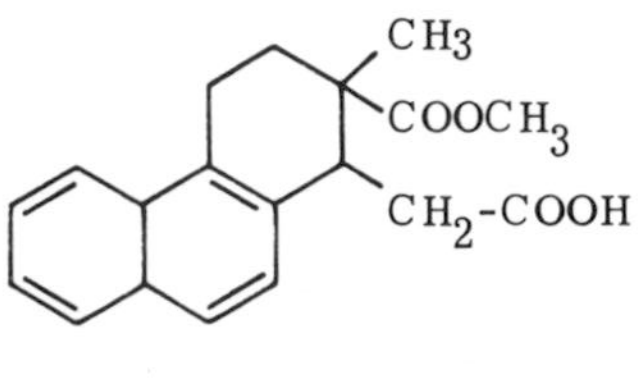

(A)

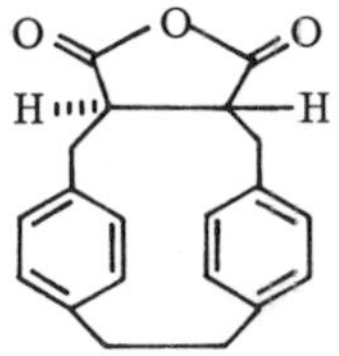

(B)

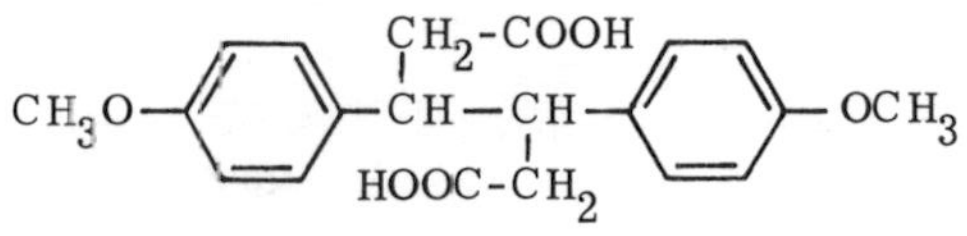

(C)

TABLE 8d (continued)

| | Substance resolved | Resolving agent and solvent | Rotation $[\alpha]_D$ of resolved compound | Reference |
|---|---|---|---|---|
| $C_{22}H_{14}O_4$ | See (A) below | Quinine<br>in EtOH-$Et_2O$ | a) - 110° (0.1 N aq. NaOH)<br>- 125.2° (c 1.023, 0.1 N aq. NaOH) (t = 22°) (546 nm)<br>b) + 124.2° (c 1.115, 0.1 N aq. NaOH) (t = 20°) (546 nm) | A 298<br>A 299<br><br>A 299 |
| $C_{23}H_{24}O_4$ | See (B) below | Brucine · 2 $H_2O$<br>in EtOH | a) - 93.7° (c 0.206, dioxane)<br>b) + 94.2° (c 0.243, dioxane) | A 300 |
| $C_{24}H_{12}O_4Br_2$ | 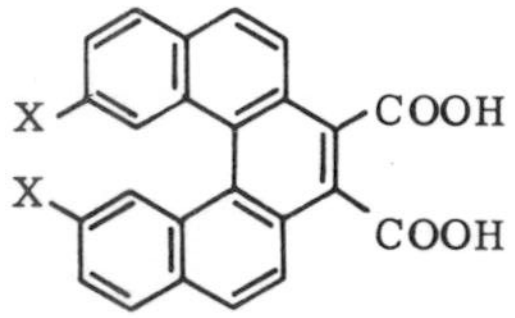 X = Br | Morphine<br>in EtOH<br>Asymmetric transformation | a) - 712° ($CHCl_3$) (t = 22°)<br>b) + 1238° ($CHCl_3$) (t = 25°) | A 301a |
| $C_{24}H_{14}O_4$ | X = H | Morphine<br>in EtOH<br>Asymmetric transformation | + 40° (EtOH) (t = 0°) | A 301b |
| $C_{24}H_{18}O_4$ | See (C) below | Strychnine<br>in acetone-$CHCl_3$ | a) - 80° (o-dichlorobenzene) (t = 30°)<br>b) + 85°<br>both on the carboxylic anhydride | A 302 |

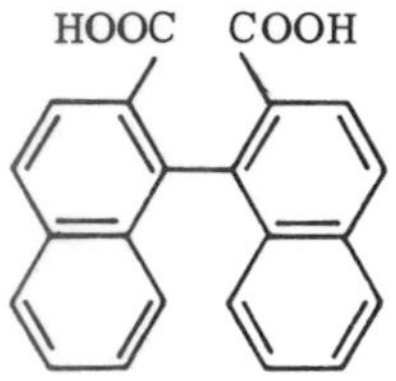

(A)

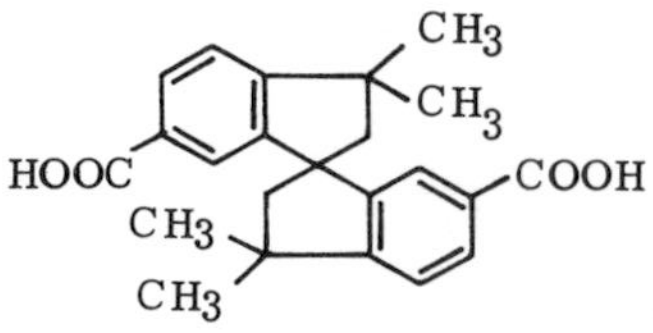

(B)

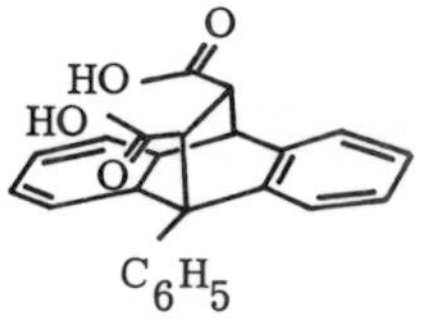

(C)

TABLE 8d (continued)

| | Substance resolved | Resolving agent and solvent | Rotation $[\alpha]_D$ of resolved compound | Reference |
|---|---|---|---|---|
| $C_{26}H_{30}O_3$ | See (A) below | Brucine<br>Asymmetric destruction<br>(partial hydrolysis in $C_6H_6$) | - 4.2° (c 4.5, $CHCl_3$) (t = 25°) | A 303 |
| $C_{28}H_{18}N_2O_4$ | 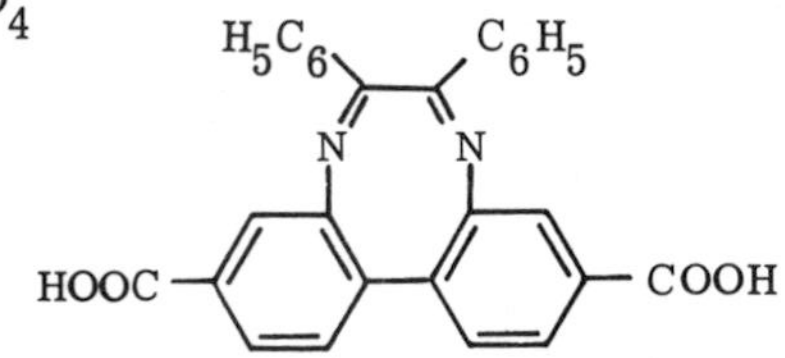 $H_5C_6$, $C_6H_5$, N, N, HOOC, COOH | Brucine<br>in EtOH | a) - 995° (c 1, 0.5 N aq. NaOH)<br>b) - 925° (c 0.91, 0.5 N aq. NaOH) | A 304 |
| $C_{30}H_{18}O_4$ | 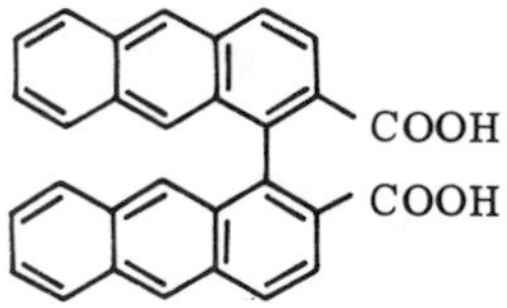 COOH, COOH | Quinidine<br>in EtOH | a) + 382° (c 0.795, acetone) (t = 21°)<br>b) - 386° (c 0.80, acetone) (t = 21°) | A 305 |
| $C_{30}H_{18}O_4$ | 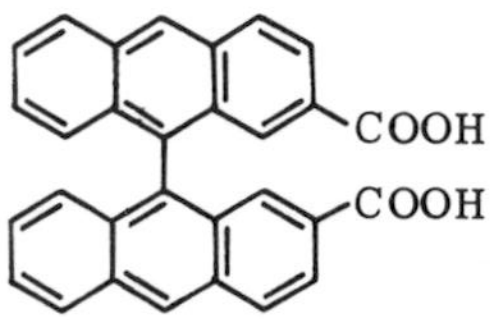 COOH, COOH | Quinidine<br>in acetone | a) + 101° (c 0.65, acetone) (t = 23.5°)<br>b) - 114° (c 0.65, acetone) (t = 22°) | A 306 |

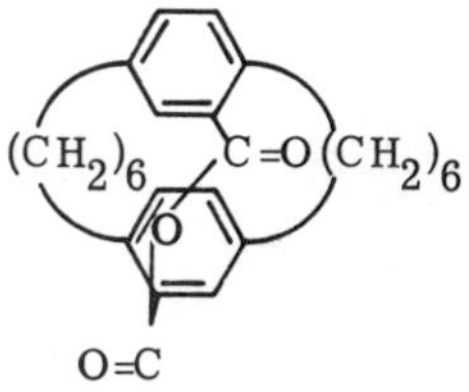

$(CH_2)_6$ C=O $(CH_2)_6$ O O=C

(A)

TABLE 8d (continued)

| | Substance resolved | Resolving agent and solvent | Rotation $[\alpha]_D$ of resolved compound | Reference |
|---|---|---|---|---|
| $C_{33}H_{28}O_4$ | $HOOC-(CH_2)_n$ … $(CH_2)_n-COOH$, n = 3 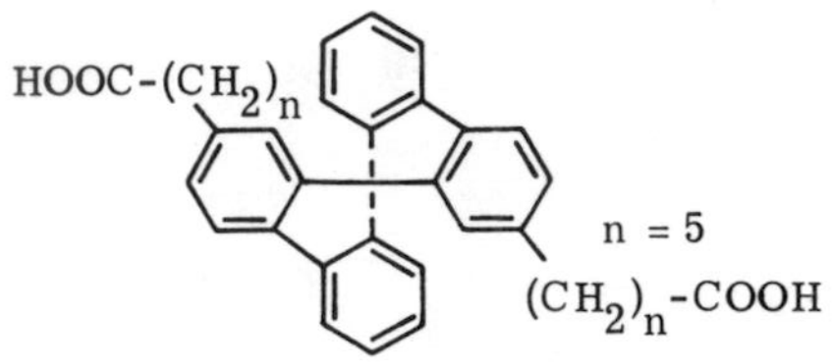 | Dehydroabietylamine<br>Amides separated by column chromatography | a) $+12.6^o$ (c 10.8, $CHCl_3$) (t = $20^o$)<br>b) $-10.1^o$ (c 6.62, $CHCl_3$) (t = $20^o$) | A 307 |
| $C_{37}H_{36}O_4$ | n = 5 | Dehydroabietylamine<br>Amides separated by column chromatography | a) $+10.1^o$ (c 1.345, $CHCl_3$) (t = $22^o$) (578 nm)<br>b) $-10.0^o$ (c 1.465, $CHCl_3$) (t = $22^o$) (578 nm) | A 307 |

TABLE 8e Sulfonic Acids.

Sulfur-containing Carboxylic Acids

| | Substance resolved | Resolving agent and solvent | Rotation $[\alpha]_D$ of resolved compound | Reference |
|---|---|---|---|---|
| $C_4H_6O_2S_2$ | S S COOH | Cinchonidine<br>in EtOH-$H_2O$ | a) + 337° (c 0.5, 96 % EtOH) (t = 25°)<br>b) - 313° (c 0.5, 96 % EtOH) (t = 25°) | A 308 |
| $C_4H_6O_4S_2$ | SH<br>HOOC-CHCH-COOH<br>SH | Brucine<br>in acetone-$H_2O$<br>Partial resolution | a) + 75° (EtOH) (t = 25°)<br>b) - 43° (EtOH) (t = 25°) | A 309 |
| $C_4H_{10}O_3S$ | α-Methylpropanesulfonic acid | See Table 3, ref. 32 | | |
| $C_5H_6O_4S_2$ | S — S<br>HOOC COOH | Brucine · 4 $H_2O$<br>in EtOH-$H_2O$ | a) + 562° (c 0.4588, EtOH) (t = 25°)<br>b) - 560° (c 0.3311, EtOH) (t = 25°) | A 310 |
| $C_5H_6O_4S_2$ | HOOC COOH<br>S S | Quinine<br>in MeOH<br>Cinchonidine<br>in EtOH-$H_2O$ (1:1) | a) - 193° (EtOH) (t = 25°)<br>b) + 194° (EtOH) (t = 25°) | A 309<br>A 309 |
| $C_5H_8O_2S$ | S COOH | Brucine<br>in 96 % EtOH<br>Cinchonidine<br>in EtOH-$H_2O$ | a) - 154.9° (c 0.4713, 96 % EtOH) (t = 25°)<br>b) + 155.7° (c 0.4271, 96 % EtOH) (t = 25°) | A 311<br>A 311 |
| $C_5H_8O_2S$ | COOH<br>S | Quinine<br>in 96 % EtOH<br>(-)-α-Methylbenzylisothio-<br>uronium chloride<br>on the sodium salt of the acid in $H_2O$ | a) + 55.7° (c 3.473, 96 % EtOH) (t = 25°)<br>b) - 53° (c 0.1883, 96 % EtOH) (t = 25°) | A 312<br>A 312 |

TABLE 8e (continued)

| Formula | Substance resolved | Resolving agent and solvent | Rotation $[\alpha]_D$ of resolved compound | Reference |
|---|---|---|---|---|
| $C_5H_8O_2S$ | COOH, S, S (ring structure) | Strychnine in 96 % EtOH | a) $-172^{\circ}$ (c 1.0, EtOH) (t = $25^{\circ}$) | A 313 |
| | | (-)-Ephedrine in EtOH | b) $+171^{\circ}$ (c 1.1, EtOH) (t = $25^{\circ}$) | A 313 |
| $C_5H_8O_2S$ | COOH, S, S (ring structure) | Quinine in 96 % EtOH | a) $-167^{\circ}$ (c 1.0, EtOH) (t = $25^{\circ}$) | A 313 |
| | | Cinchonidine in EtOH | b) $+166^{\circ}$ (c 1.1, EtOH) (t = $25^{\circ}$) | A 313 |
| $C_5H_8O_4S$ | HOOC-CH($CH_3$)-S-$CH_2$-COOH | (-)-Ephedrine in EtOH | a) $+134.2^{\circ}$ (c 0.5, EtOH) | A 314 |
| | | | b) $-134.5^{\circ}$ (c 0.5, EtOH) | |
| $C_6H_6O_3S$ | S, CH(OH)-COOH (ring structure) | Cinchonidine in EtOH-$H_2O$ | a) $-98.3^{\circ}$ (c 1.02, $H_2O$) (t = $25^{\circ}$) | A 315 |
| | | (-)-$\alpha$-Methylbenzylamine in EtOAc | b) $+99.6^{\circ}$ (c 1.010, $H_2O$) (t = $25^{\circ}$) | A 315 |
| $C_6H_6O_3S$ | CH(OH)-COOH, S (ring structure) | (+)-$\alpha$-Methylbenzylamine in EtOAc-MeOH | a) $-102.1^{\circ}$ (c 0.536, EtOH) (t = $25^{\circ}$) | A 316 |
| | | Strychnine in $H_2O$ | b) $+102.5^{\circ}$ (c 0.392, EtOH) (t = $25^{\circ}$) | A 316 |
| $C_6H_8O_4S$ | HOOC, COOH, S (ring structure) | Strychnine in EtOH | a) $-151.4^{\circ}$ (EtOH) (t = $25^{\circ}$) | A 317 |
| | | Quinine in EtOH | b) $+151.8^{\circ}$ (EtOH) (t = $25^{\circ}$) | A 317 |

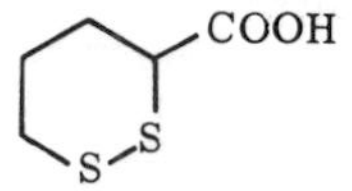

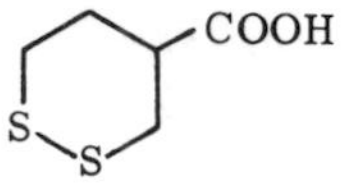

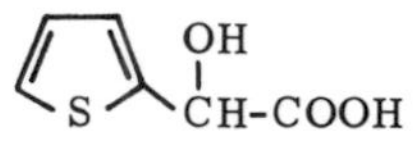

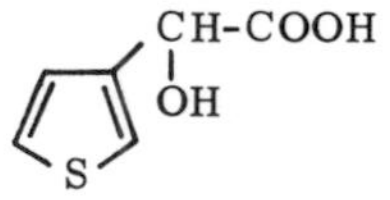

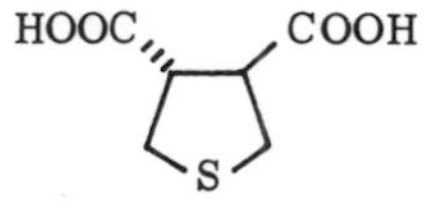

TABLE 8e (continued)

| | Substance resolved | Resolving agent and solvent | Rotation $[\alpha]_D$ of resolved compound | Reference |
|---|---|---|---|---|
| $C_6H_8O_4S_2$ | 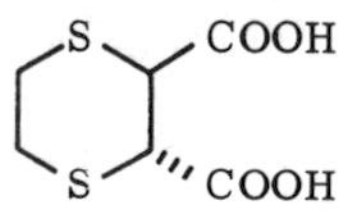 | Brucine in MeOH<br>Cinchonidine in EtOH | a) - 62.3° (EtOH) (t = 25°)<br>b) + 62.3° (EtOH) (t = 25°) | A 309<br>A 309 |
| $C_6H_{10}O_4S$ | 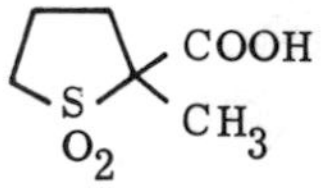 | Brucine in $H_2O$ | a) + 20.4° (c 4.7, $H_2O$) (t = 25°) (546 nm)<br>b) - 20.2° (c 4.7, $H_2O$) (t = 28°) (546 nm) | A 318 |
| $C_6H_{10}O_4S_2$ | $SCH_3$<br>HOOC-CHCH-COOH<br>$SCH_3$ | Dehydroabietylamine acetate<br>Cinchonidine<br>both in EtOH-$H_2O$ | a) - 188° (EtOH) (t = 25°)<br>b) + 188° (EtOH) (t = 25°) | A 309<br>A 309 |
| $C_7H_8O_3S$ | 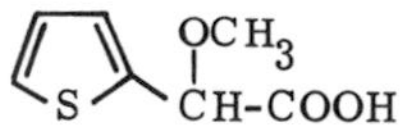 | Quinine in MeOH<br>Cinchonidine in acetone-$H_2O$<br>Spontaneous resolution | a) + 83.2° (c 1.003, acetone) (t = 25°)<br>b) - 83.1° (c 1.010, acetone) (t = 25°)<br>+ 86° (c 0.446, acetone) (t = 25°) | A 319<br>A 319<br>A 320 |
| $C_7H_8O_3S$ | 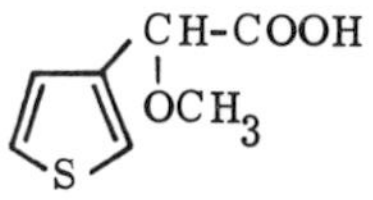 | Quinine in MeOH<br>Strychnine in acetone-$H_2O$ | a) + 94.6° (c 1.033, acetone) (t = 25°)<br>b) - 95.0° (c 1.032, acetone) (t = 25°) | A 319<br>A 319 |
| $C_7H_{10}O_4S$ | 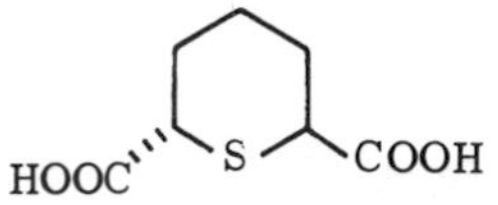 | Quinine<br>Strychnine<br>both in EtOH | a) + 91° (EtOH) (t = 25°)<br>b) - 115.5° (c 0.750, EtOH) (t = 25°) | A 321<br>A 321 |

TABLE 8e (continued)

| Formula | Substance resolved | Resolving agent and solvent | Rotation $[\alpha]_D$ of resolved compound | Reference |
|---|---|---|---|---|
| $C_7H_{10}O_4S_2$ | HOOC-(cyclic S–S, seven-membered ring)-COOH | Brucine · 4 $H_2O$ in acetone-$H_2O$ | a) + 220° (EtOH) (t = 25°) | A 322 |
| | | Brucine · 4 $H_2O$ in acetone | b) - 382° (EtOH) (t = 25°) | A 322 |
| $C_7H_{10}O_4S_3$ | HOOC-CH($CH_3$)-SC(S)S-CH($CH_3$)-COOH | Cinchonine | a) + 176.2° (c 0.7361, acetone) (t = 25°) | A 323 |
| | | Strychnine<br>both in EtOH-$H_2O$ | b) - 172.7° (c 1.903, acetone) (t = 25°) | A 323 |
| $C_7H_{12}O_4S$ | HOOC-CHCH$_2$-COOH<br>with CH$_2$CH$_2$-S-CH$_3$ on CH | Strychnine in $H_2O$ | a) + 38.8° (c 0.996, acetone) (t = 25°) | A 324 |
| | | Quinidine in acetone-$H_2O$ | b) - 38.6° (c 0.939, acetone) (t = 25°) | A 324 |
| $C_7H_{12}O_4S$ | HOOC-CHCH$_2$-COOH<br>with S-CH$_2$CH$_2$CH$_3$ on CH | Strychnine in EtOH-$H_2O$ | a) + 144.1° (c 0.878, acetone) (t = 18°) | A 324 |
| | | | b) - 131.5° (c 1.006, acetone) (t = 25°) | A 324 |
| | **Substituted Phenylsulfinylacetic Acids**<br>$C_6H_5$-S(O)-CH$_2$-COOH | | | |
| $C_8H_6O_3SCl_2$ | 2,4-Dichloro | Brucine in EtOH | a) + 340.2° (c 0.418, EtOH) (t = 25°) | A 325 |
| | | | b) - 340.1° (c 0.588, EtOH) (t = 25°) | |
| $C_8H_6O_3SCl_2$ | 3,4-Dichloro | Cinchonine | a) - 158.5° (c 1.29, EtOH) (t = 25°) | A 325 |
| | | Strychnine<br>both in EtOH | b) + 158.3° (c 0.826, EtOH) (t = 25°) | A 325 |

TABLE 8e (continued)

| | Substance resolved | Resolving agent and solvent | Rotation $[\alpha]_D$ of resolved compound | Reference |
|---|---|---|---|---|
| | Substituted Phenylsulfinylacetic Acids (continued) | | | |
| $C_8H_6O_3SCl_2$ | 3,5-Dichloro | Brucine<br>Cinchonine<br>both in EtOAc-EtOH | a) $+160.4^{o}$ (c 0.916, EtOH) (t = $25^{o}$)<br>b) $-159.7^{o}$ (c 0.608, EtOH) (t = $25^{o}$) | A 326<br>A 326 |
| $C_8H_7O_3SBr$ | 3-Bromo | Cinchonine<br>in dioxane | $-111.11^{o}$ (EtOH) (t = $20^{o}$) | A 327 |
| $C_8H_7O_3SBr$ | 4-Bromo | Cinchonidine<br>in acetone-MeOH | a) $+153.58^{o}$ (EtOH) (t = $20^{o}$)<br>b) $-155.58^{o}$ (EtOH) (t = $20^{o}$) | A 327 |
| $C_8H_8O_2S$ | CH-COOH<br>SH | Cinchonidine<br>in acetone | $-106.2^{o}$ (c 0.5, 95% EtOH) | A 328 |
| $C_8H_8O_3S$ | COOH<br>O<br>S-$CH_3$ | Brucine<br>in EtOH | a) $+232^{o}$ (c 1.2, EtOH) (t = $25^{o}$)<br>b) $-232^{o}$ (c 1.2, EtOH) (t = $25^{o}$) | A 329 |
| $C_8H_8O_4S$ | $CH_2$-COOH<br>S CH-COOH | Cinchonidine<br>in EtOH-$H_2O$ (1:1)<br>(-)-$\alpha$-Methylbenzylamine<br>in EtOAc-MeOH | a) $+103.5^{o}$ (c 1.299, EtOH) (t = $25^{o}$)<br>b) $-103.9^{o}$ (c 1.003, EtOH) (t = $25^{o}$) | A 330<br>A 330 |
| $C_8H_8O_4S$ | CH-COOH<br>$CH_2$-COOH<br>S | Cinchonidine<br>in EtOH-$H_2O$<br>(-)-$\alpha$-Methylbenzylamine<br>in EtOAc-MeOH | a) $+102.3^{o}$ (c 0.970, EtOH) (t = $25^{o}$)<br>b) $-102.5^{o}$ (c 1.0363, EtOH) (t = $25^{o}$) | A 331<br>A 331 |

TABLE 8e (continued)

| | Substance resolved | Resolving agent and solvent | Rotation $[\alpha]_D$ of resolved compound | Reference |
|---|---|---|---|---|
| $C_8H_{10}O_3S$ | α-Methylphenylmethanesulfonic acid | See Table 3, ref. 32 | | |
| $C_9H_{10}O_4S$ | 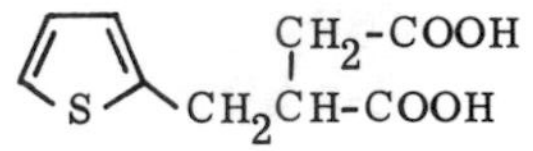 2-thienyl–$CH_2CH(COOH)$–$CH_2$-COOH | (+)-α-Methylbenzylamine in EtOAc<br>Strychnine in EtOH-$H_2O$ | a) - 14.8° (c 1.855, acetone) (t = 25°)<br>b) + 14.7° (c 1.871, acetone) (t = 25°) | A 332<br>A 332 |
| $C_9H_{10}O_4S$ | 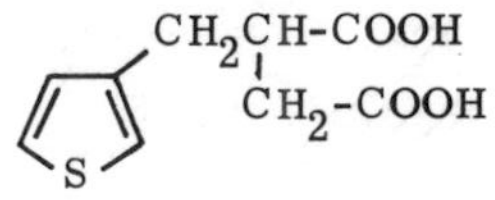 3-thienyl–$CH_2CH(COOH)$–$CH_2$-COOH | Strychnine in EtOH-$H_2O$<br>(+)-α-Methylbenzylamine in EtOAc | a) + 20.0° (c 2.102, acetone) (t = 25°)<br>b) - 20.0° (c 1.825, acetone) (t = 25°) | A 333<br>A 333 |
| $C_9H_{12}O_2S_4$ | 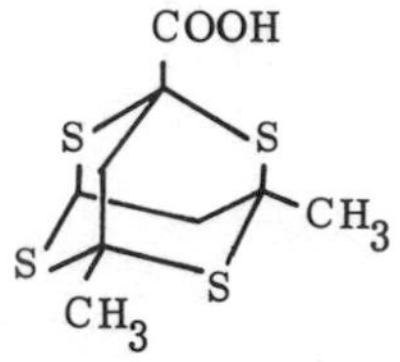 COOH, S, S, S, S, $CH_3$, $CH_3$ (tetrathiaadamantane structure) | Cinchonidine<br>Recr.: EtOH<br>Strychnine<br>Recr.: EtOH-$H_2O$ | a) + 24.6°(c 0.7, acetone) (t = 25°) (365 nm)<br>b) - 24.9°(c 0.9, acetone) (t = 25°) (365 nm) | A 334<br>A 334 |
| $C_9H_{12}O_3S$ | α-Ethylphenylmethanesulfonic acid | See Table 3, ref. 32 | | |
| $C_9H_{16}O_2S$ | 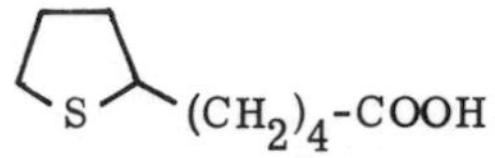 tetrahydrothien-2-yl–$(CH_2)_4$-COOH | Cinchonidine in EtOAc | a) + 73.6° (EtOH) (t = 22°)<br>b) - 72.9° (t = 25°) | A 335 |

TABLE 8e (continued)

| | Substance resolved | Resolving agent and solvent | Rotation $[\alpha]_D$ of resolved compound | Reference |
|---|---|---|---|---|
| | **Substituted 3,3'-Bithienyldicarboxylic Acids** | | | |
| | [Parent System: 3,3'-bithienyl, positions 2, 4, 5, 2', 4', 5'] | | | |
| $C_{10}H_2O_4Br_4S_2$ | 2,2',4,4'-Tetrabromo-5,5'-dicarboxy | Brucine in 96% EtOH | a) $-7.9^o$ (c 2.006, EtOH) (t = $25^o$)<br>Contains 12% $CCl_4$ crystal solvent<br>$-9.0^o$ (EtOH) (t = $25^o$) (calculated)<br>b) $+7.9^o$ (EtOH) (t = $25^o$) | A 336 |
| $C_{10}H_4N_2O_8S_2$ | 2,2'-Dicarboxy-4,4'-dinitro | Quinidine<br>Cinchonidine<br>both in EtOH | a) $-221^o$ (c 1.291, dioxane) (t = $25^o$)<br>b) $+225^o$ (c 0.805, dioxane) (t = $25^o$) | A 337<br>A 337 |
| $C_{10}H_4N_2O_8S_2$ | 2,2'-Dinitro-4,4'-dicarboxy | Brucine<br>Quinine<br>both in EtOH | a) $+98^o$ (c 0.462, EtOH) (t = $25^o$)<br>b) $-97^o$ (c 0.625, EtOH) (t = $25^o$) | A 338<br>A 338 |
| $C_{12}H_8O_4Br_2S_2$ | 2,2'-Dicarboxy-4,4'-dibromo-5,5'-dimethyl | Cinchonine in EtOH-$H_2O$ | a) $+43^o$ (c 0.836, acetone) (t = $25^o$)<br>b) $-43^o$ (c 0.26, acetone) (t = $25^o$) | A 339 |
| $C_{14}H_{14}O_4S_2$ | 2,2',5,5'-Tetramethyl-4,4'-dicarboxy | Quinine in EtOH-$H_2O$<br>Brucine · 2 $H_2O$ in 95% EtOH | a) $+61^o$ (c 0.453, dioxane) (t = $25^o$)<br>$[M]_D^{25}$ $+190^o$ (dioxane)<br>b) $[M]_D^{26}$ $-205^o$ (c 0.027 - 0.270, dioxane) | A 340<br>A 340 |
| $C_{18}H_{22}O_4S_2$ | 2,2',5,5'-Tetraethyl-4,4'-dicarboxy | Cinchonidine<br>Quinine · 3 $H_2O$<br>both in EtOH-$H_2O$ | a) $+284^o$ (c 0.173, $C_6H_6$) (t = $20^o$)<br>b) $-281^o$ (c 0.262, $C_6H_6$) (t = $20^o$) | A 292<br>A 292 |

TABLE 8e (continued)

| Formula | Substance resolved | Resolving agent and solvent | Rotation $[\alpha]_D$ of resolved compound | Reference |
|---|---|---|---|---|
| $C_{10}H_{10}N_2O_2S$ | (benzimidazol-2-yl)-S-CH($CH_3$)-COOH | Quinine in EtOH | a) - 124° (96 % EtOH) (t = 25°) | A 341 |
| | | Quinidine in EtOH-$H_2O$ | b) + 125° (96 % EtOH) (t = 25°) | A 341 |
| $C_{10}H_{10}O_4S$ | 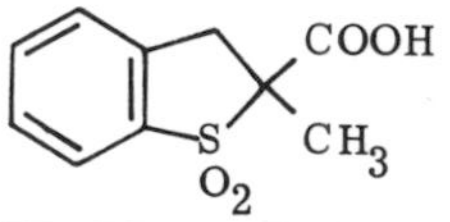 (2-methyl-2-carboxy-benzothiophene $S,S$-dioxide; COOH, $CH_3$, $O_2$) | Brucine in acetone-$H_2O$ | a) - 83.9° (c 2.9, EtOH) (t = 29°) (546 nm)<br>b) + 80.5° (c 2.9, EtOH) (t = 28°) (546 nm) | A 318 |
| $C_{10}H_{12}O_3S$ | 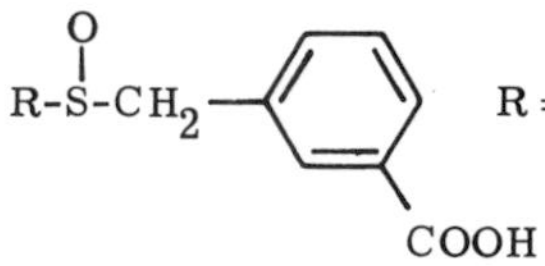 R-S(O)-$CH_2$-$C_6H_4$-COOH, R = $CH_3CH_2$ | Quinine in EtOAc | - 75° (c 1, EtOH) (t = 20°) | A 342 |
| $C_{10}H_{12}O_4S$ | 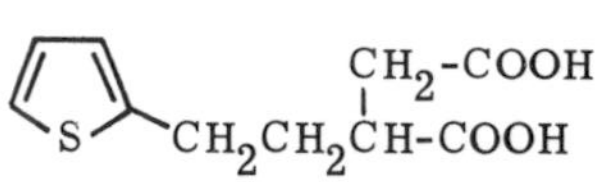 (2-thienyl)-$CH_2CH_2$CH($CH_2$-COOH)-COOH | Brucine in $H_2O$ | a) - 38.5° (EtOH) (t = 25°)<br>b) + 38.6° (EtOH) (t = 25°) | A 343 |
| $C_{10}H_{15}O_4SBr$ | 3-Bromocamphor-9-sulfonic acid | See Table 3, ref. 5 (p. 29). | | |
| $C_{10}H_{16}N_2O_3S$ | Biotin ($(CH_2)_4$-COOH) | (-)-Mandelic acid | a) + 90.7° (c 2.04, 0.1 N aq. NaOH) (t = 25°) | A 344 |
| | | (+)-Mandelic acid *via* ester formation | b) - 90.6° (c 0.5, 0.1 N aq. NaOH) (t = 25°) | A 344 |
| | | Also resolution with quinidine methohydroxide and especially with (+)-arginine | | A 344 |
| $C_{10}H_{16}O_4S$ | 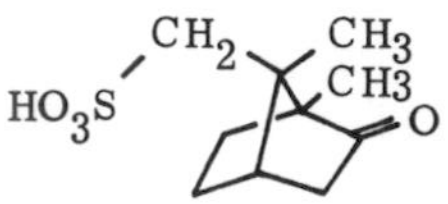 Camphor-9-sulfonic acid ($HO_3S$-$CH_2$-, $CH_3$, $CH_3$, O) | Strychnine in $H_2O$ | + 69° (c 2, $H_2O$) on the ammonium salt | A 345 |

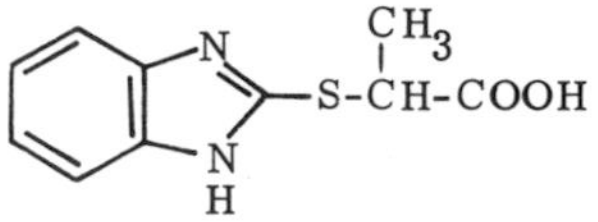

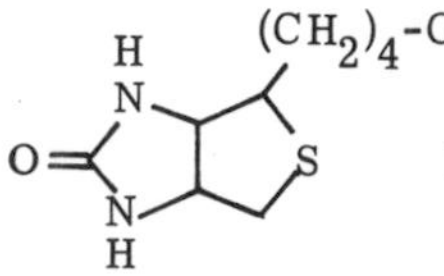

TABLE 8e (continued)

| | Substance resolved | Resolving agent and solvent | Rotation $[\alpha]_D$ of resolved compound | Reference |
|---|---|---|---|---|
| $C_{11}H_{10}O_2S$ | 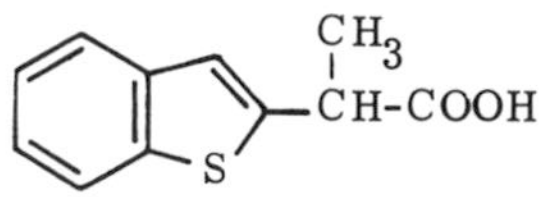 [2-benzothienyl]-CH(CH$_3$)-COOH | Morphine in EtOH-$H_2O$<br>(+)-α-(2-Naphthyl)-ethyl-amine in $Et_2O$ | a) + 31.2° (c 1.058, EtOH) (t = 25°)<br>b) - 31.4° (c 1.002, EtOH) (t = 25°) | A 346<br>A 346 |
| $C_{11}H_{10}O_2S$ | 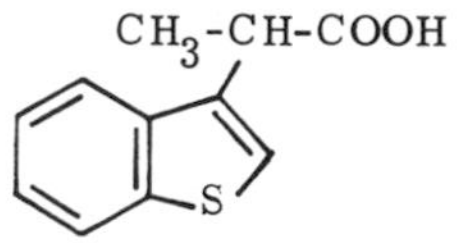 $CH_3$-CH-COOH on [3-benzothienyl] | Cinchonidine<br>Cinchonine<br>both in EtOH-$H_2O$ | a) - 80.2° (c 0.754, EtOH) (t = 25°)<br>b) + 80.0° (c 0.756, EtOH) (t = 25°) | A 347<br>A 347 |
| $C_{11}H_{10}O_2S_2$ | 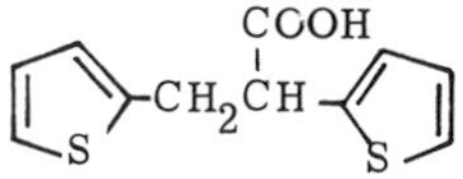 [2-thienyl]-$CH_2$CH(COOH)-[2-thienyl] | (-)-Ephedrine · 2 $H_2O$ in $C_6H_6$<br>Quinine | a) - 95.9° (c 1.269, acetone) (t = 25°)<br>b) + 70° | A 85<br>A 85 |
| $C_{11}H_{12}O_3S_2$ | 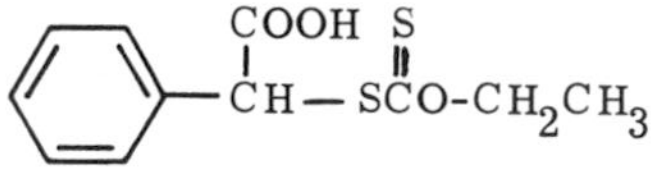 [phenyl]-CH(COOH)—SC(=S)O-$CH_2CH_3$ | Quinine in MeOH | a) + 245.2° (c 2.346, EtOAc) (t = 25°)<br>b) - 244.9° (c 2.297, EtOAc) (t = 25°) | A 348 |
| $C_{11}H_{14}O_2S$ | 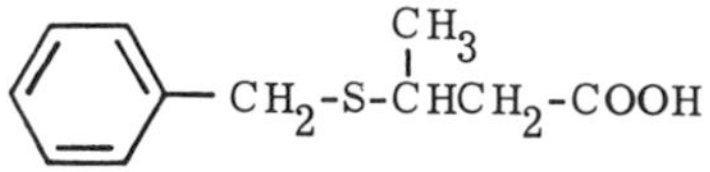 [phenyl]-$CH_2$-S-CH($CH_3$)$CH_2$-COOH | Brucine<br>Cinchonidine<br>Resolution with quinine | a) + 17.4° (c 0.68, EtOH) (t = 25°)<br>b) - 15.5° (c 0.43, EtOH) (t = 25°) | A 349<br>A 349<br>A 350 |
| $C_{11}H_{14}O_3S$ | See $C_{10}H_{12}O_3S$ above<br>R = $(CH_3)_2CH$ | Brucine in EtOH | + 134° (c 1, EtOH) (t = 20°) | A 342 |
| $C_{12}H_8O_4Br_2S_2$ | See above, under substituted 3,3'-bithienyldicarboxylic acids | | | |
| $C_{12}H_9NO_4S$ | 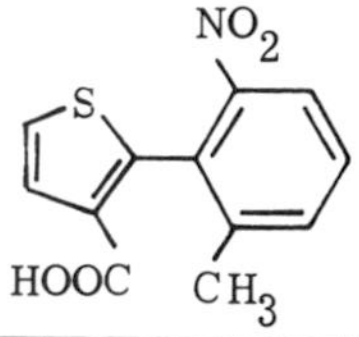 [3-HOOC-2-thienyl]-[2-$NO_2$-6-$CH_3$-phenyl] (labels: $NO_2$, S, HOOC, $CH_3$) | Brucine in EtOH-$H_2O$ | - 11.19° (c 4.002, MeOH) (t = 20°) | A 351 |

TABLE 8e (continued)

| | Substance resolved | Resolving agent and solvent | Rotation $[\alpha]_D$ of resolved compound | Reference |
|---|---|---|---|---|
| $C_{12}H_9O_3BrS$ | 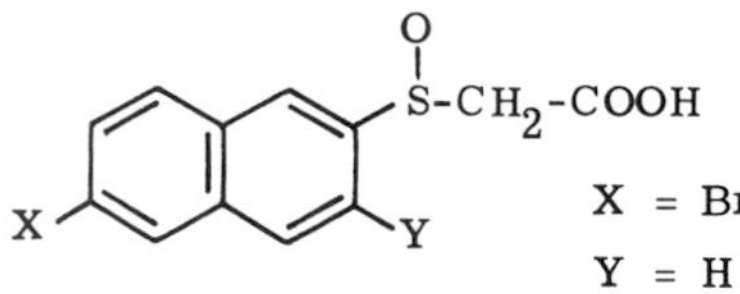 $S(O)$-$CH_2$-COOH<br>X = Br<br>Y = H | Brucine<br>Recr.: MeOH | a) - 145.45° (c 0.33, 96% EtOH) (t = 20°)<br>b) + 145.45° (c 0.33, 96% EtOH) (t = 20°) | A 352 |
| $C_{12}H_9O_3BrS$ | See $C_{12}H_9O_3BrS$ above<br>X = H, Y = Br | Cinchonidine<br>Recr.: acetone | - 179.11° (c 0.67, 96% EtOH) (t = 20°) | A 352 |
| $C_{12}H_{10}N_2O_3S_2$ | 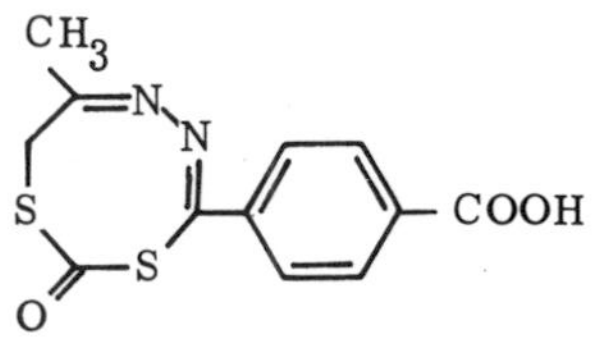 $CH_3$, N, N, S, S, O, COOH | Brucine<br>in MeOH | - 2250° (c 0.5, acetone) (t = 20°) (436 nm) | A 353 |
| $C_{12}H_{10}O_3S$ | 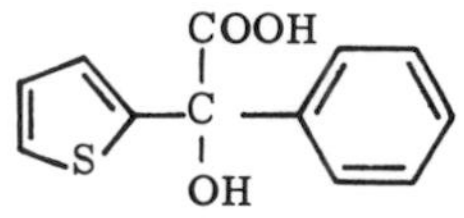 COOH, C, OH, S | (+)-α-Methylbenzylamine<br>(-)-α-Methylbenzylamine<br>both in MeOH-$H_2O$ (1:1) | a) - 28° (EtOAc)<br>b) + 25° (EtOAc) | A 354<br>A 354 |
| $C_{12}H_{12}O_5S$ | 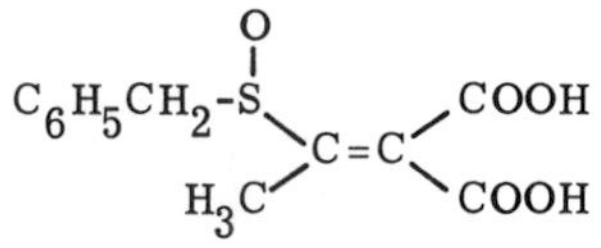 $C_6H_5CH_2$-S(O)-C(COOH)=C(COOH)-$CH_3$ | Brucine<br>in acetone-$H_2O$ (1:1) | a) - 323.0° (c 1, EtOH) (t = 25°)<br>b) + 323.1° (c 1, EtOH) (t = 25°) | A 355 |
| $C_{12}H_{14}O_4S$ | 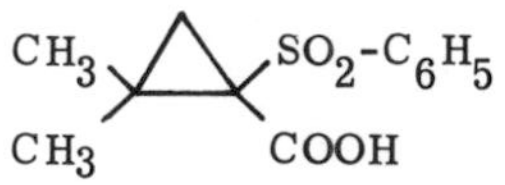 $CH_3$, $CH_3$, $SO_2$-$C_6H_5$, COOH | Quinine<br>in EtOH | a) + 20.7° (c 5, $CHCl_3$) (t = 25°) (546 nm)<br>b) - 18.2° (c 5, $CHCl_3$) (t = 25°) (546 nm) | A 356 |
| $C_{12}H_{16}O_3S$ | See $C_{10}H_{12}O_3S$ above<br>R = $(CH_3)_3C$ | Brucine<br>in EtOH | + 276° (c 0.25, EtOH) (t = 20°) | A 342 |

TABLE 8e (continued)

| | Substance resolved | Resolving agent and solvent | Rotation $[\alpha]_D$ of resolved compound | Reference |
|---|---|---|---|---|
| $C_{13}H_8O_4S_2$ | 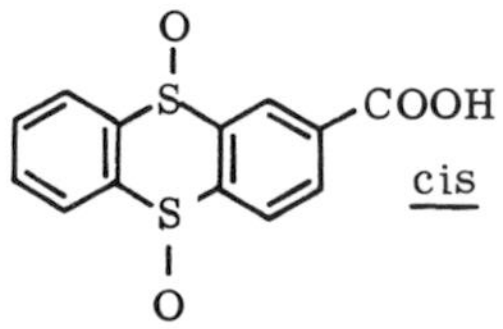 cis | Morphine<br>in EtOH<br>Hydroquinine<br>in acetone-$H_2O$ | a) - $134^o$ (1% aq. NaOH) (t = $25^o$)<br>- $126.57^o$ (c 0.47, 1% aq. NaOH) (t = $20^o$)<br>b) + $126.84^o$ (c 0.43, 1% aq. NaOH) (t = $20^o$) | A 357<br>A 358<br>A 358 |
| | trans | Cinchonine<br>in 95 % EtOH<br>Quinine<br>in MeOH | a) + $41.2^o$ (1% aq. NaOH) (t = $25^o$)<br>+ $40.17^o$ (c 0.43, 1% aq. NaOH) (t = $20^o$)<br>b) - $40.42^o$ (c 0.47, 1% aq. NaOH) (t = $20^o$) | A 357<br>A 358<br>A 358 |
| $C_{13}H_{12}O_2S$ | 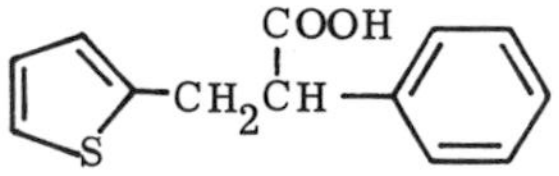 | (+)-Amphetamine<br>(+)-α-Methylbenzylamine<br>both in EtOH-$H_2O$ | a) + $126.9^o$ (c 1.07, acetone) (t = $25^o$)<br>b) - $127.9^o$ (c 1.00, acetone) (t = $25^o$) | A 359<br>A 359 |
| $C_{13}H_{12}O_2S$ | 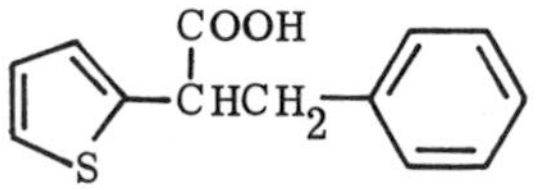 | Cinchonidine<br>(-)-Ephedrine · 2 $H_2O$<br>both in EtOH-$H_2O$ | a) + $88.3^o$ (c 0.97, acetone) (t = $25^o$)<br>b) - $87.7^o$ (c 1.1, acetone) (t = $25^o$) | A 359<br>A 359 |
| $C_{13}H_{12}O_2S$ | 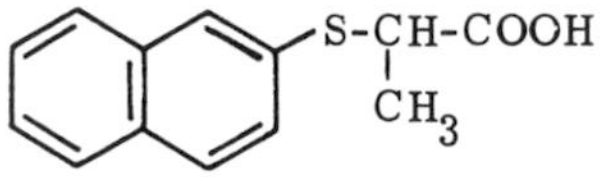 | (+)-α-Methylbenzylamine<br>(-)-α-Methylbenzylamine<br>both in EtOH-$H_2O$ | a) + $140.2^o$ (c 1.069, EtOH) (t = $25^o$)<br>b) - $139.0^o$ (c 0.952, EtOH) (t = $25^o$) | A 81<br>A 81 |
| $C_{13}H_{14}O_5S$ | 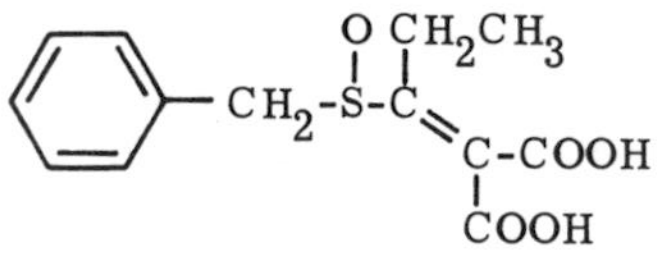 | Brucine<br>in acetone | a) - $319.1^o$ (c 1.0, EtOH) (t = $25^o$)<br>b) + $320.3^o$ (c 1.0, EtOH) (t = $25^o$) | A 360 |
| $C_{14}H_{10}O_4S$ | 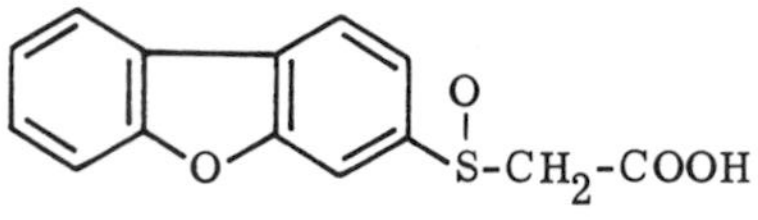 | Strychnine<br>in acetone<br>Cinchonine<br>in acetone-$H_2O$ | a) + $112.50^o$ (96% EtOH) (t = $20^o$)<br>b) - $111.25^o$ (96% EtOH) (t = $20^o$) | A 361<br>A 361 |

## TABLE 8e (continued)

| Substance resolved | | Resolving agent and solvent | Rotation $[\alpha]_D$ of resolved compound | Reference |
|---|---|---|---|---|
| $C_{14}H_{11}O_3BrS$ | 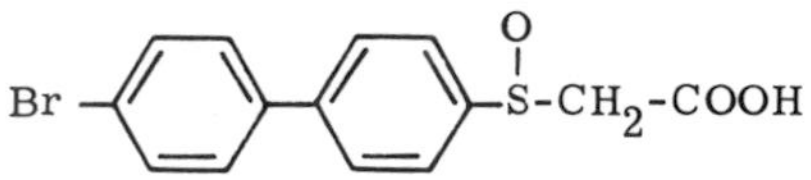 Br, O, S-$CH_2$-COOH | Cinchonine in acetone<br>Cinchonidine in acetone | a) - 125.76° (c 0.33, 96% EtOH) (t = 20°)<br>b) + 125.76° (c 0.33, 96% EtOH) (t = 20°) | A 362<br>A 362 |
| $C_{14}H_{14}O_4S_2$ | See above, under substituted 3,3'-bithienyldicarboxylic acids | | | |
| $C_{15}H_{14}O_2S$ | 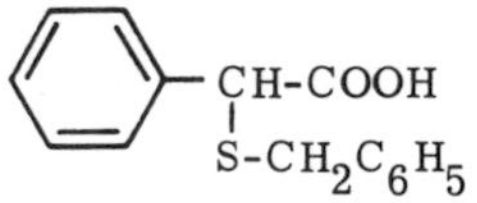 CH-COOH, S-$CH_2C_6H_5$ | (-)-α-Methylbenzylamine<br>(+)-α-Methylbenzylamine<br>both in EtOH | a) + 161° (c 0.89, EtOH) (t = 25°)<br>b) - 161° (c 1.65, EtOH) (t = 25°) | A 363<br>A 363 |
| $C_{15}H_{14}O_2S$ | 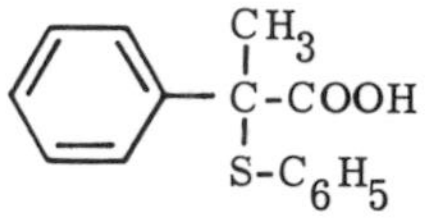 $CH_3$, C-COOH, S-$C_6H_5$ | (+)-α-Methylbenzylamine<br>(-)-α-Methylbenzylamine<br>both in EtOH-$H_2O$ | a) + 162.0° (c 1.92, EtOH) (t = 30°)<br>b) - 165.3° (c 1.903, EtOH) (t = 26°) | A 364<br>A 364 |
| $C_{15}H_{22}O_4S$ | 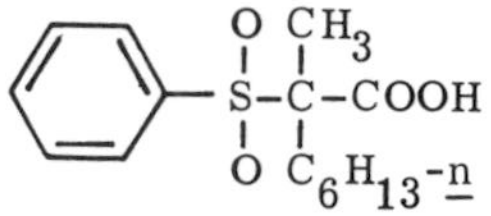 O, $CH_3$, S-C-COOH, O, $C_6H_{13}$-n | Quinine in EtOH | + 15.6° (c 8.4, $CHCl_3$) (t = 25°) (546 nm) | A 365 |
| $C_{16}H_{12}O_3S$ | 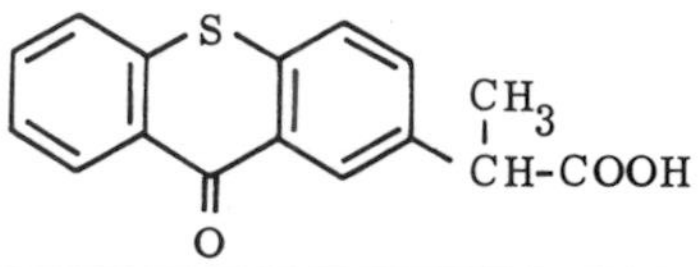 S, O, $CH_3$, CH-COOH | (-)-α-Methylbenzylamine in MeOH | - 73 ± 1.5° (t = 19.5°) | A 366 |
| $C_{16}H_{12}O_3S$ | 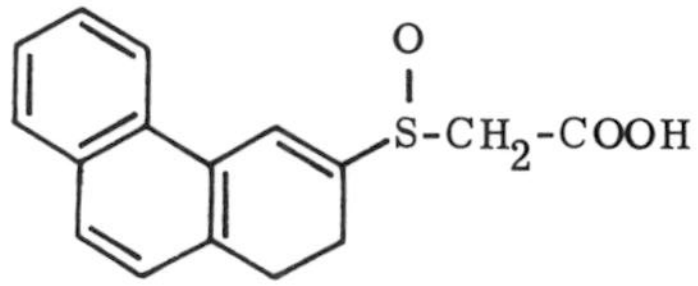 O, S-$CH_2$-COOH | Brucine<br>Cinchonidine<br>both in acetone-$H_2O$ | a) - 120.00° (c 4, 96% EtOH) (t = 20°)<br>b) + 118.75° (c 4, 96% EtOH) (t = 20°) | A 367<br>A 367 |

TABLE 8e (continued)

| | Substance resolved | Resolving agent and solvent | Rotation $[\alpha]_D$ of resolved compound | Reference |
|---|---|---|---|---|
| $C_{16}H_{12}O_5S$ | 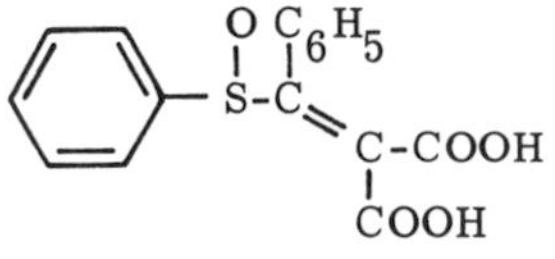 O $C_6H_5$ S-C= C-COOH COOH | Quinine in acetone<br>Cinchonidine in EtOH | a) $+61.0^o$ (c 1.0, acetone) (t = $25^o$)<br>b) $-60.7^o$ (c 1.0, acetone) (t = $25^o$) | A 360<br>A 360 |
| $C_{16}H_{14}O_4S$ | 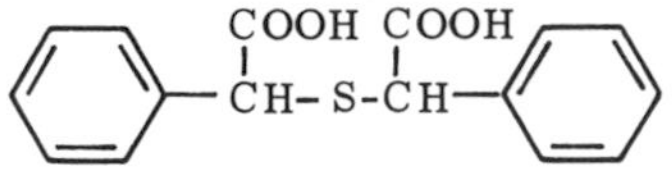 COOH COOH CH- S-CH | Brucine in MeOH<br>Strychnine in acetone-$H_2O$ | a) $-278.50^o$ (c 0.4, MeOH) (t = $20^o$)<br>b) $+281.25^o$ (c 0.4, MeOH) (t = $20^o$) | A 368<br>A 368 |
| $C_{16}H_{16}O_2S$ | 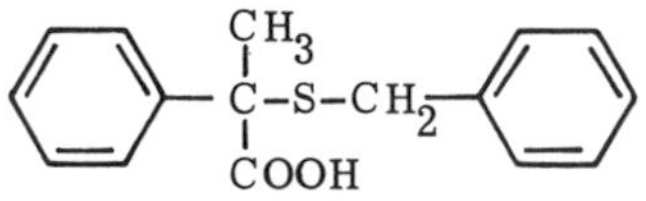 $CH_3$ C-S-$CH_2$ COOH | (+)-$\alpha$-Methylbenzylamine<br>(-)-$\alpha$-Methylbenzylamine<br>both in EtOH-$H_2O$ | a) $-46.5^o$ (c 2.2, EtOH) (t = $24^o$)<br>b) $+46.6^o$ (c 2.3, EtOH) (t = $24^o$) | A 369<br>A 369 |
| $C_{16}H_{18}O_6S_2$ | 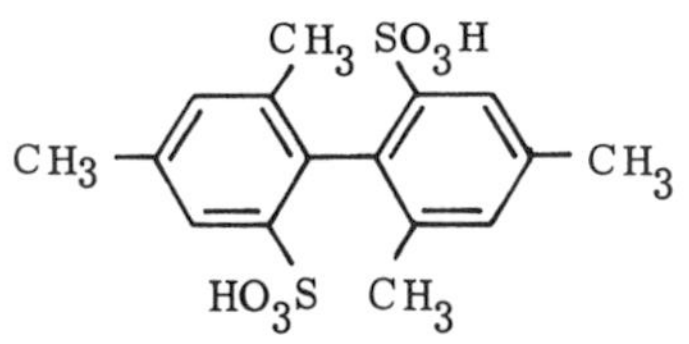 $CH_3$ $SO_3H$ $CH_3$ $CH_3$ $HO_3S$ $CH_3$ | Strychnine·HCl of the sodium salt of the acid in $H_2O$ | a) $-39.0 \pm 0.5^o$ (c 1.1015, $H_2O$) (t=$18^o$) (546 nm)<br>b) $+40.5 \pm 0.5^o$ (c 1.0870, $H_2O$) (t=$20^o$) (546 nm) | A 370 |
| $C_{17}H_{14}O_5S$ | 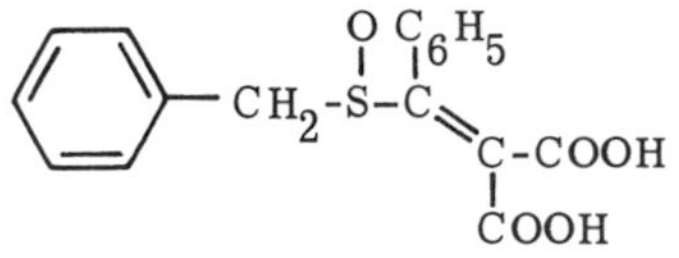 O $C_6H_5$ $CH_2$-S-C= C-COOH COOH | Brucine in acetone | a) $-306.9^o$<br>b) $+307.4^o$ | A 371 |
| $C_{17}H_{22}O_2S_2$ | 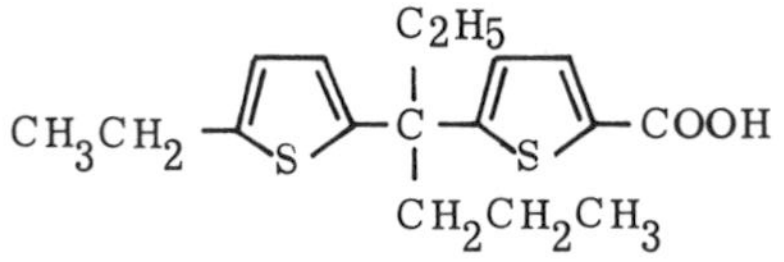 $C_2H_5$ $CH_3CH_2$ S C S COOH $CH_2CH_2CH_3$ | Cinchonidine in EtOAc | $[M]_{578} = -3.6^o$ (c 10.0, $C_6H_6$) | A 372 |

TABLE 8e (continued)

| | Substance resolved | Resolving agent and solvent | Rotation $[\alpha]_D$ of resolved compound | Reference |
|---|---|---|---|---|
| $C_{18}H_{21}NO_4S$ | $C_6H_5-SO_2-N(CH_2-COOH)-$ (2,6-$(CH_3)_2$-4-$CH_3$-3-R-phenyl); R = $CH_3$ | Cinchonine in EtOAc | $+6.9^o$ (c 1.15, EtOH) (t = $29^o$) | A 373 |
| $C_{18}H_{22}O_4S_2$ | See above, under substituted 3,3'-bithienyldicarboxylic acids | | | |
| $C_{19}H_{12}O_4S$ | See Table 8a $C_{20}H_{14}O_3$ X = $SO_2$ R = H | Brucine in acetone | a) $-250^o$ (c 1, glyme) (t = $24^o$) (436 nm)<br>b) $+250^o$ (c 1, glyme) (t = $24^o$) (436 nm) | A 177 |
| $C_{21}H_{27}NO_4S$ | See $C_{18}H_{21}NO_4S$ above R = $(CH_3)_3C$ | Cinchonidine in EtOAc | $+42.5^o$ (c 0.40, EtOH) (t = $26^o$) | A 373 |

## References to Tables 8a-e
## Resolutions of Acids


(A 1) F.Y. Edamura, E.R. Larsen and H. M. Peters, American Chemical Society, 159th National Meeting, Houston, Texas, February 1970, Organic Division, Abstract # 84.

(A 2) G. Bellucci, G. Berti, A. Borraccini and F. Macchia, Tetrahedron, 25, 2979 (1969).

(A 3) C.E. Glassick and W.E. Adcock, Ind. Eng. Chem., Prod. Res. Develop., 3, 14 (1964).

(A 4) G. Bellucci, F. Marioni and A. Marsili, Tetrahedron, 25, 4165 (1969).

(A 5) L.J. Andrews and J.E. Hardwicke, J. Amer. Chem. Soc., 74, 3582 (1952).

(A 6) R.G. Bergman, J. Amer. Chem. Soc., 91, 7405 (1969).

(A 7) T. Kaneko, H. Katsura, H. Asano and K. Wakabayashi, Chem. Ind. (London), 1960, 1187.

(A 8) R.K. Hill and W.R. Schearer, J. Org. Chem., 27, 921 (1962).

(A 9) V.E. Althouse, E. Kaufmann, P. Loeffler, K. Ueda and H.S. Mosher, J. Amer. Chem. Soc., 83, 3138 (1961).

(A 10) S. Senoh, J. Pharm. Soc. Japan, 71, 798 (1951).

(A 11) G. Odham, Ark. Kemi, 20, 507 (1963).

(A 12) J. Kenyon and W.A. Ross, J. Chem. Soc., 1951, 3407.

(A 13) G. Ställberg, Acta Chem. Scand., 11, 1430 (1957); G.I. Fray and N.J. Polgar, J. Chem. Soc., 1956, 2036.

(A 14) J.M. Walbrick, J.W. Wilson, Jr., and W.M. Jones, J. Amer. Chem. Soc., 90, 2895 (1968).

(A 15) W.G. Galetto and W. Gaffield, J. Chem. Soc. C, 1969, 2437.

(A 16) K. Mislow and I.V. Steinberg, J. Amer. Chem. Soc., 77, 3807 (1955).

(A 17) O. Ceder and B. Hansson, Acta Chem. Scand., 24, 2693 (1970).

(A 18) R.P. Zelinski, N.G. Peterson and H.R. Wallner, J. Amer. Chem. Soc., 74, 1504 (1952).

(A 19) M. Farina and E. Mostardini Peronaci, Chim. Ind. (Milan), 48, 602 (1966).

(A 20) S. Pucci, P. Pino and E. Strino, Gazz. Chim. Ital., 98, 421 (1968).

(A 21) J. Cason and R.A. Coad, J. Amer. Chem. Soc., 72, 4695 (1950).

(A 22) F.I. Carroll and R. Meck, J. Org. Chem., 34, 2676 (1969).

(A 23) J.A. Berson and D.A. Ben-Efraim, J. Amer. Chem. Soc., 81, 4083 (1959).

(A 24) J.A. Berson, J.S. Walia, A. Remanick, S. Suzuki, P. Reynolds-Warnhoff and D. Willner, J. Amer. Chem. Soc., 83, 3986 (1961).

(A 25) J.A. Berson and R.G. Bergman, J. Amer. Chem. Soc., 89, 2569 (1967).

(A 26) J.A. Berson and S. Suzuki, J. Amer. Chem. Soc., 81, 4088 (1959).

(A 27) T.L. Jacobs and W.H. Florsheim, J. Amer. Chem. Soc., 72, 256 (1950).

(A 28) A. Heymes, M. Dvolaitzky and J. Jacques, Bull. Soc. Chim. Fr., 1968, 2898.

(A 29) H. Shechter and D.K. Brain, J. Amer. Chem. Soc., 85, 1806 (1963).

(A 30) C. Aaron, D. Dull, J.L. Schmiegel, D. Jaeger, Y. Ohashi and H.S. Mosher, J. Org. Chem., 32, 2797 (1967).

(A 31) A. Fredga and C. Vázquez de Castro y Sarmiento, Ark. Kemi, 7, 387 (1954).

(A 32) (a) A. Fredga, Ark. Kemi, 7, 241 (1954); (b) F. Nerdel and H. Härter, Justus Liebigs Ann. Chem., 621, 22 (1959).

(A 33) F.A. Abd Elhafez and D.J. Cram, J. Amer. Chem. Soc., 74, 5846 (1952).

(A 34) R. Roger and D.G. Neilson, J. Chem. Soc., 1960, 627.

(A 35) K. Pettersson, Ark. Kemi, 10, 283 (1956).

(A 36) J.A. Berson, J.A. Hammons, A.W. Mc Rowe, R.G. Bergman, A. Remanick and D. Houston, J. Amer. Chem. Soc., 89, 2590 (1967).

(A 37) J.H. Wotiz and R.J. Palchak, J. Amer. Chem. Soc., 73, 1971 (1951).

(A 38) W. von E. Doering and K.B. Wiberg, J. Amer. Chem. Soc., 72, 2608 (1950).

(A 39) T.L. Jacobs, R. Reed and E. Pacovska, J. Amer. Chem. Soc., 73, 4505 (1951).

(A 40) H. L. Goering and F.H. McCarron, J. Amer. Chem. Soc., 80, 2287 (1958).

(A 41) H.M. Walborsky and L. Plonsker, J. Amer. Chem. Soc., 83, 2138 (1961).

(A 42) Y. Inouye, T. Sugita and H.M. Walborsky, Tetrahedron, 20, 1695 (1964).

(A 43) A. Fredga, Chem. Ber., 89, 322 (1956); compare D. Battail and D. Gagnaire, Bull. Soc. Chim. Fr., 1964, 3076.

(A 44) D.J. Cram and A.S. Wingrove, J. Amer. Chem. Soc., 86, 5490 (1964).

(A 45) A.W. Schrecker, J. Org. Chem., 22, 33 (1957).

(A 46) D.F. De Tar and C. Weiss, J. Amer. Chem. Soc., 79, 3045 (1957).

(A 47) J.S. Birtwistle, K. Lee, J.D. Morrison, W.A. Sanderson and H.S. Mosher, J. Org. Chem., 29, 37 (1964). See also M. Delepine and F. Larèze, Bull. Soc. Chim. Fr., 22, 104 (1955).

(A 48) W.A. Bonner and T.W. Greenlee, J. Amer. Chem. Soc., 81, 3336 (1959).

(A 49) A.-M. Weidler and G. Bergson, Acta Chem. Scand., 18, 1483 (1964).

(A 50) E.L. Eliel, P.H. Wilken and F.T. Fang, J. Org. Chem., 22, 231 (1957).

(A 51) D.J. Cram, J. Amer. Chem. Soc., 74, 2137 (1952).

(A 52) J. Almy and D.J. Cram, J. Amer. Chem. Soc., 91, 4459 (1969).

(A 53) I. G. M. Campbell and S. M. Harper, J. Sci. Food Agr., 3, 189 (1952).

(A 54) (a) British Patent 1,178,423 (Jan. 21, 1970) [Chem. Abstr., 72, 90671 h (1970)]; (b) M. Matsui and F. Horiuchi, Ger. Offen. 2,043,173 (Apr. 29, 1971) [Chem. Abstr., 75, 19772 y (1971)].

(A 55) E.W. Yankee and D. J. Cram, J. Amer. Chem. Soc., 92, 6329 (1970).

(A 56) D. J. Cram and P. Haberfield, J. Amer. Chem. Soc., 83, 2354 (1961).

(A 57) H. Veldstra and C. van de Westeringh, Rec. Trav. Chim. Pays-Bas, 70, 1113 (1951).

(A 58) A. Fredga and L. Westman, Ark. Kemi, 7, 193 (1954).

(A 59) C.H. De Puy, F.W. Breitbeil and K.R. De Bruin, J. Amer. Chem. Soc., 88, 3347 (1966).

(A 60) L. Westman, Ark. Kemi, 12, 161 (1958).

(A 61) D. J. Cram and J.D. Knight, J. Amer. Chem. Soc., 74, 5835 (1952).

(A 62) R. K. Hill and G. R. Newkome, Tetrahedron, 25, 1249 (1969).

(A 63) K. Pettersson and G. Willdeck, Ark. Kemi, 9, 333 (1956); compare K. Mislow and K. Hamermesh, J. Amer. Chem. Soc., 77, 1590 (1955).

(A 64) G. Sörlin and G. Bergson, Ark. Kemi, 29, 593 (1968).

(A 65) M.S. Silver and T. Sone, J. Amer. Chem. Soc. 90, 6193 (1968).

(A 66) A. Fredga, Ark. Kemi, 12, 547 (1958).

(A 67) D. J. Cram, J. Amer. Chem. Soc., 74, 2152 (1952).

(A 68) J. Applequist, P. Rivers and D.E. Applequist, J. Amer. Chem. Soc., 91, 5705 (1969).

(A 69) H. Hamill and M. A. McKervey, Chem. Commun., 1969, 864.

(A 70) J. Martel and Huynh Chanh, French Patent Addition 94,309 (Aug. 1, 1969) [Chem. Abstr., 72, 100141b (1970)].

(A 71) G. Büchi, E. Koller and C.W. Perry, J. Amer. Chem. Soc., 86, 5646 (1964).

(A 72) A. Fredga and T. Svensson, Ark. Kemi, 25, 81 (1966).

(A 73) M. Janczewski and K. Kut, Rocz. Chem., 42, 1159 (1968) [Chem. Abstr., 70, 3618 p (1969)]; M. Janczewski, K. Kut and T. Jablonska, ibid., 44, 563 (1970) [Chem. Abstr., 74, 12876 f (1971)].

(A 74) A. Khawam and E.V. Brown, J. Amer. Chem. Soc., 74, 5603 (1952).

(A 75) A. Fredga, Ark. Kemi, 8, 463 (1956).

(A 76) B. Sjöberg, Ark. Kemi, 9, 295 (1956).

(A 77) J. Almy, R.T. Uyeda and D.J. Cram, J. Amer. Chem. Soc., 89, 6768 (1967).

(A 78) C.L. Arcus, J. Kenyon and S. Levin, J. Chem. Soc., 1951, 407.

(A 79) D.F. DeTar and J.C. Howard, J. Amer. Chem. Soc., 77, 4393 (1955).

(A 80) B. Sjöberg, Ark. Kemi, 12, 573 (1958).

(A 81) M. Matell, Ark. Kemi, 5, 187 (1953).

(A 82) H.E. Smith and T.C. Willis, Chem. Commun., 1969, 873.

(A 83) D.J. Cram and L. Gosser, J. Amer. Chem. Soc., 86, 5445 (1964).

(A 84) D.J. Cram, W.T. Ford and L. Gosser, J. Amer. Chem. Soc., 90, 2598 (1968).

(A 85) K. Pettersson, Ark. Kemi, 7, 339 (1954).

(A 86) H.L. Goering and H. Hopf, J. Amer. Chem. Soc., 93, 1224 (1971).

(A 87) H.M. Walborsky, L. Barash, A.E. Young and F.J. Impastato, J. Amer. Chem. Soc., 83, 2517 (1961).

(A 88) H.M. Walborsky and A.E. Young, J. Amer. Chem. Soc., 86, 3288 (1964).

(A 89) I.A. D'yakonov, M.I. Komendantov, K.H. Fu, G.L. Korichev, Zh. Obshch. Khim., 32, 928 (1962) [Engl. transl., p. 917].

(A 90) M. Janczewski and T. Matynia, Rocz. Chem., 40, 2029 (1966).

(A 91) H.M. Walborsky and C.G. Pitt, J. Amer. Chem. Soc., 84, 4831 (1962); H.M. Walborsky and J.M. Motes, ibid., 92, 2445 (1970).

(A 92) W. J. Chambers, W. R. Brasen and C. R. Hauser, J. Amer. Chem. Soc., 79, 879 (1957).

(A 93) F. Bell and G. A. Dinsmore, J. Chem. Soc., 1950, 3691.

(A 94) E. W. Yankee and D. J. Cram, J. Amer. Chem. Soc., 92, 6328 (1970).

(A 95) H. E. Zimmerman, S. S. Hixson and E. F. McBride, J. Amer. Chem. Soc., 92, 2000 (1970).

(A 96) M. H. Delton and D. J. Cram, J. Amer. Chem. Soc., 92, 7623 (1970).

(A 97) D. J. Cram and N. L. Allinger, J. Amer. Chem. Soc., 77, 6289 (1955).

(A 98) H. Falk, P. Reich-Rohrwig and K. Schlögl, Tetrahedron, 26, 511 (1970). See also H. Falk and K. Schlögl, Angew. Chem., 80, 405 (1968); Angew. Chem. Int. Ed. Engl., 7, 383 (1968).

(A 99) D. T. Hefelfinger and D. J. Cram, J. Amer. Chem. Soc., 92, 1073 (1970).

(A 100) E. Wenkert, A. Afonso, J. B-son Bredenberg, C. Kaneko and A. Tahara, J. Amer. Chem. Soc., 86, 2038 (1964).

(A 101) A. T. Blomquist, R. E. Stahl, Y. C. Meinwald and B. H. Smith, J. Org. Chem., 26, 1687 (1961).

(A 102) A. G. Brook, J. Amer. Chem. Soc., 85, 3051 (1963).

(A 103) W. Klötzer, Monatsh. Chem., 87, 346 (1956).

(A 104) K. Mislow, S. Heyden and H. Schaeffer, J. Amer. Chem. Soc., 84, 1449 (1962).

(A 105) W. Tochtermann and C. Franke, Angew. Chem., 81, 32 (1969); Angew. Chem. Int. Ed. Engl., 8, 68 (1969).

(A 106) V. Boekelheide and E. Sturm, J. Amer. Chem. Soc., 91, 902 (1969).

(A 107) D. J. Cram, W. J. Wechter and R. W. Kierstead, J. Amer. Chem. Soc., 80, 3126 (1958).

(A 108) W. Tochtermann, H. Küppers and C. Franke, Chem. Ber., 101, 3808 (1968).

(A 109) M. S. Newman and R. M. Wise, J. Amer. Chem. Soc., 78, 450 (1956).

(A 110) J.H. Weisburger, E.K. Weisburger and F.E. Earl, J. Amer. Chem. Soc., 72, 4250 (1950).

(A 111) F.W. Bachelor and G.A. Miana, Can. J. Chem., 47, 4089 (1969).

(A 112) J.I. Brauman and A.J. Pandell, J. Amer. Chem. Soc., 89, 5421 (1967). See also E.J. Corey and W.R. Hertler, ibid., 82, 1657 (1960).

(A 113) G.S. Myers, P. Morozovitch, W.L. Glen, R. Barber, G. Papineau-Couture and G.A. Grant, J. Amer. Chem. Soc., 77, 3348 (1955).

(A 114) J.R. Sjolander, K. Folkers, E.A. Adelberg and L. Tatum, J. Amer. Chem. Soc., 76, 1085 (1954).

(A 115) E.S. Zhdanovich, V.N. Krilova, G.S. Kozlova and N.A. Preobrazhenskii, Zh. Org. Khim., 3, 826 (1967) [Engl. transl., p. 794]. See also E.S. Zhdanovich, G.S. Kozlova, T.D. Marieva, T.V. Mel'nikova and N.A. Preobrazenskii, Zh. Org. Khim., 4, 1359 (1968) [Engl. transl., p. 1310].

(A 116) W. Ozegowski and H. Haering, Pharmazie, 12, 254 (1957).

(A 117) S. Bauer and S. Országh, Czech. Patent 88,066 (Dec. 15, 1966) [Chem. Abstr., 54, 15250h (1960)]; F.D. Pickel, J.I. Fass and S. Chodroff, U.S. Patents 2,460,239 and 2,460,240 (Jan. 25, 1949) [Chem. Abstr., 43, 3448g (1949)].

(A 118) British Patent 1,183,387 (Mar. 4, 1970) [Chem. Abstr., 72, 132087e (1970)].

(A 119) L.A. Paquette and J.P. Freeman, J. Org. Chem., 35, 2249 (1970).

(A 120) P. Block, Jr., J. Org. Chem., 30, 1307 (1965).

(A 121) P. Block, Jr., and M.S. Newman, Org. Syn., 48, 120 (1968).

(A 122) T. Tanabe, S. Yajima and M. Imaida, Bull. Chem. Soc. Japan, 41, 2178 (1968).

(A 123) R. Winterbottom and W.M. Ziegler, U.S. Patent 3,541,139 (Nov. 17, 1970) [Chem. Abstr., 74, 31559w (1971)].

(A 124) L.D. Bergel'son, E.V. Dyatlovitskaya, M. Tichy and V.V. Voronkova, Izv. Akad. Nauk S.S.S.R., Otd. Khim. Nauk, 1962, 1612 [Engl. transl., p. 1525].

(A 125) C.H. Shunk, B.O. Linn, J.W. Huff, J.L. Gilfillan, H.R. Skeggs and K. Folkers, J. Amer. Chem. Soc., 79, 3294 (1957); C.H. Shunk and K.A. Folkers, U.S. Patent 2,945,059 (July 12, 1960) [Chem. Abstr., 55, 5350h (1961)].

(A 126) K. Chilina, U. Thomas, A.F. Tucci, K.D. McMichael and C.M. Stevens, Biochemistry, 8, 2846 (1969).

(A 127) J. Sánchez Real and J. Pascual, An. Real Soc. Espan. Fis. Quim., 49 B, 445 (1953).

(A 128) J.B. Kay and J.B. Robinson, J. Chem. Soc. C, 1969, 248.

(A 129) P. Gomis Torne, Rev. Real Acad. Cienc. Exactas, Fis. Natur. Madrid, 60, 419 (1966) [Chem. Abstr., 66, 55082 w (1967)].

(A 130) R. Grewe and S. Kersten, Chem. Ber., 100, 2546 (1967). See also E.E. Smissman, J.T. Suh, M. Oxman and R. Daniels, J. Amer. Chem. Soc., 84, 1040 (1962).

(A 131) D.S. Noyce and D.B. Denney, J. Amer. Chem. Soc., 74, 5912 (1952).

(A 132) R. Sandberg, Acta Chem. Scand., 16, 1124 (1962).

(A 133) J. Kenyon and M.C.R. Symons, J. Chem. Soc., 1953, 3580.

(A 134) (a) A.M. Likhosherstov, V.N. Kulakov and N.K. Kochetkov, Zh. Obshch. Khim., 37, 1012 (1967) [Engl. transl., p. 956]; (b) R. Adams and B.L. Van Duuren, J. Amer. Chem. Soc., 72, 5349 (1952). See also L.J. Dry and F.L. Warren, J. Chem. Soc., 1952, 3445; (c) V.N. Kulakov, A.M. Likhosherstov and N.K. Kochetkov, Zh. Obshch. Khim., 37, 146 (1967) [Engl. transl., p. 132].

(A 135) D.G. Neilson, U. Zakir and C.M. Scrimgeour, J. Chem. Soc. C, 1971, 898.

(A 136) D.G. Neilson, I.A. Khan and R.S. Whitehead, J. Chem. Soc. C, 1968, 1853.

(A 137) G.F. Grillot and R.I. Bashford, Jr., J. Amer. Chem. Soc., 73, 5598 (1951).

(A 138) J.L. Norula and J. Kenyon, Current Sci. (India), 32, 260 (1963).

(A 139) K. Mislow and M. Heffler, J. Amer. Chem. Soc., 74, 3668 (1952).

(A 140) W.A. Bonner, J. Amer. Chem. Soc., 73, 3126 (1951).

(A 141) J. Jacobus, M. Raban and K. Mislow, J. Org. Chem., 33, 1142 (1968).

(A 142) C.A. McKenzie, Experimental Organic Chemistry, 3rd. ed., Prentice Hall, New York, 1967, p. 257.

(A 143) E.T. Kaiser and F.W. Carson, J. Amer. Chem. Soc., 86, 2922 (1964).

(A 144) G. Nicolau and A. Greff, Analele Univ. "C.I. Parhon," Ser. Stiint. Nat., 9, 211 (1960) [Chem. Abstr., 58, 11261a (1963)].

(A 145) K. Mislow, J. Amer. Chem. Soc., 73, 3954 (1951).

(A 146) A. Deljac, Z. Cibilić and K. Balenović, Bull. Sci., Cons. Acad. Sci. Arts, RSF Yougoslavie, Sect. A, 12, 181 (1967).

(A 147) T.M. Dawson, J. Dixon, B. Lythgoe and I.A. Siddiqui, Chem. Commun., 1970, 992.

(A 148) R.M. Lukes, G.I. Poos and L.H. Sarett, J. Amer. Chem. Soc., 74, 1401 (1952).

(A 149) P. Blattman and J. Retey, Chem. Commun., 1970, 1393.

(A 150) J.A. Dale, D.L. Dull and H.S. Mosher, J. Org. Chem., 34, 2543 (1969).

(A 151) S.G. Cohen and S.Y. Weinstein, J. Amer. Chem. Soc., 86, 725 (1964).

(A 152) D.S. Noyce and C.A. Lane, J. Amer. Chem. Soc., 84, 1635 (1962).

(A 153) D.S. Noyce and C.A. Lane, J. Amer. Chem. Soc., 84, 1641 (1962).

(A 154) W.A. Bonner, J.A. Zderic and G.A. Casaletto, J. Amer. Chem. Soc., 74, 5086 (1952).

(A 155) E.J. Corey, T.K. Schaaf, W. Huber, U. Koelliker and N.M. Weinshenker, J. Amer. Chem. Soc., 92, 397 (1970).

(A 156) A.G. Davies, F.M. Ebeid and J. Kenyon, J. Chem. Soc., 1957, 3154.

(A 157) J.A. Reid and E.E. Turner, J. Chem. Soc., 1950, 3694.

(A 158) J.F. Bunnett and J.L. Marks, J. Amer. Chem. Soc., 74, 5893 (1952).

(A 159) (a) L. Novák and M. Protiva, Collect. Czech. Chem. Commun., 27, 2702 (1962); (b) L. Novák, J.O. Jilek, B. Kakáč, I. Ernest and M. Protiva, ibid., 25, 2196 (1960).

(A 160) M. Matell, Ark. Kemi, 1, 455 (1950).

(A 161) D.S. Noyce, L. Gortler, M.J. Jorgensen, F.B. Kirby and E.C. McGoran, J. Amer. Chem. Soc., 87, 4329 (1965).

(A 162) (a) G. Muller, A. Gaston, A. Poittevin and T. Vesperto, French Patent 1,481,978 (May 26, 1967) [Chem. Abstr., 69, 2698g (1968)]; (b) G. Nominé, G. Amiard and V. Torelli, Bull. Soc. Chim. Fr., 1968, 3664.

(A 163) T. Svensson, Ark. Kemi, 26, 27 (1967).

(A 164) French Patent 1,329,640 (June 14, 1963) [Chem. Abstr., 60, 462b (1964)].

(A 165) R.P. Zelinsky, B.W. Turnquist and E.C. Martin, J. Amer. Chem. Soc., 73, 5521 (1951).

(A 166) K. Mislow and M. Siegel, J. Amer. Chem. Soc., 74, 1060 (1952).

(A 167) A.G. Pinkus, J.I. Riggs, Jr., and S.M. Broughton, J. Amer. Chem. Soc., 90, 5043 (1968).

(A 168) R.H. Baker and S.H. Jenkins, Jr., J. Amer. Chem. Soc., 71, 3969 (1949).

(A 169) K. Sisido, K. Kumazawa and H. Nozaki, J. Amer. Chem. Soc., 82, 125 (1960).

(A 170) British Patent 914,731 (Jan. 2, 1963) [Chem. Abstr., 59, 529d (1963)].

(A 171) Belgian Patent 619,892 (Jan. 7, 1963) [Chem. Abstr., 59, 11381e (1963)].

(A 172) W. Nagata, T. Terasawa and T. Sugasawa, S. African Patent 67 06,757 (Apr. 9, 1968) [Chem. Abstr., 70, 68004e (1969)].

(A 173) R.L. Clarke and S.J. Daum, J. Med. Chem., 13, 320 (1970).

(A 174) E. Wyrzykiewicz, M. Kielczewski and J. Bartz, Zeszyty Nauk. Uniw. Poznaniu, Mat., Fiz., Chem. No. 9, 29 (1965) [Chem. Abstr., 65, 3781c (1966)].

(A 175) M. Dvolaitzky, M.J. Luche-Ronteix and A. Marquet, C.R. Acad. Sci., Paris, Ser. C, 266, 1797 (1968).

(A 176) F.M. Dean, P. Halewood, S. Mongkolsuk, A. Robertson and W.B. Whalley, J. Chem. Soc., 1953, 1250.

(A 177) W. Tochtermann, C. Franke and D. Schäfer, Chem. Ber., 101, 3122 (1968).

(A 178) K.V. Narayanan, R. Selvarajan and S. Swaminathan, J. Chem. Soc. C, 1968, 540.

(A 179) W.J. Gensler, C.M. Samour, S.Y. Wang and F. Johnson, J. Amer. Chem. Soc., 82, 1714 (1960).

(A 180) D. Shapiro and H.M. Flowers, J. Amer. Chem. Soc., 83, 3327 (1961).

(A 181) K. Harada and J. Oh-hashi, Bull. Chem. Soc. Jap., 39, 2311 (1966).

(A 182) C. Wasielewski, Rocz. Chem., 42, 1479 (1968).

(A 183) J.E. Faixat, A. Febrer and J. Pascual, An. Real Soc. Espan. Fis. Quim., 57B, 705 (1961).

(A 184) K. Harada, J. Org. Chem., 31, 1407 (1966).

(A 185) J. Jacobus and M. Raban, J. Chem. Educ., 46, 351 (1969).

(A 186) D.G. Neilson and D.A.V. Peters, J. Chem. Soc., 1962, 1519.

(A 187) W.J. Gottstein and L.C. Cheney, J. Org. Chem., 30, 2072 (1965); W.J. Gottstein, U.S. Patent 3,454,626 (July 8, 1969) [Chem. Abstr., 72, 12917d (1970)].

(A 188) B. Sjöberg and S. Sjöberg, Ark. Kemi, 22, 447 (1964).

(A 189) Y.G. Perron, W.F. Minor, C.T. Holdrege, W.J. Gottstein, J.C. Godfrey, L.B. Crast, R.B. Babel and L.C. Cheney, J. Amer. Chem. Soc., 82, 3934 (1960).

(A 190) A. Fredga and M. Matell, Ark. Kemi, 4, 325 (1952).

(A 191) British Patent 899,023 (June 20, 1962) [Chem. Abstr., 58, 10125b (1963)].

(A 192) A. Fredga and T. Raźnikiewicz, Ark. Kemi, 14, 11 (1959).

(A 193) M. Matell, Ark. Kemi, 9, 157 (1956).

(A 194) A. Fredga and G. Ekstedt, Ark. Kemi, 23, 123 (1965).

(A 195) A. Fredga, Croat. Chem. Acta, 29, 313 (1957).

(A 196) M. Matell, Ark. Kemi, 4, 473 (1952).

(A 197) A. Fredga, E. Gamstedt and L. Ekermo, Ark. Kemi, 29, 515 (1968).

(A 198) A. Fredga and M. Andersson, Ark. Kemi, 25, 223 (1966).

(A 199) A. Fredga and M. Andersson, Ark. Kemi, 21, 555 (1964).

(A 200) A. Fredga, Ark. Kemi, 18, 507 (1962).

(A 201) A. Fredga, A.M. Weidler and C. Grönwall, Ark. Kemi, 17, 265 (1961).

(A 202) M. Matell, Ark. Kemi, 7, 437 (1954).

(A 203) A. Fredga, Ark. Kemi, 25, 87 (1966).

(A 204) A. Fredga, Acta Chem. Scand., 23, 2216 (1969).

(A 205) A. Fredga and E. Gamstedt, Ark. Kemi, 28, 109 (1968).

(A 206) E. Gamstedt, Ark. Kemi, 32, 151 (1969).

(A 207) A. Fredga and K.I. Sändström, Ark. Kemi, 23, 245 (1965).

(A 208) A. Fredga, A. Kjellqvist and E. Tornqvist, Ark. Kemi, 32, 301 (1970).

(A 209) A. Fredga, E. Thimon and K. Rosberg, Ark. Kemi, 32, 369 (1970).

(A 210) M. Matell, Ark. Kemi, 6, 365 (1953).

(A 211) A. Fredga and R. Bäckström, Ark. Kemi, 25, 455 (1966).

(A 212) A. Fredga, I. Kiriks and C. Lundström, Ark. Kemi, 25, 249 (1966).

(A 213) A. Fredga and I. Avalaht, Ark. Kemi, 24, 425 (1965).

(A 214) G.R. Clemo, W.A. Cummings and R. Raper, J. Chem. Soc., 1949, 1923.

(A 215) A. Fredga and K. Olsson, Ark. Kemi, 30, 409 (1969).

(A 216) A. Fredga and F. Plénat, Ark. Kemi, 24, 577 (1965).

(A 217) A. Fredga and E. Gamstedt, Ark. Kemi, 28, 211 (1968).

(A 218) A. Fredga and U. Löfroth, Ark. Kemi, 23, 239 (1965).

(A 219) M. Andersson, Ark. Kemi, 26, 335 (1967).

(A 220) L. Angiolini, P. Costa Bizzarri and M. Tramontini, Tetrahedron, 25, 4211 (1969).

(A 221) D.J. Cram and K.R. Kopecky, J. Amer. Chem. Soc., 81, 2748 (1959).

(A 222) M. Matell, Ark. Kemi, 6, 251 (1953).

(A 223) J.W. Clark-Lewis and R.W. Jemison, Aust. J. Chem., 18, 1791 (1965).

(A 224) T.J. Leitereg and D.J. Cram, J. Amer. Chem. Soc., 90, 4019 (1968).

(A 225) M. Matell, Ark. Kemi, 8, 79 (1955).

(A 226) M. Matell, Ark. Kemi, 6, 375 (1954).

(A 227) M. Matell and S. Larsson, Ark. Kemi, 5, 379 (1953).

(A 228) A. Fredga and M. Matell, Ark. Kemi, 3, 429 (1952).

(A 229) E. Grabowska, J. Arct and Z. Eckstein, Rocz. Chem., 43, 715 (1969).

(A 230) C. van der Stelt, W.J. Heus and W. Th. Nauta, Arzneim.-Forsch., 19, 2010 (1969).

(A 231) D.J. Collins and J.J. Hobbs, Aust. J. Chem., 23, 1605 (1970).

(A 232) T.Y. Jen, G.A. Hughes and H. Smith, J. Amer. Chem. Soc., 89, 4551 (1967).

(A 233) J. Oh-hashi and K. Harada, Bull. Chem. Soc. Jap., 40, 2977 (1967).

(A 234) (a) N.K. Richtmyer, in Methods of Carbohydrate Chemistry, Vol. 2, R.L. Whistler and M.L. Wolfrom, Eds., Academic Press, New York, 1963, p. 49; W.T. Haskins and C.S. Hudson, J. Amer. Chem. Soc., 61, 1266 (1939); (b) J. Read and

W.G. Reid, J. Soc. Chem. Ind., 47, 9 T (1928) and N.B. Chapman and J.F.A. Williams, J. Chem. Soc., 1953, 2797.

(A 235) W.C. Agosta, J. Amer. Chem. Soc., 86, 2638 (1964).

(A 236) J.F. Tocanne and C. Asselineau, Bull. Soc. Chim. Fr., 1965, 3346.

(A 237) P.E. Wilcox, C. Heidelberger and V.R. Potter, J. Amer. Chem. Soc., 72, 5019 (1950).

(A 238) O. Gawron, A.J. Glaid, T.P. Fondy and M.M. Bechtold, J. Amer. Chem. Soc., 84, 3877 (1962).

(A 239) S. Tatsumi, Y. Izumi, M. Imaida, Y. Fukuda and S. Akabori, Bull. Chem. Soc. Jap., 39, 602 (1966).

(A 240) R. Adams and D. Fleš, J. Amer. Chem. Soc., 81, 4946 (1959).

(A 241) R. Adams and F.B. Hauserman, J. Amer. Chem. Soc., 74, 694 (1952).

(A 242) G. Ställberg, Ark. Kemi, 12, 79 (1958).

(A 243) A. Fredga and Å. Sikström, Ark. Kemi, 8, 433 (1956).

(A 244) L. Schotte, Ark. Kemi, 9, 407 (1956).

(A 245) R.P. Linstead, J.C. Lunt and B.C.L. Weedon, J. Chem. Soc., 1950, 3333.

(A 246) A. Lardon and T. Reichstein, Helv. Chim. Acta, 32, 1613 (1949).

(A 247) J.A. Berson and R. Swidler, J. Amer. Chem. Soc., 75, 4366 (1953).

(A 248) G.P. Mueller, J. Amer. Chem. Soc., 72, 4092 (1950).

(A 249) D.E. Applequist and N.D. Werner, J. Org. Chem., 28, 48 (1963).

(A 250) R. Adams, B.L. Van Duuren and B.H. Braun, J. Amer. Chem. Soc., 74, 5608 (1952).

(A 251) K. Serck-Hanssen, Ark. Kemi, 10, 135 (1956).

(A 252) S. Tatsumi, M. Imaida and Y. Izumi, Bull. Chem. Soc. Jap., 39, 2543 (1966).

(A 253) J. Kenyon and W.A. Ross, J. Chem. Soc., 1951, 3407.

(A 254) A. Ställberg-Stenhagen, Ark. Kemi, 3, 273 (1951).

(A 255) A. Fredga and U. Sahlberg, Ark. Kemi, Min., Geol., 18A, No. 16 (1944).

(A 256) A. Fredga, Ark. Kemi, Min., Geol., 23 B, No. 2 (1946).

(A 257) R.E. Pincock, M.-M. Tong and K.R. Wilson, J. Amer. Chem. Soc., 93, 1669 (1961).

(A 258) H. Wynberg and J.P.M. Houbiers, J. Org. Chem., 36, 835 (1971).

(A 259) A. Campbell and H.N. Rydon, J. Chem. Soc., 1953, 3002.

(A 260) T.L. Dawson and G.R. Ramage, J. Chem. Soc., 1950, 3523.

(A 261) R. Sandberg, Ark. Kemi, 16, 255 (1960).

(A 262) A. Fredga, Ark. Kemi, 6, 277 (1953).

(A 263) A. Fredga and M. Matell, Bull. Soc. Chim. Belges, 62, 47 (1953).

(A 264) M. Tichý and J. Sicher, Tetrahedron Lett., 1969, 4609.

(A 265) C.C.J. Culvenor and T.A. Geissman, J. Amer. Chem. Soc., 83, 1647 (1961).

(A 266) L. Ahlquist, J. Asselineau, C. Asselineau, K. Serck-Hanssen, S. Ställberg-Stenhagen and E. Stenhagen, Ark. Kemi, 14, 171 (1959).

(A 267) (a) S. Kushner, J. Morton, II, J.H. Boothe and J.H. Williams, J. Amer. Chem. Soc., 75, 1097 (1953); (b) S. Kushner, J.H. Boothe, J. Morton, II, J. Petisi and J.H. Williams, ibid., 74, 3710 (1952).

(A 268) A. Fredga, Ark. Kemi, Min., Geol., 26 B, No. 11 (1948).

(A 269) L. Westman, Ark. Kemi, 11, 431 (1957). See also K. Kawazu, T. Fujita and T. Mitsui, J. Amer. Chem. Soc., 81, 932 (1959).

(A 270) E. Selegny and M. Vert, Bull. Soc. Chim. Fr., 1968, 2549.

(A 271) J. Martel, Ger. Offen., 1,935,320 (Jan. 15, 1970) [Chem. Abstr., 72, 121078b (1970)].

(A 272) F. Gautschi, O. Jeger, V. Prelog and R.B. Woodward, Helv. Chim. Acta, 38, 296 (1955).

(A 273) J. Porath, Ark. Kemi, Min., Geol., 26 B, No. 16 (1948).

(A 274) L. Westman, Ark. Kemi, 12, 167 (1958).

(A 275) W.S. Briggs, M. Suchý and C. Djerassi, Tetrahedron Lett., 1968, 1097.

(A 276) C.G. Overgerger and D.A. Labianca, J. Org. Chem., 35, 1762 (1970).

(A 277) G.S. Saharia and M.P. Tyagi, J. Indian Chem. Soc., 44, 601 (1967).

(A 278) T.D. Hoffman and D.J. Cram, J. Amer. Chem. Soc., 91, 1000 (1969).

(A 279) A. Hofmann and P. Stadler, Swiss Patent 454,826 (June 28, 1968) [Chem. Abstr., 70, 3569y (1969)].

(A 280) G. Stork and F.H. Clarke, Jr., J. Amer. Chem. Soc., 83, 3114 (1961).

(A 281) M. Rieger and F.H. Westheimer, J. Amer. Chem. Soc., 72, 28 (1950).

(A 282) D.C. Iffland and H. Siegel, J. Amer. Chem. Soc., 80, 1947 (1958).

(A 283) K. Takeda, S. Hagishita, M. Sugiura, K. Kitahonoki, I. Ban, S. Miyazaki and K. Kuriyama, Tetrahedron, 26, 1435 (1970).

(A 284) H. Stetter and O.-E. Bänder, Chem. Ber., 88, 1535 (1955).

(A 285) K. Mislow, R. Graeve, A.J. Gordon and G.H. Wahl, Jr., J. Amer. Chem. Soc., 86, 1733 (1964).

(A 286) J. Suszko and M. Kielczewski, Rocz. Chem., 41, 1291 (1967).

(A 287) J. Suszko and M. Kielczewski, Rocz. Chem., 41, 1565 (1967).

(A 288) Y. Abe, T. Harukawa, H. Ishikawa, T. Miki, M. Sumi and T. Toga, J. Amer. Chem. Soc., 78, 1422 (1956).

(A 289) K. Mislow and H.D. Perlmutter, J. Amer. Chem. Soc., 84, 3591 (1962).

(A 290) L.V. Dvorken, R.B. Smyth and K. Mislow, J. Amer. Chem. Soc., 80, 486 (1958).

(A 291) T. Sato, S. Akabori, M. Kainosho and K. Hata, Bull. Chem. Soc. Jap., 41, 218 (1968).

(A 292) S. Gronowitz and J.E. Skramstad, Ark. Kemi, 28, 115 (1968).

(A 293) W.J. Gensler and H.N. Schlein, J. Amer. Chem. Soc., 78, 169 (1956).

(A 294) W.E. Bachmann, A.S. Dreiding and E.F.M. Stephenson, J. Amer. Chem. Soc., 73, 2765 (1951).

(A 295) H. J. Reich and D. J. Cram, J. Amer. Chem. Soc., 91, 3517 (1969).

(A 296) A. L. Wilds and R. E. Sutton, J. Org. Chem., 16, 1371 (1951).

(A 297) T. Ozawa, J. Pharm. Soc. Japan, 72, 285 (1952) [Chem. Abstr., 47, 2746 b (1953)].

(A 298) A. K. Colter and L. M. Clemens, J. Amer. Chem. Soc., 87, 847 (1965).

(A 299) D. M. Hall and E. E. Turner, J. Chem. Soc., 1955, 1242.

(A 300) S. Hagishita, K. Kuriyama, M. Hayashi, Y. Nakano, K. Shingy and M. Nakagawa, Bull. Chem. Soc. Jap., 44, 496 (1971).

(A 301) (a) M. Crawford, R. A. M. Mackinnon and V. R. Supanekar, J. Chem. Soc., 1959, 2807; (b) A. Brown, C. M. Kemp and S. F. Mason, J. Chem. Soc. A, 1971, 751.

(A 302) J. A. Berson and W. A. Mueller, J. Amer. Chem. Soc., 83, 4940 (1961).

(A 303) D. J. Cram and J. Abell, J. Amer. Chem. Soc., 77, 1179 (1955).

(A 304) F. Bell, J. Chem. Soc., 1952, 1527.

(A 305) F. Bell and D. H. Waring, J. Chem. Soc., 1949, 1579.

(A 306) C. Koukotas and L. H. Schwartz, Chem. Commun., 1969, 1400; C. Koukotas, Ph. D. Dissertation, The City University of New York, 1968.

(A 307) G. Haas and V. Prelog, Helv. Chim. Acta, 52, 1202 (1969).

(A 308) G. Claeson, Ark. Kemi, 30, 277 (1969).

(A 309) M.-O. Hedblom, Ark. Kemi, 31, 489 (1969).

(A 310) L. Schotte, Ark. Kemi, 9, 429 (1956).

(A 311) G. Claeson and H.-G. Jonsson, Ark. Kemi, 26, 247 (1967).

(A 312) G. Claeson and H.-G. Jonsson, Ark. Kemi, 28, 167 (1968).

(A 313) G. Claeson, Ark. Kemi, 30, 511 (1969).

(A 314) M. S. Rabinovitch and G. N. Kulikova, Zh. Obshch. Khim., 35, 235 (1965) [Engl. transl., p. 237].

(A 315) S. Gronowitz, Ark. Kemi, 13, 87 (1958).

(A 316) S. Gronowitz, Ark. Kemi, 13, 231 (1958).

(A 317) E. Jonsson, Acta Chem. Scand., 19, 2247 (1965).

(A 318) D. J. Cram and T. A. Whiting, J. Amer. Chem. Soc., 89, 4651 (1967).

(A 319) T. Raźnikiewicz, Ark. Kemi, 18, 467 (1962).

(A 320) T. Raźnikiewicz, Acta Chem. Scand., 16, 1097 (1962).

(A 321) J.E. Némorin, E. Jonsson and A. Fredga, Ark. Kemi, 30, 403 (1969).

(A 322) L. Schotte, Ark. Kemi, 9, 413 (1956).

(A 323) A. Fredga and A. Björn, Ark. Kemi, 23, 91 (1965).

(A 324) M. Matell, Ark. Kemi, 3, 129 (1951).

(A 325) B. Stridsberg, Ark. Kemi, 32, 9 (1970).

(A 326) B. Stridsberg, Ark. Kemi, 32, 295 (1970).

(A 327) M. Janczewski and T. Rak-Najda, Rocz. Chem., 44, 1599 (1970).

(A 328) W.A. Bonner, J. Org. Chem., 33, 1831 (1968).

(A 329) G. Barbieri, V. Davoli, I. Moretti, F. Montanari and G. Torre, J. Chem. Soc. C, 1969, 731; S. Allenmark and C.-E. Hagberg, Acta Chem. Scand., 24, 2225 (1970).

(A 330) K. Pettersson, Ark. Kemi, 7, 39 (1954).

(A 331) S. Gronowitz, Ark. Kemi, 11, 361 (1957).

(A 332) A. Fredga and O. Palm, Ark. Kemi, Min., Geol., 26 A, No. 26 (1949).

(A 333) S. Gronowitz and S. Larsson, Ark. Kemi, 8, 567 (1956).

(A 334) M.-O. Hedblom and K. Olsson, unpublished work; personal communication from A. Fredga.

(A 335) G. Claeson and H.G. Jonsson, Ark. Kemi, 31, 83 (1969) [Chem. Abstr., 71, 91191t (1969)].

(A 336) R. Hakansson and E. Wiklund, Ark. Kemi, 31, 101 (1969).

(A 337) S. Gronowitz, Ark. Kemi, 23, 307 (1965).

(A 338) S. Gronowitz and P. Gustafson, Ark. Kemi, 20, 289 (1963).

(A 339) S. Gronowitz and H. Frostling, Acta Chem. Scand., 16, 1127 (1962).

(A 340) S. Gronowitz and R. Beselin, Ark. Kemi, 21, 349 (1963).

(A 341) A. Fredga and K. Stadell, unpublished work; personal communication from A. Fredga.

(A 342) U. Folli and D. Iarossi, Gazz. Chim. Ital., 99, 1306 (1969).

(A 343) A. Fredga, unpublished work; personal communication from A. Fredga.

(A 344) D.E. Wolf, R. Mozingo, S.A. Harris, R.C. Anderson and K. Folkers, J. Amer. Chem. Soc., 67, 2100 (1945).

(A 345) A.M.T. Finch, Jr. and W.R. Vaughan, J. Amer. Chem. Soc., 91, 1416 (1969).

(A 346) B. Sjöberg, Ark. Kemi, 12, 565 (1958).

(A 347) B. Sjöberg, Ark. Kemi, 11, 439 (1957).

(A 348) A. Fredga, Ark. Kemi, Min., Geol., 24B, No. 15 (1947).

(A 349) S. Allenmark and C.-E. Hagberg, Acta Chem. Scand., 22, 1461 (1968).

(A 350) H. Schulz and V. du Vigneaud, J. Amer. Chem. Soc., 88, 5015 (1966).

(A 351) L.J. Owen and F.F. Nord, J. Org. Chem., 16, 1864 (1951).

(A 352) M. Janczewski and S.E. Sadowska, Rocz. Chem., 41, 945 (1967).

(A 353) G. Ege and W. Planer, Angew. Chem., 81, 749 (1969); Angew. Chem. Int. Ed. Engl., 8, 758 (1969).

(A 354) A. Fredga, K. Aejmelaeus and B. Tollander, Ark. Kemi, 3, 331 (1951).

(A 355) S. Allenmark, O. Bohman and C.E. Hagberg, Ark. Kemi, 30, 273 (1969).

(A 356) D.J. Cram and A. Ratajczak, J. Amer. Chem. Soc., 90, 2198 (1968).

(A 357) J. Chickos and K. Mislow, J. Amer. Chem. Soc., 89, 4815 (1967).

(A 358) M. Janczewski, M. Dec and W. Charmas, Rocz. Chem., 40, 1021 (1966).

(A 359) K. Pettersson, Ark. Kemi, 7, 279 (1954).

(A 360) O. Bohman and S. Allenmark, Ark. Kemi, 31, 299 (1969).

(A 361) M. Janczewski, H. Maziarczyk and E. Ulanowska, Rocz. Chem., 43, 237 (1969).

(A 362) M. Janczewski and W. Charmas, Ann. Univ. Mariae Curie-Sklodowska, Lublin-Polonia, Sect. AA, 18, 81 (1963). [Chem. Abstr., 66, 37570z (1967)].

(A 363) W.A. Bonner, J. Org. Chem., 32, 2496 (1967).

(A 364) W.A. Bonner, J. Amer. Chem. Soc., 74, 1034 (1952).

(A 365) D.J. Cram and A.S. Wingrove, J. Amer. Chem. Soc., 85, 1100 (1963).

(A 366) D. Farge and M.N. Messer, South African Patent 68 00,524 (Jun. 25, 1968) [Chem. Abstr., 70, 77597c (1969)].

(A 367) M. Janczewski and M. Podgorski, Rocz. Chem., 43, 1479 (1969).

(A 368) M. Janczewski, J. Pytlarz and J. Poplawski, Rocz. Chem., 37, 1153 (1963).

(A 369) W.A. Bonner and R.A. Grimm, J. Org. Chem., 32, 3022 (1967).

(A 370) W.L.F. Armarego and E.E. Turner, J. Chem. Soc., 1956, 3668.

(A 371) S. Allenmark, C.E. Hagberg and O. Bohman, Ark. Kemi, 30, 269 (1969).

(A 372) H. Wynberg, G.L. Hekkert, J.P.M. Houbiers and H.W. Bosch, J. Amer. Chem. Soc., 87, 2635 (1965).

(A 373) R. Adams and J.S. Dix, J. Amer. Chem. Soc., 80, 4579 (1958).

TABLE 9a  Simple Amines and Onium Salts *)

| | Substance resolved | Resolving agent and solvent | Rotation $[\alpha]_D$ of resolved compound | Reference |
|---|---|---|---|---|
| $C_4H_{11}N$ | $CH_3CH_2CH(NH_2)CH_3$ | (+)-Tartaric acid in $H_2O$ | a) $+ 8.1^o$ (neat) ($t = 25^o$) | B 1 |
| | | Improved procedure | $+ 7.48^o$ (neat) ($t = 20^o$) | B 2 |
| | | Resolution in MeOH | $\alpha^{20}_D + 5.38^o$ (l 1 dm, neat) | B 3 |
| | | (-)-Tartaric acid in $H_2O$ | b) $- 7.64^o$ (neat) ($t = 19^o$) | B 2 |
| $C_5H_{11}N$ | cyclopropyl-$CH(NH_2)-CH_3$ | (+)-Tartaric acid in $H_2O$ | $\alpha^{25}_D + 21.80^o$ (l 1 dm, neat) | Al 27 (p.218) |
| $C_5H_{13}N$ | $CH_3CH(CH_3)CH(NH_2)CH_3$ | (+)-Tartaric acid in MeOH | $- 3.45^o$ (neat) ($t = 24^o$)<br>Optical purity $\geq 95\%$ | B 4 |
| $C_5H_{13}N$ | $CH_3CH_2CH_2CH(NH_2)CH_3$ | (+)-Tartaric acid in MeOH | $\alpha^{25}_D - 6.04^o$ (l 1 dm, neat)<br>Optical purity 100 % | B 3 |
| $C_6H_{15}N$ | $CH_3CH(CH_3)CH_2CH(NH_2)CH_3$ | (+)-Tartaric acid in MeOH | $- 11.2^o$ (neat) ($t = 25^o$)<br>$+ 3.5^o$ (c 1, MeOH) ($t = 25^o$) | B 5, B 4 |
| | | Resolution with (-)-O,O'-dibenzoyltartaric acid | | B 6 |
| $C_6H_{15}N$ | $CH_3CH_2CH_2CH_2CH(NH_2)CH_3$ | (+)-Tartaric acid in MeOH | $\alpha^{25}_D - 4.52^o$ (l 1 dm, neat)<br>Optical purity 77.4 % | B 3 |

*) References to Tables 9a-d begin on p. 182

TABLE 9a (continued)

| Formula | Substance resolved | Resolving agent and solvent | Rotation $[\alpha]_D$ of resolved compound | Reference |
|---|---|---|---|---|
| $C_7H_{17}N$ | $CH_3CH(CH_3)CH_2CH_2CH(NH_2)CH_3$ | (+)-Tartaric acid in MeOH | - 7.2° (neat) (t = 25°)<br>- 0.6° (c 10, MeOH) (t = 25°) | B 5 |
| $C_7H_{17}N$ | $CH_3CH_2C(CH_3)(NH_2)CH_2CH_2CH_3$ | (+)-Tartaric acid in $H_2O$ | - 0.096° (l 4 dm, neat) (t = 25°) | B 7 |
| $C_7H_{17}N$ | $CH_3CH_2CH_2CH_2CH_2CH(NH_2)CH_3$ | (+)-Tartaric acid in MeOH | $\alpha_D^{24}$ - 5.02° (l 1 dm, neat)<br>Optical purity 69.2 % | B 3, B 5 |
| | | Resolution with (-)-2-pyrrolidone-5-carboxylic acid | | B 8 |
| $C_8H_{10}N_2O_2$ | $O_2N$-$C_6H_4$-$CH(NH_2)CH_3$ m-isomer | (+)-Camphor-10-sulfonic acid in $H_2O$ | a) + 2.5° (c 8.054, $H_2O$) (t = 20°)<br>- 3.8° (c 2.248, EtOH) (t = 20°)<br>b) - 2.1° (c 6.34, $H_2O$) (t = 20°)<br>all on the amine hydrochlorides | B 9 |
| $C_8H_{10}N_2O_2$ | p-isomer | (+)-Camphor-10-sulfonic acid in $H_2O$ | + 16.7° (neat) (t = 22°) | B 10, B 11 |
| | | Resolution with (+)-tartaric acid | | B 12 |
| $C_8H_{11}N$ | $C_6H_5$-$CH(NH_2)CH_3$ | (+)-Tartaric acid in MeOH | a) - 40.3° (neat) (t = 22°)<br>- 40.1° (neat) (t = 25°)<br>- 39.4° (neat) (t = 29°) | B 13<br>B 10<br>B 14 |
| | | For convenient procedures and modifications, see references (B 14), (B 10) and (B 15) | | |
| | | (-)-Malic acid in $H_2O$ | b) + 39.7° (neat) (t = 29°)<br>+ 40.1° (neat) (t = 25°) | B 14, B 13<br>B 10, B 16 |
| | | (-)-(Hydroxymethylene)-camphor in MeOH | + 40.5° (neat) (t = 20°) | B 17 |

Also, resolution with (-)-N-acetyl-L-leucine (B 18, B 1), (+)-N-acetyl-3,5-dibromo-L-tyrosine (B 18), (-)-barium bornyl sulfate (B 19), (-)-diisopropylidene-2-oxo-L-gulonic acid (see Chapter II, ref. 4), (+)- and (-)-N-($\alpha$-methylbenzyl)-succinamic acid (B 20), (-)-2-pyrrolidone-5-carboxylic acid (B 8), and (-)-2,3,4,6-tetra-O-acetyl-D-glucose (B 21, B 22).

TABLE 9a (continued)

| Formula | Substance resolved | Resolving agent and solvent | Rotation $[\alpha]_D$ of resolved compound | Reference |
|---|---|---|---|---|
| $C_8H_{19}N$ | n-$C_6H_{13}CH(NH_2)CH_3$ | (+)-Tartaric acid in MeOH | $\alpha_D^{19}$ - 5.32° (1 1 dm, neat) | B 23, B 3 |
| | | Resolution with (-)-2-pyrrolidone-5-carboxylic acid | | B 8 |
| $C_9H_{11}N$ | $H_5C_6$, H, H, $NH_2$ (cyclopropane) | (+)-Tartaric acid in EtOH-$Et_2O$ | a) - 30.5° (c 1%, $H_2O$) (t = 25°)<br>b) + 31.0° (c 1%, $H_2O$) (t = 25°)<br>both evidently on tartrate salts | B 24 |
| $C_9H_{13}N$ | $CH_3$-$C_6H_4$-$CH(NH_2)CH_3$ | (-)-N-Acetyl-L-leucine in $H_2O$ | - 34.0° (neat) (t = 25°) | B 18 |
| $C_9H_{13}N$ | $C_6H_5$-$CH(NH_2)CH_2CH_3$ | (-)-Malic acid in EtOH | a) + 21.2° (neat) (t = 24°)<br>+ 8.1° (c 7.9, EtOH) (t = 27°) | B 25, B 26 |
| | | (+)-N-Acetyl-3,5-dibromo-L-tyrosine in $H_2O$ | b) - 12.2° (c 4, EtOH) (t = 25°) | B 18 |
| | | Partial resolution with (+)-tartaric acid in EtOH | | B 27 |
| $C_9H_{13}N$ | $C_6H_5$-$CH_2CH(NH_2)CH_3$<br>Amphetamine | (+)-Tartaric acid in EtOH | a) + 36.3° (neat) (t = 19°)<br>+ 40.2° (c 8.7, $C_6H_6$) (t = 15°) | B 28, B 29<br>B 30 |
| | | See also, Table 2, ref. 4a | | |
| | | Menthyl hydrogen sulfate | b) - 26° (neat) (t = 20°) | B 28 |
| | | (-)-Tartaric acid | - 40.1° (c 8.83, $C_6H_6$) (t = 15°) | B 30 |
| | | Resolution with (+)- and (-)-N-(α-methylbenzyl)-phthalamic acid in 2-butanone (B 20) | | |
| $C_9H_{15}N$ | $N(CH_3)_2$, H (cycloheptadiene)<br>α-Methyltropidine | (+)-O,O'-Dibenzoyltartaric acid in EtOH | - 141° ($CHCl_3$) (t = 25°) | B 31 |

TABLE 9a (continued)

| | Substance resolved | Resolving agent and solvent | Rotation $[\alpha]_D$ of resolved compound | Reference |
|---|---|---|---|---|
| $C_9H_{17}N$ | Camphenilylamine ($H_3C$, $NH_2$, H, $CH_3$ on bicyclic ring) | (+)-Camphoric acid in dioxane or (+)-tartaric acid in $Et_2O$-EtOH-$H_2O$ | + 76° (c 1.5-2.5 %, $C_6H_6$) (t = 26°) on camphenilone obtained from camphorate salt | B 32 |
| $C_{10}H_{15}N$ | Y, Z, $CH_3$-substituted benzene with $-CH(NH_2)CH_3$; Y = $CH_3$ Z = H | (-)-Barium menthyl sulfate | + 36.1° (neat)<br>+ 25.5° ($Et_2O$) (t = 20°) | B 12<br>B 19 |
| | Y = H Z = $CH_3$ | (-)-Barium menthyl sulfate | + 43.8° (neat)<br>+ 32.0° ($Et_2O$) (t = 18°) | B 12<br>B 19 |
| $C_{10}H_{15}N$ | $C_6H_5-CH(NH_2)CH(CH_3)_2$ | (+)-Tartaric acid<br><br>(-)-Malic acid in EtOH-$Et_2O$ (3:2) | + 1.14° (neat)<br>$\alpha_D^{19.5}$ - 1.40° (1 1 dm, neat)<br>+ 1.04° (EtOH) | B 33<br>B 27<br>B 34 |
| $C_{10}H_{15}N$ | $C_6H_5-CH_2CH(NH(CH_3))CH_3$<br>Deoxyephedrine | (+)-Tartaric acid in MeOH<br>See also Table 2, ref. 8 | a) - 2.21° (neat) (t = 24°)<br>b) + 2.48° (neat) (t = 24°) | A 2 (p. 121) |
| $C_{10}H_{15}N$ | $C_6H_5-CH_2CH(NH_2)CH_2CH_3$ | (+)-Tartaric acid in EtOH | - 32° (neat) (t = 20°) | B 17 |
| | | Partial resolution with (-)-(hydroxymethylene)-camphor | | B 17 |
| $C_{10}H_{15}N$ | $C_6H_5-C(CH_3)(NH_2)CH_2CH_3$ | (-)-Malic acid<br>Recr.: $H_2O$<br>(+)-Tartaric acid | a) - 17.0° (neat)<br><br>+ 19.2° (c 7.0, $CHCl_3$) (t = 25°)<br>b) - 19.2° (c 7.0, $CHCl_3$) (t = 25°)<br>both on the N-benzoyl derivatives | B 35<br><br>B 36<br>B 36 |

TABLE 9a (continued)

| Formula | Substance resolved | Resolving agent and solvent | Rotation $[\alpha]_D$ of resolved compound | Reference |
|---|---|---|---|---|
| $C_{10}H_{15}N$ | $C_6H_5-CH_2CH_2CH(NH_2)CH_3$ | (+)-Tartaric acid in EtOH | a) - 4.47° (neat)<br>b) + 3.12° (neat) | B 29 |
| $C_{10}H_{19}N$ | Fenchylamine ($\alpha$-isomer) | (-)-N-Acetyl-L-leucine<br>See also Table 2, ref. 11 | a) + 25.5° (c 5, 95% EtOH) (t = 25°)<br>b) - 25.4° (c 4, 95% EtOH) (t = 25°) | B 37 |
| $C_{10}H_{19}N$ | cyclooctenyl-$N(CH_3)_2$ | (+)-Camphor-10-sulfonic acid. Recr.: diisobutyl ketone | a) - 59.5° (c 7.0, acetone) (t = 25°)<br>b) + 61.5° (c 2.4, acetone) (t = 25°) | B 38 |
| $C_{10}H_{21}N$ | cyclohexyl(H)-N(H)-$CH(CH_3)CH_2CH_3$ | Methyl hydrogen O,O'-dibenzoyl-D-tartrate in DMF | a) $[\alpha]^{25}_{Hg}$ + 5.2°<br>b) $[\alpha]^{25}_{Hg}$ - 5.0°<br>on the hydrochloride salts | B 39 |
| $C_{10}H_{23}N$ | $\underline{n}$-$C_8H_{17}CH(NH_2)CH_3$ | (+)-Tartaric acid in MeOH | $\alpha^{20}_D$ - 3.98° (1 1 dm, neat)<br>Optical purity 76.8% | B 3 |
| $C_{11}H_{17}N$ | $C_6H_5-CH(NH_2)C(CH_3)_3$ | (+)-Tartaric acid in EtOH | a) $\alpha^{19}_D$ - 1.65°(1 1 dm, neat)<br>b) $\alpha^{20}_D$ + 0.99°(1 1 dm, neat) | B 27 |
| | | (-)-N-Acetyl-L-leucine in $H_2O$ | + 5.6° (neat) (t = 21°) | B 25 |
| | | Partial resolution with (+)-camphoric acid | | B 34 |

TABLE 9a (continued)

| Formula | Substance resolved | Resolving agent and solvent | Rotation $[\alpha]_D$ of resolved compound | Reference |
|---|---|---|---|---|
| $C_{11}H_{17}N$ | $C_6H_5CH_2CH(N(CH_3)_2)CH_3$ | (+)-Tartaric acid in MeOH | a) + 3.65° (neat) (t = 22°)<br>+ 13.1° (c 10.0, $H_2O$) (t = 16°) on the hydrochloride<br>b) - 13.1° (c 10.0, $H_2O$) (t = 16°) on the hydrochloride | B 40 |
| $C_{12}H_{13}N$ | 1-naphthyl-$CH(CH_3)NH_2$ | (-)-Diisopropylidene-2-oxo-L-gulonic acid in acetone<br>See also Table 2, ref. 15a | - 79.57° (neat) (t = 25°) | Chapter II, ref. 4 (p.14) |
| $C_{12}H_{13}N$ | 2-naphthyl-$CH(NH_2)CH_3$ | (+)-Tartaric acid in $H_2O$<br>See also Table 2, ref. 15a | a) + 19° (EtOH) | B 41 |
| $C_{12}H_{16}NF_3$ | 3-$CF_3C_6H_4CH_2CH(NHCH_2CH_3)CH_3$ | (+)-O,O'-Dibenzoyltartaric acid in EtOH<br>(+)-Camphoric acid in EtOH | a) - 9.6° (c 8, EtOH) (t = 25°)<br>b) + 9.5° (c 8, EtOH) (t = 25°) | B 42<br>B 42 |
| $C_{12}H_{19}N$ | $C_6H_5C(CH_3)(NH_2)CH_2CH_2CH_2CH_3$ | (+)-Tartaric acid in isoPrOH-EtOH-acetone | + 12.1° (c 2.81, $CHCl_3$) (t = 27°) | B 43 |
| $C_{13}H_{12}NCl$ | $C_6H_5CH(NH_2)C_6H_4$-4-Cl | (+)-Tartaric acid<br>(-)-Tartaric acid<br>both in $H_2O$ | a) - 10.9° (c 5.054, EtOH)<br>b) + 10.8° (c 2.184, EtOH) | B 44<br>B 44<br>Also, M106 (p. 266) |
| $C_{13}H_{17}N$ | bicyclic (norbornane) with H, $C_6H_5$, $NH_2$, H | (+)-Tartaric acid<br>Recr.: EtOH | + 47.6° (c 1, EtOH) (t = 25°) on the hydrochloride salt | B 45 |

TABLE 9a (continued)

| | Substance resolved | Resolving agent and solvent | Rotation $[\alpha]_D$ of resolved compound | Reference |
|---|---|---|---|---|
| $C_{14}H_{15}N$ | 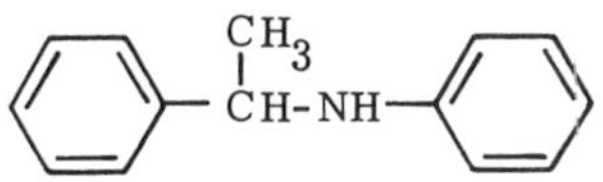 $CH_3$ CH-NH | (+)-Camphor-10-sulfonic acid. Recr.: $C_6H_6$ Partial resolution | $-26.1 \pm 0.9^o$ (c 2.15, EtOH) ($t = 25^o$) | B 46 |
| $C_{14}H_{15}N$ | 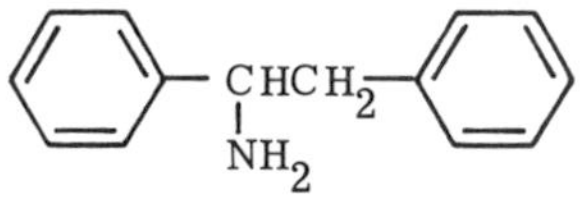 $CHCH_2$ $NH_2$ | (+)-Tartaric acid in MeOH | a) $-11.9^o$ (presumably neat) ($t = 23^o$)<br>b) $+11.9^o$ (presumably neat) ($t = 24^o$) | B 47 |
| $C_{14}H_{21}N$ | 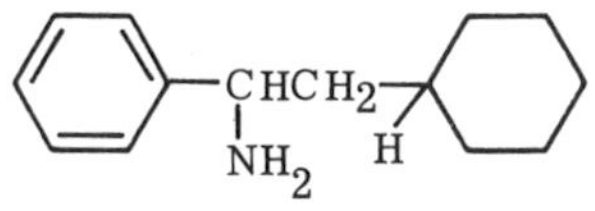 $CHCH_2$ $NH_2$ H | (+)-N-Acetyl-3,5-dibromo-L-tyrosine in MeOH-$H_2O$ | a) $-18.8^o$ (EtOH) ($t = 22^o$) (on the base)<br>$+42.5^o$ (EtOH) ($t = 19^o$) (HCl salt)<br>b) $-42.3^o$ (EtOH) ($t = 18^o$) (HCl salt) | B 48 |
| $C_{14}H_{21}N$ | 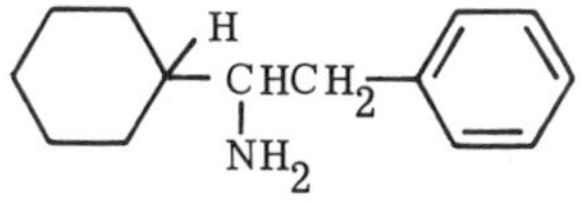 H $CHCH_2$ $NH_2$ | (+)-N-Acetyl-3,5-dibromo-L-tyrosine in MeOH-$H_2O$ | a) $-23.1^o$ (MeOH) ($t = 21^o$)<br>b) $+23.1^o$ (MeOH) ($t = 19^o$)<br>both on hydrochloride salts | B 48 |
| $C_{15}H_{17}N$ | 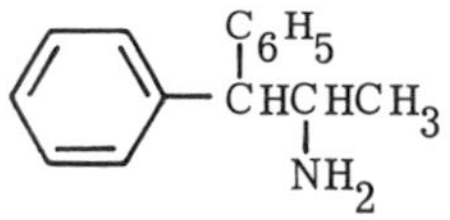 $C_6H_5$ $CHCHCH_3$ $NH_2$ | (+)-Camphor-10-sulfonic acid in $H_2O$<br>(+)-Camphoric acid in EtOH-$H_2O$ | a) $+10.7^o$ (EtOH) ($t = 25^o$)<br>b) $-10.8^o$ (EtOH) ($t = 24^o$) | B 49<br>B 49 |
| $C_{15}H_{23}N$ | 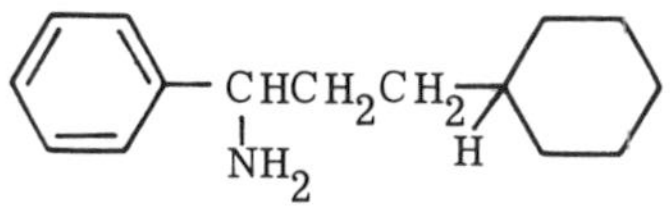 $CHCH_2CH_2$ $NH_2$ H | (+)-N-Acetyl-3,5-dibromo-L-tyrosine in MeOH-$H_2O$ | a) $-22.2^o$ (EtOH) ($t = 22^o$)<br>b) $+23.0^o$ (EtOH) ($t = 20^o$)<br>both on hydrochloride salts | B 48 |
| $C_{15}H_{23}N$ | 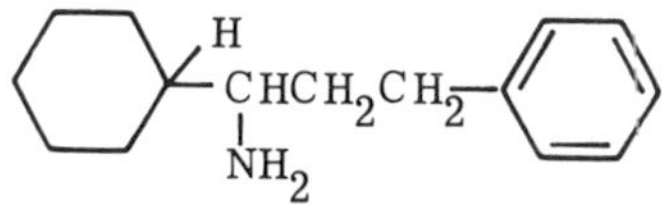 H $CHCH_2CH_2$ $NH_2$ | (-)-N-Acetyl-L-leucine in MeOH-$H_2O$ | a) $-13.1^o$ (MeOH) ($t = 18^o$)<br>b) $+13.8^o$ (MeOH) ($t = 23^o$)<br>both on hydrochloride salts | B 48 |

TABLE 9a (continued)

| | Substance resolved | Resolving agent and solvent | Rotation $[\alpha]_D$ of resolved compound | Reference |
|---|---|---|---|---|
| $C_{17}H_{37}NO$ | cyclooctyl–$\overset{+}{N}(CH_3)(CH_2CH_2CH_2CH_3)(CH_2CH(CH_3)CH_3)$ $OH^-$ | (+)-Camphor-10-sulfonic acid in $H_2O$ | a) $- 1.65^o$ (c 1.73, EtOH) ($t = 25^o$) Optical purity 99.6 % b) $+ 0.81^o$ (c 1.76, EtOH) ($t = 25^o$) both on perchlorate salts | B 50 |
| $C_{18}H_{19}NBrCl$ | $CH_3$, Br, $CH_3$, $CH_3$-substituted phenyl–C(Cl)=C($CH_3$)–C$_6$H$_4$–$NH_2$ | (+)-Camphor-10-sulfonic acid in $H_2O$-acetone | $+ 56.1^o$ (c 1.67, EtOH) ($t = 20^o$) on the N-benzoyl derivative | B 51 |
| $C_{19}H_{17}N$ | $C_6H_5$–$C_6H_4$–CH($NH_2$)–$C_6H_5$ | (+)-Tartaric acid in EtOH | $+ 12.0^o$ (EtOH) | B 34 |
| $C_{20}H_{19}N$ | $C_6H_5$–CH($C_6H_5$)–CH($NH_2$)–$C_6H_5$ | (+)-Camphor-10-sulfonic acid in EtOH-$H_2O$ (+)-Tartaric acid in EtOH-$H_2O$ | a) $- 69.5^o$ (c 3.7, EtOH) ($t = 25^o$) b) $+ 74.6^o$ (EtOH) ($t = 25^o$) | B 52 B 52 |
| $C_{20}H_{23}N$ | bicyclic amine with $N(CH_3)_2$, H, H, $H_5C_6$, $H_5C_6$ | (-)-Malic acid in $Et_2O$-EtOH | $+ 52.6^o$ (c 0.00095, EtOH) ($t = 24.7^o$) | B 53 |
| $C_{20}H_{25}N$ | cyclopentane with $N(CH_3)_2$, $C_6H_5CH_2$, $C_6H_5$ | (-)-Malic acid in $Et_2O$-MeOH | $+ 17.0^o$ (c 0.0030, EtOH) ($t = 25.7^o$) | B 53 |
| $C_{24}H_{21}N$ | 1-naphthyl–C($C_6H_5$)($NH_2$)–$C_6H_4$–$CH_3$ | (+)-Camphor-10-sulfonic acid in EtOH (-)-Camphor-10-sulfonic acid in EtOH | a) $- 1.3^o$ (c 3.7, $CS_2$) ($t = 17^o$) b) $+ 1.3^o$ (c 4.7, $CS_2$) ($t = 17^o$) | A 78 (p.125) A 78 |

TABLE 9b Hydroxy, Alkoxy, Keto, Mercapto, and Sulfinyl Amines. Amine Oxides

| | Substance resolved | Resolving agent and solvent | Rotation $[\alpha]_D$ of resolved compound | Reference |
|---|---|---|---|---|
| $C_3H_9NO$ | $CH_3CH(OH)-CH_2(NH_2)$ | (+)-Tartaric acid in $H_2O$ Partial resolution | $+16.75^o$ (t = $25^o$) | B 54 |
| | | (+)-Tartaric acid in $EtOH-Et_2O$ on the N-benzyl O-p-nitrobenzoate derivative | a) $-31.5^o$ (c 1, MeOH) (t = $25^o$)<br>b) $+35^o$ (c 1, MeOH) (t = $25^o$)<br>both on the hydrochloride salts | B 55 |
| $C_3H_9NS$ | $CH_3CH(SH)-CH_2(NH_2)$ | (+)-Glucose in MeOH *via* the thiazolidine derivative | a) $+35.3 \pm 0.9^o$ (c 1.46, $H_2O$) (t = $25^o$)<br>b) $-36.4 \pm 1.1^o$ (c 1.27, $H_2O$) (t = $25^o$)<br>both on the hydrochloride salts | B 56 |
| $C_4H_{11}NO$ | $CH_3CH_2CH(H_2N)CH_2(OH)$ | (+)-Tartaric acid in $H_2O$ | a) $-10.1^o$ (neat) (t = $25^o$)<br>b) $+10.1^o$ (neat) (t = $20^o$) | B 57, B 58, B 59 |
| | | (+)-Glutamic acid in $H_2O$ or EtOH | $+9.7 \pm 0.0^o$ (neat) (t = $25^o$) | B 58, B 60 |
| | | For variants of the resolution with (+)-tartaric acid, see refs. (B 61) (B 62) (B 63). Resolution with (-)- and (+)-N-($\alpha$-methylbenzyl)-succinamic acid (B 20). | | |
| $C_4H_{11}NO$ | $CH_3CH(OH)CH(NH_2)CH_3$ *threo* isomer | (+)-Tartaric acid in $EtOH-MeOH-H_2O$ | a) $-17.05^o$ (neat) (t = $25^o$) | B 64 |
| | | (-)-Malic acid Recr.: EtOH | b) $+16.95^o$ (neat) (t = $25^o$) | B 64 |
| $C_4H_{11}NOS$<br>$C_9H_{13}NOS$ | See Table 12 Sulfur Compounds | | | |
| $C_6H_{13}NS$ | 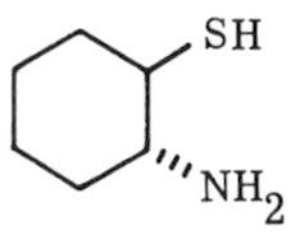 SH, $NH_2$ | (+)-Glucose in $H_2O$ *via* the thiazolidine derivative | a) $-49.1^o$ (c 1.85, $H_2O$) (t = $17^o$)<br>b) $+49.1^o$ (c 1.9, $H_2O$) (t = $17^o$)<br>both on the hydrochloride salts | B 65 |
| $C_7H_9NOS$ | 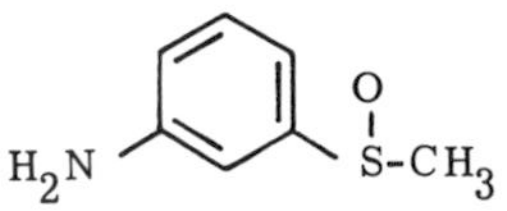 $H_2N$, $S(O)-CH_3$ | (+)-Camphor-10-sulfonic acid in isoPrOH Partial resolution | a) $-23^o$ (c 1, $CHCl_3$) (t = $20^o$)<br>b) $+30^o$ (c 1, $CHCl_3$) (t = $20^o$) | B 66 |

TABLE 9b (continued)

| Formula | Substance resolved | Resolving agent and solvent | Rotation $[\alpha]_D$ of resolved compound | Reference |
|---|---|---|---|---|
| $C_7H_{11}NO$ | $NH_2$, O | (+)-Camphor-10-sulfonic acid in $H_2O$ on the amino ketone hydrochloride | + 23.9° (MeOH) (578 nm) on the camphorsulfonate salt | B 67 |
| $C_7H_{15}NO$ | $CH_3C(O)CH(CH_3)CH_2N(CH_3)CH_3$ | (+)-Camphor-10-sulfonic acid in EtOAc | + 32.35° (c ca. $10^{-2}$, $H_2O$) (t = 30°) on the camphorsulfonate salt | B 68 |
| $C_7H_{15}NO$ | $NHCH_3$, OH | (+)-Tartaric acid<br>(-)-Tartaric acid<br>both in EtOH | a) + 86.3° (c 5, $H_2O$) (t = 20°)<br>b) - 84.4° (c 5, $H_2O$) (t = 20°) | A 128 (p.128) |
| $C_8H_{11}NO_2$ | HO, $CHCH_2-NH_2$, OH | (-)-O, O'-Dibenzyltartaric acid in $H_2O$ or MeOH | a) + 39° (c 3 %, $H_2O$) (t = 20°) (HCl salt)<br>+ 45° (c 0.4, MeOH) (t = 19°) (HCl salt)<br>b) - 33° ($H_2O$) (t = 20°) (HCl salt) | B 69<br>B 70<br>B 69 |
| $C_8H_{11}NO_2$ | HO, $CHCH_2-NH_2$, OH<br>Octopamine | (+)-Camphor-10-sulfonic acid in EtOH-EtOAc | a) - 37.4° (c 1, $H_2O$) (t = 25°)<br>b) + 37.2° (c 1, $H_2O$) (t = 25°) | B 71 |
| $C_8H_{11}NO_3$ | HO, HO, $CHCH_2-NH_2$, OH<br>Norepinephrine; Arterenol | (+)-Tartaric acid in $H_2O$<br>See also Table 2, ref. 18 | a) - 37.3° (c 5%, $H_2O$ containing 1 equivalent HCl) (t = 25°)<br>b) + 37.4° (c 5%, $H_2O$ containing 1 equivalent HCl) (t = 25°) | B 72 |
| $C_8H_{11}NO_3$ | HO, HO, $CHCH_2-NH_2$, OH | (+)-Camphor-10-sulfonic acid in MeOH-isoPrOH | b) - 30.3° (t = 21°) | B 73 |

TABLE 9b (continued)

| Formula | Substance resolved | Resolving agent and solvent | Rotation $[\alpha]_D$ of resolved compound | Reference |
|---|---|---|---|---|
| $C_9H_{11}NO$ | $C_6H_5$-CH(NH$_2$)-C(=O)-CH$_3$ | (-)-Camphor-10-sulfonic acid in $H_2O$ | a) - 360° (c 0.5, EtOH) (t = 25°)<br>b) + 360° (c 0.5, EtOH) (t = 25°)<br>both on the hydrochloride salts | B 74 |
| $C_9H_{12}N_2O_4$ | $O_2N$-$C_6H_4$-CH(OH)-CH(NH$_2$)-CH$_2$OH<br>threo isomer | (-)-O,O'-Dibenzoyltartaric acid in $H_2O$ | a) - 23.5° (c 1.316, MeOH) (t = 20°)<br>b) + 23.4° (c 1.321, MeOH) (t = 20°) | B 75, B 76 |
| | | Also, resolutions with (+)-camphor-10-sulfonic acid (B 77, B 78), (+)-tartaric acid (B 79, B 80, B 81), (+)-mandelic acid (B 78), (-)-diisopropylidene-2-oxo-L-gulonic acid (see Chapter II, ref. 4), (-)- and (+)-N-($\alpha$-methylbenzyl)-succinamic acid (B 20), and (-)-quinic acid (B 75). | | |
| $C_9H_{13}NO$ | $CH_3O$-$C_6H_4$-CH(NH$_2$)-CH$_3$ | (+)-Tartaric acid<br>Recr.: EtOH | a) + 31.7° (c 5.32, $CHCl_3$) (t = 26°) on the N-benzoyl derivative of the (+)-amine<br>b) - 29.7° (c 7.89, $C_6H_6$) (t = 27°) | B 82 |
| $C_9H_{13}NO$ | $C_6H_5$-CH(OH)-CH$_2$NHCH$_3$ | (+)-Tartaric acid<br>(-)-Tartaric acid<br>both in EtOH | a) - 52.46° (c 0.88, $H_2O$)<br>b) + 52.78° (c 0.78, $H_2O$)<br>both on the hydrochloride salts | B 83 |
| $C_9H_{13}NOS$ | $H_2N$-$C_6H_4$-CH$_2$-S(=O)-R (meta)<br>R = CH$_3$CH$_2$ | (+)-Camphor-10-sulfonic acid in $CH_3CN$<br>Partial resolution | a) + 10° (c 2, $CHCl_3$) (t = 20°)<br>b) - 6° (c 2, $CHCl_3$) (t = 20°) | B 66 |
| $C_9H_{13}NO_2$ | $C_6H_5$-CH(OH)-CH(NH$_2$)-CH$_2$OH<br>threo isomer | (+)-Tartaric acid in MeOH-isoPrOH | a) + 18° (c 2%, $H_2O$) (t = 27°)<br>b) - 18° (c 2%, $H_2O$) (t = 27°) | B 84 |
| | | Also, resolution with (-)-glutamic acid (B 85). | | |

TABLE 9b (continued)

| Formula | Substance resolved | Resolving agent and solvent | Rotation $[\alpha]_D$ of resolved compound | Reference |
|---|---|---|---|---|
| $C_{10}H_{13}NO$ | 2-(1-Hydroxyethyl)-indoline | See Table 10c | | |
| $C_{10}H_{15}NO$ | $CH_3CH(NHCH_2C_6H_5)CH_2OH$ | (-)-Diisopropylidene-2-oxo-L-gulonic acid in EtOAc | + 39.8° (c 1.02, MeOH) (t = 25°) | Chapter II, ref. 4 (p. 14) |
| $C_{10}H_{15}NO$ | $C_6H_5-CH(OH)-CH(NHCH_3)CH_3$ erythro isomer Ephedrine | (+)-Arabonic acid in MeOH or $H_2O$ | a) + 34.6° ($H_2O$)<br>b) - 34.8° ($H_2O$)<br>both on the hydrochloride salts | B 86 |
| | | (-)-N-Carbobenzoxy-L-alanine in EtOAc | + 34.59° (c 4, $H_2O$) (t = 25°) on the hydrochloride salts | B 87 |
| | | Also, resolution with (+)- and (-)-mandelic acid (B 88). See also p. 19. | | |
| $C_{10}H_{15}NO$ | threo isomer Pseudoephedrine | (+)-Tartaric acid in $H_2O$ | a) + 52.9° (c 4.0, EtOH) (t = 22.5°)<br>b) - 52.50°(c 4.0, EtOH) (t = 22.5°) | B 89 |
| $C_{10}H_{15}NO_2$ | $3-CH_3O-C_6H_4-CH(OCH_3)CH_2-NH_2$ | (+)-Tartaric acid in acetone | a) + 77.1° (c 1.07%, $H_2O$) (t = 21°) | B 90 |
| | | (-)-Tartaric acid in MeOH | b) - 77.9° (c 1.04%, $H_2O$) (t = 21°) | B 90 |
| $C_{10}H_{15}NO_2S$ | $4-CH_3Y-C_6H_4-CH(OH)CH(NH_2)CH_2OH$<br>Y = S threo isomer | (+)-Tartaric acid in MeOH | a) - 21° (c 1%, EtOH) (t = 25°)<br>b) + 21° (c 1%, EtOH) (t = 25°) | B 91 |
| $C_{10}H_{15}NO_4S$ | Y = $SO_2$ threo isomer | (+)-Mandelic acid in EtOH | + 20° (c 2%, EtOH) (t = 22°) | B 92 |

TABLE 9b (continued)

| | Substance resolved | Resolving agent and solvent | Rotation $[\alpha]_D$ of resolved compound | Reference |
|---|---|---|---|---|
| $C_{10}H_{23}NO$ | $CH_3(CH_2)_5CH(OH)CH_2N(CH_3)_2$ | (-)-O, O'-Dibenzoyltartaric acid in EtOH | $-15.3^o$ (neat) ($t = 21^o$) | B 93 |
| $C_{11}H_{17}NO$ | $CH_2CH(NHCH_3)CH_3$ <br> $OCH_3$ | (+)-Tartaric acid in MeOH-acetone | a) $-14.25^o$ (c 2%, $H_2O$) ($t = 28^o$) <br> b) $+13.80^o$ (c 2%, $H_2O$) ($t = 28^o$) <br> both on the hydrochloride salts | B 94 |
| $C_{11}H_{17}NOS$ | See $C_9H_{13}NOS$ above <br> $R = (CH_3)_3C$ | (+)-Camphor-10-sulfonic acid in $CH_3CN$ <br> Partial resolution | a) $+102^o$ (c 2, $CHCl_3$) ($t = 20^o$) <br> $+126^o$ (c 2, EtOH) ($t = 20^o$) <br> b) $-80^o$ (c 1, EtOH) ($t = 20^o$) | B 66 |
| $C_{11}H_{17}NO_3$ | HO, HO <br> $CH(OH)CH_2-NHCH(CH_3)_2$ | (+)-Tartaric acid in MeOH | a) $+45.0^o$ (0.5 N aqueous HCl) ($t = 19^o$) <br> b) $-45.0^o$ (0.5 N aqueous HCl) ($t = 19^o$) | B 95 |
| $C_{11}H_{21}NO$ | $(CH_3)_2N$, HO, H, H | (+)-Tartaric acid in acetone-EtOH | $+25.6^o$ (EtOH) ($t = 25^o$) | B 96 |
| $C_{11}H_{21}NO_2$ | $O\;NCH_2CH_2C(=O)CH(CH_3)CH_2CH_3$ | (+)-Camphor-10-sulfonic acid in EtOAc | $+12.47^o$ (c ca. $10^{-2}$, $H_2O$) ($t = 30^o$) <br> on the camphorsulfonate salt | B 68 |
| $C_{12}H_{17}NO$ | $C(=O)CH(CH_3)CH_2-N(CH_3)_2$ | (-)-O, O'-Dibenzoyltartaric acid in acetone <br> (+)-O, O'-Dibenzoyltartaric acid in acetone | a) $-49^o$ (c 1, $H_2O$) ($t = 25^o$) <br> $+48^o$ (c 1.0, EtOH) ($t = 20^o$) <br> b) $+47^o$ (c 1, $H_2O$) ($t = 25^o$) <br> both on the hydrochloride salts | B 97, B 68 <br> B 98 <br> B 97 |

TABLE 9b (continued)

| Substance resolved | | | Resolving agent and solvent | Rotation $[\alpha]_D$ of resolved compound | Reference |
|---|---|---|---|---|---|
| $C_{12}H_{23}NO$ | 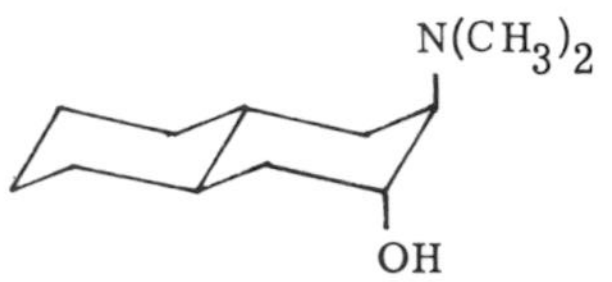 | | (+)-Tartaric acid | a) $-21.5^o$ (c ca. 0.1 M, $CHCl_3$) ($t = 25^o$) | B 99 |
| | | | (-)-Tartaric acid | b) $+21.5^o$ (c ca. 0.1 M, $CHCl_3$) ($t = 25^o$) | B 99 |
| | | | both in acetone- EtOH | | |
| $C_{13}H_{27}NO$ | 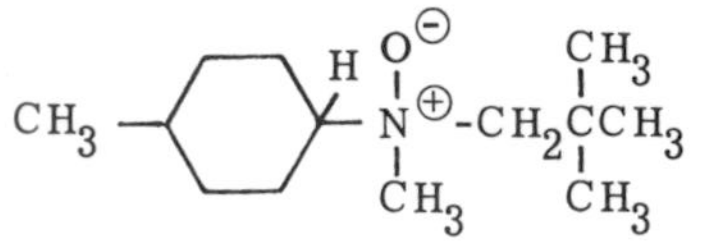 | cis isomer | (-)-O,O'-Dibenzoyltartaric acid in EtOH | a) $+2.90 \pm 0.19^o$ (c 8.07, $CHCl_3$) ($t = 22^o$) | B 100 |
| | | | | b) $-2.87 \pm 0.20^o$ (c 14.19, $CHCl_3$) ($t = 26^o$) | |
| | | trans isomer | (-)-O,O'-Dibenzoyltartaric acid in EtOH | a) $-0.70 \pm 0.05^o$ (c 11.2, $CHCl_3$) ($t = 29^o$) | B 100 |
| | | | | b) $+2.35 \pm 0.28^o$ (c 5.33, $CHCl_3$) ($t = 27^o$) | |
| $C_{14}H_{15}NO$ | 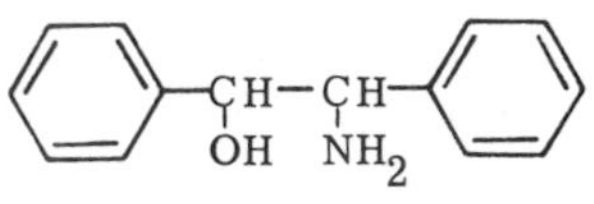 | erythro isomer | (-)-Glutamic acid in $EtOH-H_2O$ | a) $-10.1^o$ (c 0.59, EtOH) ($t = 25^o$) | B 101a, B 101b |
| | | | | b) $+10.2^o$ (c 0.61, EtOH) ($t = 25^o$) | |
| | | threo isomer | (+)-Tartaric acid | a) $-121.2^o$ (c 1.3, EtOH) ($t = 17^o$) | B 101b |
| | | | (+)-Camphor-10-sulfonic acid | b) $+124.4^o$ (c 1.31, EtOH) ($t = 20^o$) | B 101b |
| $C_{14}H_{21}NO$ | 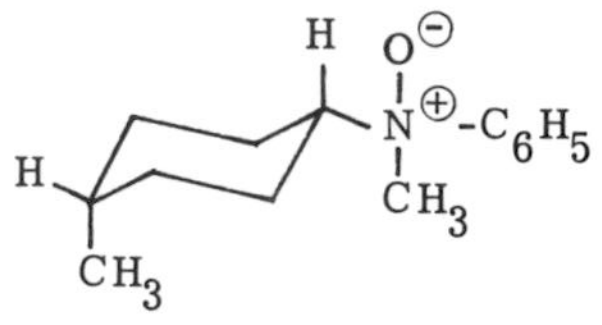 | | (-)-O,O'-Dibenzoyltartaric acid | $+17.4^o$ (c 0.5, $CHCl_3$) ($t = 18^o$) (578 nm) | B 102 |
| $C_{15}H_{17}NO$ | 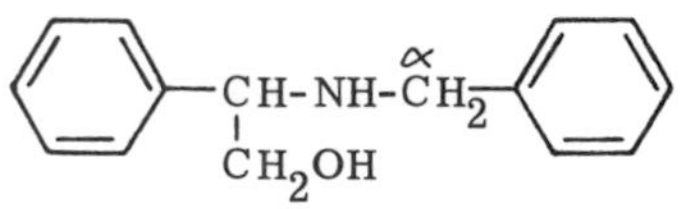 | | (+)-O,O'-Di-p-toluoyltartaric acid in $EtOH-H_2O$ | $+80^o$ (c 2, EtOH) ($t = 25^o$) | B 103 |
| $C_{15}H_{16}NOD$ | $\alpha$-deuterio analog | | (-)-O,O'-Di-p-toluoyltartaric acid in EtOAc | a) $-65^o$ ($C_6H_6$) | B 104 |
| | | | (+)-O,O'-Di-p-toluoyltartaric acid in EtOAc | b) $+59^o$ ($C_6H_6$) | B 104 |

TABLE 9b (continued)

| Formula | Substance resolved | Resolving agent and solvent | Rotation $[\alpha]_D$ of resolved compound | Reference |
|---|---|---|---|---|
| $C_{15}H_{17}NO$ | $C_6H_5$-CH(OH)-CH($NHCH_3$)-$C_6H_5$ <br> erythro isomer | (+)-Potassium benzylpenicillinate on the HCl salt of the amine in $H_2O$ | a) - 40.0° (c 1, MeOH) (t = 25°) <br> b) + 38.2° (c 1, MeOH) (t = 25°) | B 105 |
| $C_{15}H_{17}NO$ | $C_6H_5$-C($C_6H_5$)(OH)-CH($NH_2$)$CH_3$ | (+)-Camphor-10-sulfonic acid in EtOH-$H_2O$ (1:1) | - 59.3° (EtOH) (t = 24°) | B 106 |
| $C_{15}H_{17}NO$ | $C_6H_5$-CH($NH_2$)-C($CH_3$)(OH)-$C_6H_5$ <br> erythro isomer | (+)-Tartaric acid <br> (+)-Camphor-10-sulfonic acid <br> both in EtOH-$H_2O$ | a) - 52.3° (EtOH) (t = 25°) <br> b) + 52.5° (EtOH) (t = 25°) <br> both on the hydrochloride salts | B 49 <br> B 49 |
| $C_{15}H_{17}NO$ | $C_6H_5$-CH(OH)$CH_2$CH($NH_2$)-$C_6H_5$ <br> isomer m.p. 127° | (-)-Mandelic acid in MeOH-$Et_2O$-Pet E <br> (+)-Camphor-10-sulfonic acid in MeOH-$Et_2O$ | a) - 23.5° (c 5.221, MeOH) (t = 17°) <br> b) + 25.3° (c 4.6540, MeOH) (t = 17°) | B 107 <br> B 107 |
| | isomer m.p. 121-122° | (+)-3-Bromocamphor-9-sulfonic acid in MeOH <br> Ammonium (-)-3-bromo-camphor-9-sulfonate in MeOH | a) + 3.8° (c 0.660, MeOH) (t = 17°) <br> b) - 3.8° (c 0.660, MeOH) (t = 17°) | B 107 <br> B 107 |
| $C_{15}H_{17}NO_2$ | $C_6H_5$-$C_6H_4$-CH(OH)-CH($NH_2$)$CH_2OH$ <br> threo isomer | (-)-Phenylethylsuccinic acid in BuOH <br> (+)-Phenylethylsuccinic acid in BuOH | a) + 28° (c 5%, DMA) (t = 25°) <br> b) - 28° (c 5%, DMA) (t = 25°) | B 108 <br> B 108 |

TABLE 9b (continued)

| | Substance resolved | Resolving agent and solvent | Rotation $[\alpha]_D$ of resolved compound | Reference |
|---|---|---|---|---|
| $C_{15}H_{19}NO$ | 2-naphthyl-$CH(OH)CH_2$-$NHCH(CH_3)_2$<br>and analogous amino-alcohols | (-)-O,O'-Di-p-toluoyltartaric acid in $MeOH-H_2O$ | a) $+28.4^{o}$ (c 0.99%, EtOH) (t = $21^{o}$) | B 109 |
| | | (+)-O,O'-Di-p-toluoyltartaric acid in $MeOH-H_2O$ | b) $-29.0^{o}$ (c 1.3%, EtOH) (t = $21^{o}$) | B 109 |
| $C_{16}H_{19}NO$ | $C_6H_5$-$CH(NH_2)$-$C(CH_3)(OH)$-$C_6H_4$-$CH_3$ erythro isomer | (+)-Tartaric acid in $H_2O$ | a) $+15.3^{o}$ (EtOH) (t = $24^{o}$) | B 74 |
| | | (+)-Camphoric acid in $EtOH-H_2O$ | b) $-15.5^{o}$ (EtOH) (t = $24^{o}$) | B 74 |
| | threo isomer | (+)-Tartaric acid in $H_2O$ | a) $-64.5^{o}$ (c 1, EtOH) (t = $25^{o}$)<br>b) $+58^{o}$ | B 74 |
| $C_{18}H_{23}NO$ | $C_6H_5$-$C(C_6H_5)(OH)$-$CH(CH_3)CH_2N(CH_3)_2$ | (+)-Camphor-10-sulfonic acid in acetone | $+53.4^{o}$ (c 1.2, $C_6H_6$) (t = $20^{o}$) | B 98 |
| $C_{18}H_{37}NO_2$ | $CH_3(CH_2)_{12}CH=CH$-$CH(OH)$-$CH(NH_2)$-$CH_2OH$<br>Sphingosine<br>See also $C_{25}H_{41}NO_3$ below | (-)-Glutamic acid in $EtOH-H_2O$ | $-12.8^{o}$ (t = $24^{o}$)<br>on the triacetyl derivative | B 110 |
| $C_{18}H_{39}NO_2$ | n-$C_{15}H_{31}$$CH(OH)$$CH(NH_2)$$CH_2OH$<br>Dihydrosphingosine<br>See also $C_{25}H_{43}NO_3$ below | (+)-Glutamic acid in EtOH or $EtOH-H_2O$ | a) $+13.5 \pm 2^{o}$ (c 0.74, $CHCl_3$) (t = $28^{o}$)<br>and $-19.35^{o}$ (c 0.66, $CHCl_3$) (t = $22^{o}$)<br>on the triacetyl derivative | B 111<br>B 112 |
| | | (-)-Glutamic acid in EtOH | b) $-14.1 \pm 2^{o}$ (c 0.74, $CHCl_3$) (t = $27^{o}$)<br>$+19.2^{o}$ ($CHCl_3$) (t = $22^{o}$) on the triacetyl derivative | B 111<br>B 112 |

TABLE 9b (continued)

| Substance resolved | | Resolving agent and solvent | Rotation $[\alpha]_D$ of resolved compound | Reference |
|---|---|---|---|---|
| $C_{19}H_{21}NO_5$ | Deacetylcolchiceine | (+)-Camphor-10-sulfonic acid in MeOH | + 185° (c 1, $CHCl_3$) | B 113 |
| $C_{19}H_{25}NO$ | $C_6H_5-CH_2C(C_6H_5)(OH)-CH(CH_3)CH_2-N(CH_3)_2$ | (+)-Camphor-10-sulfonic acid in EtOH | a) + 54.9° (c 0.7, $H_2O$) (t = 25°)<br>b) - 54.7° (c 0.7, $H_2O$) (t = 25°)<br>both on the hydrochloride salts | B 114 |
| $C_{20}H_{36}N_2O_2$ | | (-)-3-Bromocamphor-9-sulfonic acid in EtOAc | a) - 21° (c 1.7, MeOH) (578 nm) | B 115 |
| | | (+)-3-Bromocamphor-9-sulfonic acid in EtOAc | b) + 21° (c 1, MeOH) (578 nm) | B 115 |
| $C_{21}H_{27}NO$ | $CH_3CH_2C(=O)-C(C_6H_5)(C_6H_5)-CH(CH_3)CH_2-N(CH_3)_2$<br>Isomethadone | (+)-Tartaric acid in $H_2O$ | a) - 20° (c 1.5, EtOH) (t = 20°)<br>b) + 21° (c 1.5, EtOH) (t = 20°) | B 116 |
| | | Also, resolution with (+)-p-nitrobenzoyl-L-glutamic acid (B 117). | | |
| $C_{21}H_{27}NO$ | $CH_3CH_2C(=O)-C(C_6H_5)(C_6H_5)-CH_2CH(CH_3)-N(CH_3)_2$<br>Methadone | (+)-Tartaric acid in acetone | a) - 29.91° (c 2.66, EtOH) (t = 22°) | B 118 |
| | | and in PrOH | b) + 29.51° (EtOH) (t = 26°) | B 116 |
| | | Also, resolution with (+)-ammonium 3-bromocamphor-9-sulfonate (B 117) and with (+)-p-nitrobenzoyl-L-glutamic acid (B 119) | | |

TABLE 9b (continued)

| | Substance resolved | Resolving agent and solvent | Rotation $[\alpha]_D$ of resolved compound | Reference |
|---|---|---|---|---|
| $C_{23}H_{25}NO_3$ | $CH_3O$, $OCH_3$-substituted phenyl–$CH(CH_2NH_2)$–$C_6H_4$–$OCH_2C_6H_5$ | (-)-Diisopropylidene-2-oxo-L-gulonic acid in isoPrOH<br>(-)-O,O'-Di-p-toluoyltartaric acid in acetone | a) $+6.5^o$ (c 1%, MeOH) (t = $25^o$)<br>b) $-6.5^o$ (c 1%, MeOH) (t = $25^o$) | Chapter II, ref. 5 (p. 14) |
| $C_{23}H_{27}NO_2$ | 1-Naphthyl–$O$-$CH_2CH(OH)CH_2$-$N(CH(CH_3)_2)CH_2C_6H_5$ | (-)-O,O'-Di-p-toluoyltartaric acid in MeOH | $-30^o$ (c 1.961, MeOH) (t = $25^o$) | B 120 |
| $C_{25}H_{41}NO_3$ | 3-O-Benzoylsphingosine see $C_{18}H_{37}NO_2$ above | Barium L-tartrate on the hemisulfate salt<br>Barium D-tartrate on the hemisulfate salt<br>both in THF-$H_2O$ | a) $+8^o$ (c 1, $CHCl_3$) (t = $26^o$) on the tribenzoyl derivative<br>b) $-10.8^o$ on the triacetyl derivative | A 180 (p. 131) |
| $C_{25}H_{43}NO_3$ | 3-O-Benzoyldihydrosphingosine see $C_{18}H_{39}NO_2$ above | Barium L-tartrate on the hemisulfate salt in EtOH | $-26^o$ on the tribenzoyl derivative | A 180 |

TABLE 9c  Diamines, Amidines, Hydrazines and their Salts

| | Substance resolved | Resolving agent and solvent | Rotation $[\alpha]_D$ of resolved compound | Reference |
|---|---|---|---|---|
| $C_3H_9N_2OCl$ | $CH_3CH(OH)-C(=NH_2^{\oplus})-NH_2$ $Cl^{\ominus}$ | Sodium (+)-mandelate<br>Sodium (-)-mandelate<br>both in $H_2O$ | a) $-16.5^o$ (c 0.95, $H_2O$) (t = $18^o$) (546 nm)<br>b) $+16.8^o$ (c 1.5, $H_2O$) (t = $22^o$) (546 nm) | B 121 |
| $C_3H_{10}N_2$ | $CH_3-CH(NH_2)CH_2-NH_2$ | (+)-Tartaric acid<br>in HOAc - $H_2O$ | a) $-34.8 \pm 0.4^o$ ($C_6H_6$) (t = $20^o$)<br>b) $+34.8 \pm 0.6^o$ ($C_6H_6$) (t = $20^o$) | B 122 |
| $C_4H_{12}N_2$ | $CH_2(NH_2)CH_2CH(NH_2)CH_3$ | (+)-Tartaric acid<br>in $H_2O$ | $\alpha_D^{25}$ + 1.80$^o$ (1 2 cm, neat) | B 123 |
| $C_4H_{12}N_2$ | $CH_3CH(NH_2)CH(NH_2)CH_3$ | (+)-Tartaric acid<br>in EtOH | a) $+29.48^o$ (neat) (t = $25^o$) ($\alpha_D^{25}$ + 25.18$^o$)<br>b) $\alpha_D^{25}$ - 11.43$^o$ | B 64 |
| $C_6H_{14}N_2$ | $NH_2$<br>$NH_2$ | (+)-Tartaric acid<br>in $H_2O$ | a) $-15.8^o$ (c 20, $H_2O$) (t = $25^o$)<br>on the hydrochloride salt<br>$-42.6^o$ ($C_6H_6$) (t = $21^o$)<br>b) $+15.8^o$ (c 20, $H_2O$) (t = $25^o$)<br>on the hydrochloride salt | B 124<br><br>B 125<br>B 124 |
| $C_7H_{18}N_2S_2$ | $CH_2(NH_2)CH_2SCH(CH_3)CH_2SCH_2CH_2(NH_2)$ | (+)-O,O'-Dibenzoyltartaric acid in $H_2O$ + NaOAc<br>(-)-O,O'-Dibenzoyltartaric acid in $H_2O$ + NaOAc | a) $-50^o$ ($C_6H_6$)<br>b) $+50^o$ ($C_6H_6$) | B 126<br>B 126 |
| $C_8H_9N_2F_3$ | $C_6H_5-CH(NHNH_2)CF_3$ | 5$\alpha$-Androstan-17-one<br>in HOAC-EtOH + NaOAc<br>via hydrazone formation | $+36.2^o$ (c 0.915, EtOH) (t = $27^o$) | B 127 |

TABLE 9c (continued)

| | Substance resolved | Resolving agent and solvent | Rotation $[\alpha]_D$ of resolved compound | Reference |
|---|---|---|---|---|
| $C_8H_{10}N_2OCl_2$ | OH; CH-C-$NH_2$; ‖⊕ $NH_2$ $Cl^{\ominus}$; Cl | Sodium (-)-mandelate | a) + 56.3° (c 0.75, $H_2O$)<br>b) - 54.1° (c 0.45, $H_2O$) | B 128 |
| Also resolved, 2-bromo and 2,4-dichloro derivatives | | | | |
| $C_8H_{12}N_2$ | CH-$CH_2$-$NH_2$; $NH_2$ | (+)-Tartaric acid | - 34.6° (neat) | B 129 |
| $C_8H_{20}N_2$ | 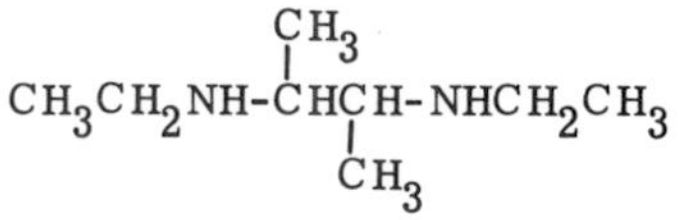 $CH_3CH_2NH$-$\underset{CH_3}{}$CHCH($CH_3$)-$NHCH_2CH_3$ (CH<sub></sub>$_3$ above, $CH_3$ below)<br>*threo* isomer | (-)-O, O'-Dibenzoyltartaric acid in EtOH | a) + 106.0° (presumably neat) (t = 25°)<br>b) - 103.7° (presumably neat) (t = 25°) | B 130 |
| $C_9H_{13}N_2OCl$ | 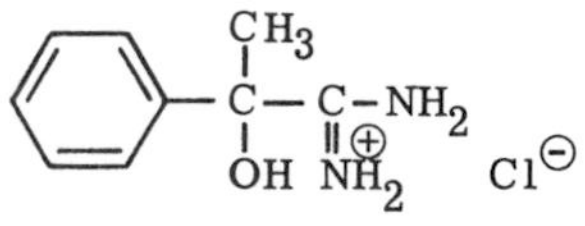 $CH_3$; C — C-$NH_2$; OH ‖⊕ $NH_2$ $Cl^{\ominus}$ | Sodium (+)-mandelate<br>Sodium (-)-mandelate<br>both in MeOH | a) - 55.6° (c 0.54, $H_2O$) (t = 15°) (546 nm)<br>b) + 54.8° (c 0.58, $H_2O$) (t = 15°) (546 nm) | B 131 |
| $C_9H_{13}N_2O_2Cl$ | 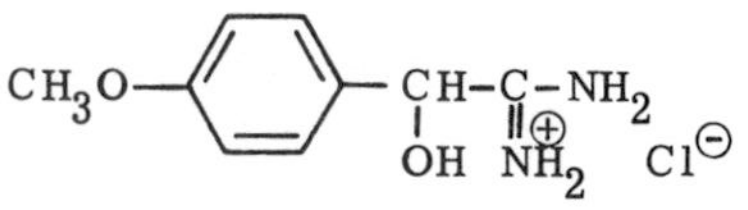 $CH_3O$—; CH-C-$NH_2$; OH ‖⊕ $NH_2$ $Cl^{\ominus}$ | Sodium (+)-mandelate<br>Sodium (-)-mandelate<br>both in $H_2O$ | a) + 87.8° (c 0.56, $H_2O$) (t = 22°) (546 nm)<br>b) - 87.9° (c 0.44, $H_2O$) (t = 22°) (546 nm) | B 121 |
| $C_{10}H_{15}N_2OCl$ | 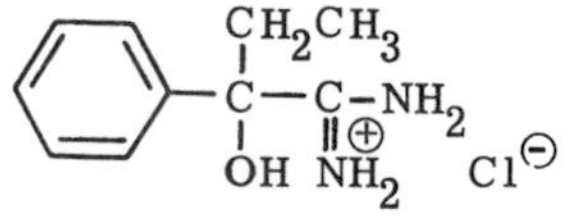 $CH_2CH_3$; C — C-$NH_2$; OH ‖⊕ $NH_2$ $Cl^{\ominus}$ | Sodium (-)-mandelate<br>Sodium (+)-mandelate<br>both in $H_2O$ | a) + 48.9° (c 0.44, $H_2O$) (t = 19°) (546 nm)<br>b) - 49° (c 0.68, $H_2O$) (t = 17°) (546 nm) | B 132 |
| $C_{10}H_{15}N_2OCl$ | 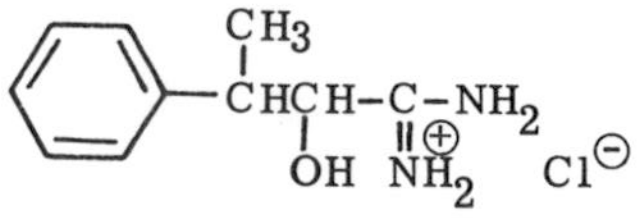 $CH_3$; CHCH-C-$NH_2$; OH ‖⊕ $NH_2$ $Cl^{\ominus}$ | Sodium (-)-mandelate in $H_2O$ | - 43.6° (c 0.86, $H_2O$) (t = 13°) (546 nm) | B 132 |

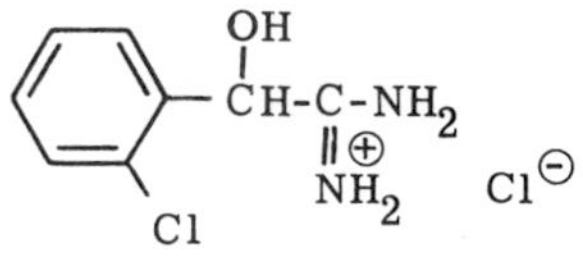

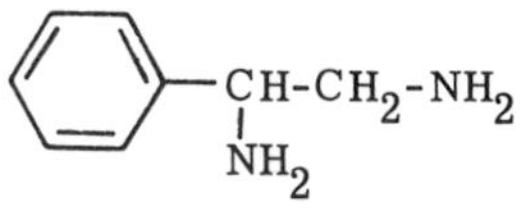

TABLE 9c (continued)

| Substance resolved | Resolving agent and solvent | Rotation $[\alpha]_D$ of resolved compound | Reference |
|---|---|---|---|
| $C_{10}H_{16}N_2$ — $C_6H_5$-C($CH_3$)($CH_2CH_3$)-$NHNH_2$ | (+)-O, O'-Dibenzoyltartaric acid in EtOH-$H_2O$ | $\alpha^{25}_{546}$ + 11.8° (l 1 dm, neat) | B 133 |
| $C_{10}H_{24}N_2$ — $CH_3CH_2C(CH_3)(NH_2)CH_2CH_2C(CH_3)(NH_2)CH_2CH_3$ | (+)-Tartaric acid | + 316° (c 0.71, MeOH) on the derivative [3,6-diethyl-3,6-dimethyl-tetrahydropyridazine: $CH_3$, $CH_3CH_2$ / N=N / $CH_2CH_3$, $CH_3$] | B 134 |
| $C_{12}H_{17}N_3O_3$ — $O_2N$-$C_6H_4$-$CH_2$-$\overset{\oplus}{N}$($CH_3$)($CH_2CH_3$)-$\overset{\ominus}{N}$-C(=O)$CH_3$ | (+)-O, O'-Dibenzoyltartaric acid in EtOH | a) + 98.3° (95% EtOH) (t = 23°)<br>b) - 80.3° | B 135 |
| $C_{14}H_{16}N_2$ — $C_6H_5$-CH($NH_2$)-CH($NH_2$)-$C_6H_5$ | (+)-Tartaric acid in 96% EtOH | a) - 87° ($Et_2O$)<br>- 108° (MeOH) (t = 20°)<br>b) + 86° ($Et_2O$) | B 136 |
| $C_{14}H_{16}N_2$ — biphenyl with $CH_3$, $NH_2$, $NH_2$, $CH_3$ | (+)-Tartaric acid<br>(-)-Tartaric acid<br>both in EtOH | a) + 48° (c 2.5, EtOH) (t = 31°)<br>b) - 47° (c 3.0, EtOH) (t = 26°) | B 137 |
| $C_{15}H_{17}N_3O_2S$ — $C_6H_5$-CH(C(=NH)$NH_2$)-NH-$SO_2$-$C_6H_4$-$CH_3$ | (-)-Mandelic acid in EtOH-acetone | a) - 38.2° (c 0.8, MeOH) (t = 18-20°) (546 nm)<br>b) + 36.4° (c 0.4, MeOH) (t = 18-20°) (546 nm)<br>both on the hydrochloride salts | B 138 |
| $C_{15}H_{24}N_3O_3I$ — $O_2N$-$C_6H_4$-N(C(=O)$CH_3$)-$\overset{\oplus}{N}$($CH_2CH_2CH_2CH_3$)($CH_2CH_3$)($CH_3$) $I^{\ominus}$ | (+)-Camphor-10-sulfonic acid in EtOH<br>on the ammonium hydroxide salt | + 2.78° (MeOH) (t = 25°) | B 135 |

TABLE 9c (continued)

| Substance resolved | Resolving agent and solvent | Rotation $[\alpha]_D$ of resolved compound | Reference |
|---|---|---|---|
| $C_{16}H_{18}N_2D_2$ | 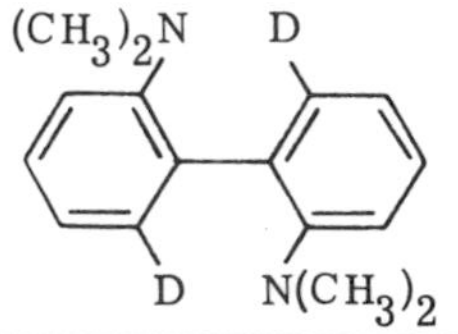 Silver (+)-camphor-10-sulfonate in EtOH-$H_2O$ on the amine mono methiodide salt | $+106^{o}$ ($H_2O$) (t = $98^{o}$) (436 nm) on the (+)-camphor-10-sulfonate salt | B 139 |
| $C_{16}H_{20}N_2$ | 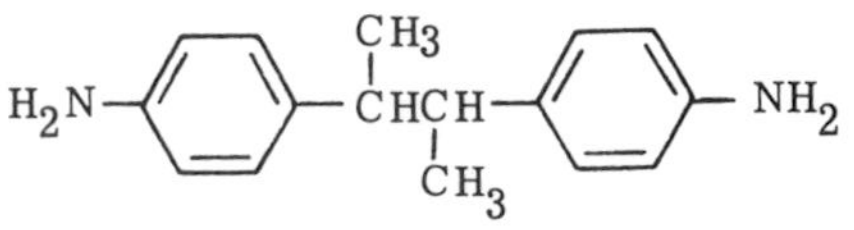 (+)-Tartaric acid in MeOH | $+103^{o}$ (t = $20^{o}$) | B 140 |
| $C_{16}H_{20}N_2$ | 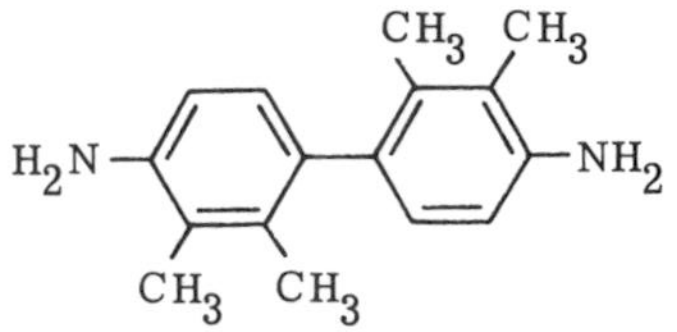 Ammonium (+)-3-bromo-camphor-9-sulfonate in $H_2O$-acetone | a) ca. $+10^{o}$ (c 1.8-2.2, EtOAc)<br>b) $-9.5^{o}$ (c 0.97, EtOAc) | B 141 |
| $C_{16}H_{20}N_2O_2$ | 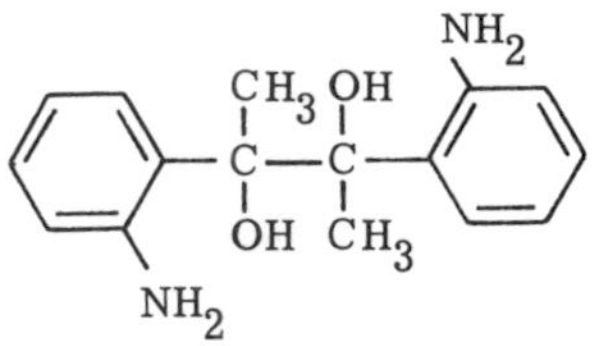 D-Camphoric acid in dioxane | a) $+152 \pm 5^{o}$ (EtOH) (t = $23.5^{o}$)<br>b) $-157 \pm 5^{o}$ (EtOH) (t = $23.5^{o}$) | B 142 |
| $C_{20}H_{16}N_2$ | 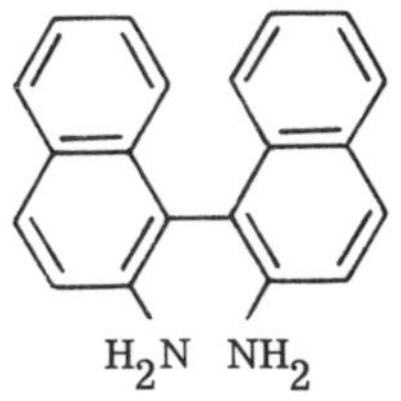 (+)-Camphor-10-sulfonic acid in $C_6H_5Cl$-EtOH | $+154^{o}$ (c 3.58, py) (t = $30^{o}$) | B 143 |

TABLE 9c (continued)

| Formula | Substance resolved | Resolving agent and solvent | Rotation $[\alpha]_D$ of resolved compound | Reference |
|---|---|---|---|---|
| $C_{20}H_{16}N_2$ | $H_2N$, $NH_2$ | Ammonium (+)-3-bromo-camphor-9-sulfonate in $H_2O$-acetone | a) $+42 \pm 1^o$ (c 0.30, acetone) (t = 22$^o$) | B 141 |
| | | | $+46.5^o$ (t = 20$^o$) (579 nm) | B 144 |
| | | | b) $-16^o$ (c 0.34, acetone) | B 141 |
| $C_{26}H_{24}N_2$ | 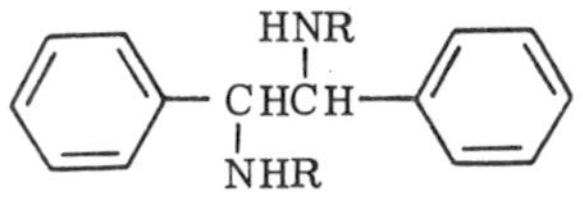 HNR, CHCH, NHR; R = $C_6H_5$ | (+)-3-Bromocamphor-10-sulfonic acid in THF-$Et_2O$ | a) $-204^o$ (c 1.60, $C_6H_6$) (t = 22$^o$) | B 145 |
| | | (+)-Camphor-10-sulfonic acid in THF-$Et_2O$ | b) $+200^o$ (c 1.525, $C_6H_6$) (t = 18$^o$) | B 145 |
| $C_{26}H_{36}N_2$ | R 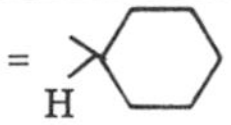 = H | (+)-3-Bromocamphor-9-sulfonic acid in $H_2O$ | $-19^o$ (c 0.865, $C_6H_6$) (t = 18$^o$) | B 145 |
| $C_{28}H_{28}N_2$ | R 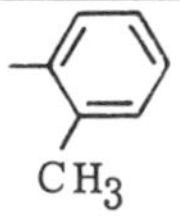 = $CH_3$ | (+)-Camphor-10-sulfonic acid in THF-$Et_2O$ | a) $-181^o$ (c 1.325, $C_6H_6$) (t = 18$^o$) | B 145 |
| | | (-)-Camphor-10-sulfonic acid in THF-$Et_2O$ | b) $+186.0^o$ (c 0.9225, $C_6H_6$) (t = 21$^o$) | B 145 |
| $C_{28}H_{28}N_2$ | R 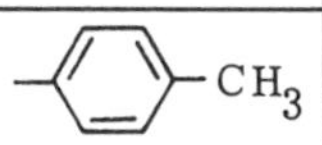 = $CH_3$ | (+)-Camphor-10-sulfonic acid in THF-$Et_2O$ | $+155^o$ (c 0.6075, $C_6H_6$) (t = 19$^o$) | B 145 |

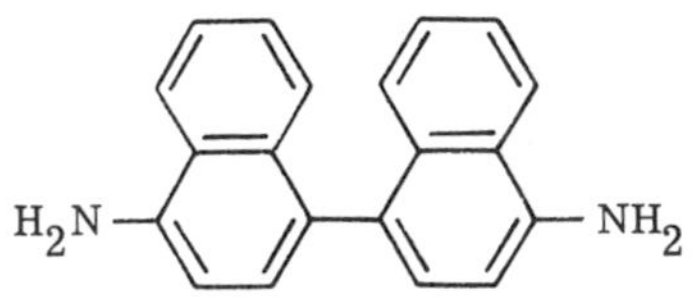

TABLE 9d Heterocyclic Amines

| | Substance resolved | Resolving agent and solvent | Rotation $[\alpha]_D$ of resolved compound | Reference |
|---|---|---|---|---|
| $C_3H_6N_2O_2$ | 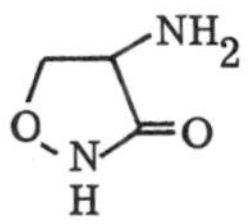 Cycloserine<br>See Table 13 | (+)-Tartaric acid<br>(-)-Tartaric acid | a) $+115^{\circ}$ (c 1.0, $H_2O$) (t = $22^{\circ}$)<br>b) $-115^{\circ}$ (c 1.0, $H_2O$) (t = $22^{\circ}$) | B 146 |
| $C_5H_{10}N_2$<br>and<br>$C_5H_{12}N_2$ | 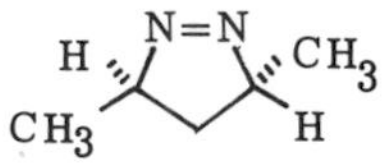 | (+)-Camphor-10-sulfonic acid in $Et_2O$ on the pyrazolidine $C_5H_{12}N_2$ | $+605^{\circ}$ (c 0.75, MeOH) (t = $25^{\circ}$) | B 147 |
| $C_6H_9NS$ | 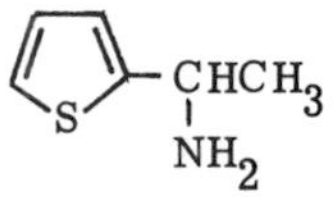 | (+)-2-Nitrotartranilic acid | $-2.96^{\circ}$ (neat) (t = $20^{\circ}$) | B 148 |
| $C_6H_{12}N_2O$ | 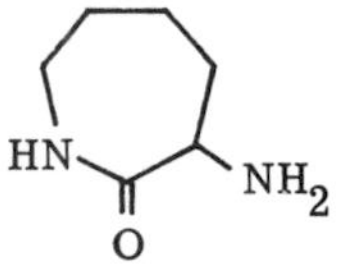 | N-Carbamoyl-L-valine<br>N-Carbamoyl-D-valine<br>both in $H_2O$ | a) $+28.5^{\circ}$ (c 2, $H_2O$) (t = $20^{\circ}$)<br>b) $-28.5^{\circ}$ (c 2, $H_2O$) (t = $20^{\circ}$)<br>both on the hydrochloride salts | B 149<br>B 149 |
| | | Also, resolution with (-)-2-pyrrolidone-5-carboxylic acid (B 150); see also Table 13. | | B 150 |
| $C_6H_{13}NO$ | 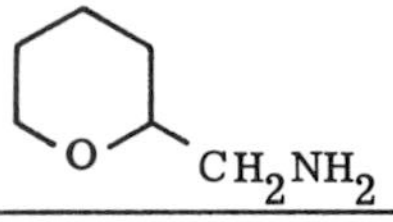 | (+)-Tartaric acid in MeOH | $\alpha_D^{24}$ $+8.3^{\circ}$ (l 1 dm, neat) | A 18 (p. 121) |
| $C_7H_{13}NO$ | 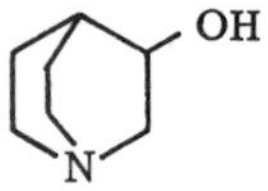 | (+)-Camphor-10-sulfonic acid in isoPrOH - $H_2O$ | a) $-2.0^{\circ}$ (c 6.5, $H_2O$) (t = $25^{\circ}$)<br>b) $+10.5^{\circ}$ (c 3.3, 0.5 N HCl) (t = $25^{\circ}$) on the diphenylacetate ester | B 151a<br>B 151b |
| $C_7H_{15}NO$ | 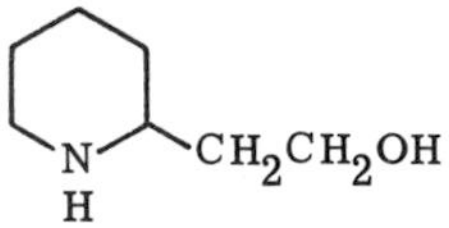 | (+)-Camphor-10-sulfonic acid in EtOH | a) $+11.0^{\circ}$ (c 3%, $CHCl_3$)<br>b) $-11.0^{\circ}$ (c 3%, $CHCl_3$) | B 152 |

TABLE 9d (continued)

| | Substance resolved | Resolving agent and solvent | Rotation $[\alpha]_D$ of resolved compound | Reference |
|---|---|---|---|---|
| $C_7H_{16}N_2$ | 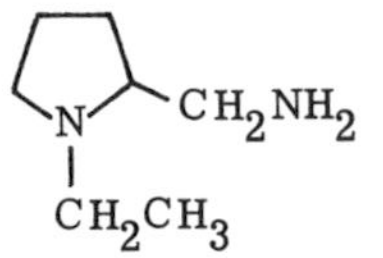 $CH_2NH_2$; $CH_2CH_3$ | (+)-Tartaric acid in $H_2O$ | $\alpha_{589}$ + 146.5° | B 153 |
| $C_8H_8N_2O_3S$ | 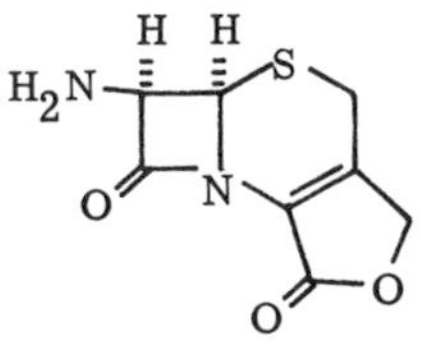 H H; $H_2N$; S; N; O; O; O | Sodium (+)-bitartrate in $H_2O$ on the amine hydrochloride | a) + 248 ± 6° ($H_2O$) (t = 24°)<br>b) - 248 ± 6° ($H_2O$) (t = 24°) | B 154 |
| $C_8H_{11}NO$ | 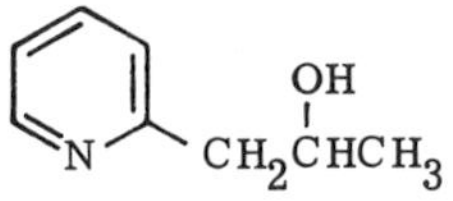 N; OH; $CH_2CHCH_3$ | (+)-Tartaric acid in acetone<br>(-)-Tartaric acid in acetone both on the tosylate esters | a) + 54.7 ± 0.5° (c 1.0, MeOH) (t = 18°)<br>b) - 54.9 ± 0.5° (c 1.0, MeOH) (t = 15°) both on the tosylate esters | B 155<br>B 155 |
| $C_8H_{14}N_2O$ | 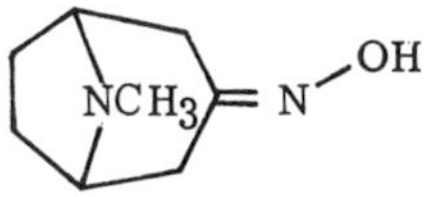 $NCH_3$; = N; OH | (+)-Camphor-10-sulfonic acid in EtOH-$Et_2O$ | - 21.65° (c 1.38, MeOH) (t = 20°) | B 156 |
| $C_8H_{15}N$ | 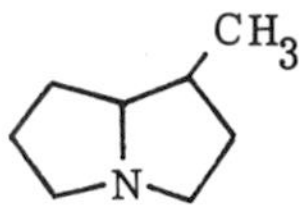 $CH_3$; N<br>Pseudoheliotridane | (+)-Tartaric acid in EtOAc-95% EtOH | + 6.94° (neat) (t = 25°) | B 157 |
| $C_8H_{15}N$ | 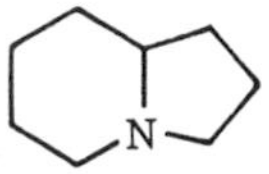 N | (-)-O,O'-Dibenzoyltartaric acid in MeOH | - 7.89° (t = 27°) | B 158 |

TABLE 9d (continued)

| Formula | Substance resolved | Resolving agent and solvent | Rotation $[\alpha]_D$ of resolved compound | Reference |
|---|---|---|---|---|
| $C_8H_{15}NO_2$ | 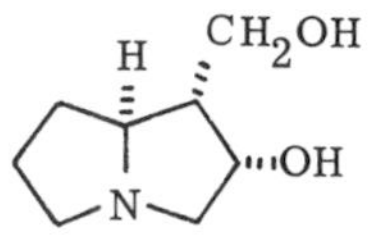 Macronecine | Ammonium (+)-3-bromo-camphor-9-sulfonate in EtOH | a) $+42.6^{o}$ (c 1.1, EtOH) ($t = 18^{o}$)<br>b) $-42.1^{o}$ (c 1.0, EtOH) ($t = 18^{o}$) | B 159 |
| $C_8H_{17}NO$ | Conhydrine, isomer m.p. $100^{o}$ 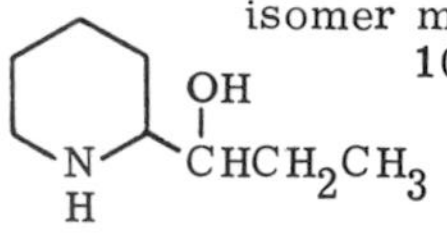 | (+)-6,6'-Dinitro-2,2'-diphenic acid<br>(-)-6,6'-Dinitro-2,2'-diphenic acid<br>both in MeOH | a) $+9.60^{o}$ (c 7.29, EtOH) ($t = 18^{o}$)<br>b) $-9.76^{o}$ (c 4.1, EtOH) ($t = 18^{o}$) | B 160<br>B 160 |
| $C_8H_{17}NO$ | 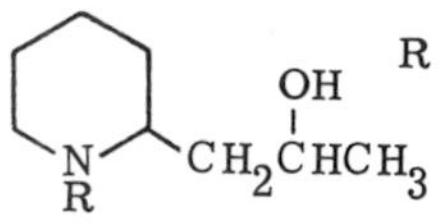 Allosedridine, R = H | (+)-O,O'-Dibenzoyltartaric acid in $H_2O$-acetone<br>(-)-O,O'-Dibenzoyltartaric acid in MeOH-acetone | a) $+16.2 \pm 0.5^{o}$ (c 4.01, MeOH) ($t = 29^{o}$)<br>b) $-16.4 \pm 0.5^{o}$ (c 4.06, MeOH) ($t = 29^{o}$) | B 161<br>B 161 |
| $C_9H_{11}NO$ | 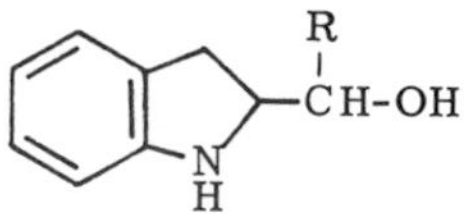 R = H | (+)-Mandelic acid in $C_6H_6$-MeOH | $+60.5^{o}$ (c 0.89, 95% EtOH) ($t = 27^{o}$) | B 162 |
| $C_9H_{13}N$ | 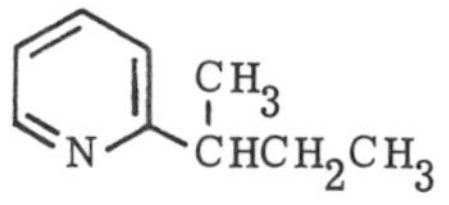 | (-)-O,O'-Dibenzoyltartaric acid<br>Recr.: EtOH | a) $-29.7^{o}$ ($t = 25^{o}$)<br>b) $+14.3^{o}$ ($t = 25^{o}$) | B 163 |
| $C_9H_{15}NS_2$ | 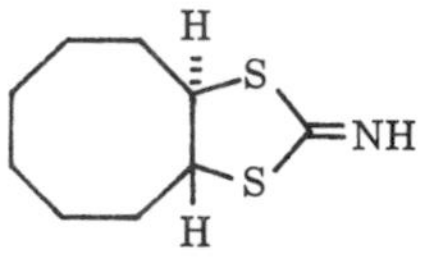 | (-)-$\alpha$-Methylphenylmethane-sulfonic acid<br>Recr.: 2-butanone | $+173.6^{o}$ (c 1.00, EtOH) ($t = 25^{o}$) (578 nm) | B 164 |

TABLE 9d (continued)

| | Substance resolved | Resolving agent and solvent | Rotation $[\alpha]_D$ of resolved compound | Reference |
|---|---|---|---|---|
| $C_9H_{17}NO$ | | (-)-O,O'-Dibenzoyltartaric acid in $Et_2O$ | $\alpha_D^{24}$ + 0.79 ± 0.01° (1 1 dm, neat) | B 165 |
| $C_9H_{17}NO_2S$ | | (+)-Camphor-10-sulfonic acid in acetone | + 70.0 ± 0.3° (c 3.100, EtOH) (t = 22°) | A 119 (p.127) |
| $C_{10}H_{13}NO$ | See $C_9H_{11}NO$ above R = $CH_3$ | see Table 10c | | |
| $C_{10}H_{14}N_2O$ | | (+)-Camphor-10-sulfonic acid in EtOH-$Et_2O$ | - 28.06° (c 0.588, MeOH) (t = 27°) | A177 (p.221) |
| $C_{10}H_{15}N_3$ | | (+)-Tartaric acid | a) + 69.75° (c 0.5, $H_2O$) (t = 11°)<br>b) - 56.08° (t = 12°) | B 166 |
| $C_{10}H_{19}N$ | | (+)-Tartaric acid in EtOH | + 1.6° (c 6.6, $H_2O$)<br>on the hydrochloride salt | B 167 |
| $C_{10}H_{19}NO$ | 3-Lupinine | (+)-Tartaric acid in EtOH | a) - 19.6 ± 0.5° (c 10, MeOH) (t = 20°)<br>b) + 19.5 ± 0.5° (c 10, MeOH) (t = 20°) | B 168 |

TABLE 9d (continued)

| Formula | Substance resolved | Resolving agent and solvent | Rotation $[\alpha]_D$ of resolved compound | Reference |
|---|---|---|---|---|
| $C_{10}H_{20}N_2$ | | (-)-6,6'-Dinitro-2,2'-diphenic acid in EtOH-$H_2O$ | a) + 1.2° (c 3.40%, 96% EtOH) (t = 18°) | B 169 |
| | | (+)-6,6'-Dinitro-2,2'-diphenic acid in EtOH-$H_2O$ | b) - 1.3° (c 3.20%, 96% EtOH) (t = 19°) | B 169 |
| $C_{10}H_{20}N_2$ | | (-)-6,6'-Dinitro-2,2'-diphenic acid in 96% EtOH | a) + 3.3° (c 5.56, 96% EtOH) (t = 17°)<br>b) - 3.3° (c 5.43, 96% EtOH) (t = 18°) | B 169 |
| $C_{11}H_{12}N_2S$ | Tetramisole | (+)-Camphor-10-sulfonic acid in $CHCl_3$ | a) + 122.5° (c 10, $H_2O$) (t = 25°)<br>b) - 120° (t = 25°)<br>both on the hydrochloride salts | B 170<br>B 170 |
| | | Sodium (+)-N-(p-toluenesulfonyl)-glutamate in $H_2O$ on the amine hydrochloride | - 85.3° (c 10, $CHCl_3$) (t = 20°) | B 171 |
| | | Also, resolution with sodium (-)-O,O'-dibenzoyltartrate | | B 172 |
| $C_{11}H_{13}N_2Br$ | Y = Br | (-)-Mandelic acid in EtOH-EtOAc | + 54.2° (c 1-1.5, MeOH) (546 nm) on the hydrochloride salt | A 136 (p.128) |
| $C_{11}H_{13}N_2Cl$ | Y = Cl | (-)-Mandelic acid in EtOH-EtOAc | + 61.1° (c 1-1.5, MeOH) (546 nm) on the hydrochloride salt | A 136 |
| $C_{11}H_{14}N_2O$ | Cytisine | (+)-Camphor-10-sulfonic acid in MeOH-acetone | - 188.5° ($H_2O$) (t = 22°) | B 173 |

TABLE 9d (continued)

| Formula | Substance resolved | Resolving agent and solvent | Rotation $[\alpha]_D$ of resolved compound | Reference |
|---|---|---|---|---|
| $C_{11}H_{14}N_2O_3$ | 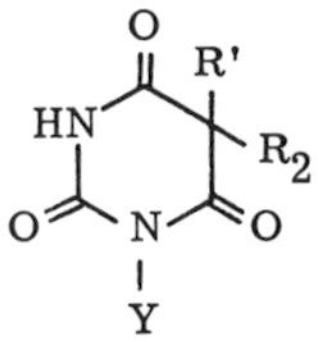 (barbiturate ring: HN, O, R', $R_2$, N–Y) $Y = CH_3$, $R_1 = H$, $R_2 = -CH(CH_3)C{\equiv}CCH_2CH_3$ | Brucine in EtOH | a) $-5.2^o$ (c 5%, EtOH) (t = $28^o$)<br>b) $+5.0^o$ (c 5%, EtOH) (t = $30^o$) | B 174 |
| $C_{11}H_{15}N_2O_3Br$ | $Y = CH_3$ $R_1 = -CH(CH_3)CH_3$<br>$R_2 = -CH_2CBr = CH_2$<br>Pronarcon | Quinine methohydroxide in MeOH<br>May be only partial resolution | a) $+5.7^o$ (c 9.435, EtOH) (t = $20^o$)<br>b) $-5.5^o$ (c 9.020, EtOH) (t = $20^o$) | B 175 |
| $C_{11}H_{16}N_2O_3$ | $Y = H$ $R_1 = -CH{=}CH_2$<br>$R_2 = -CH(CH_3)CH_2CH_2CH_3$<br>Vinylbital | Quinine methohydroxide in MeOH<br>May be only partial resolution | a) $-9.4^o$ (c 4.275, EtOH) (t = $20^o$)<br>b) $+9.3^o$ (c 3.460, EtCH) (t = $20^o$) | B 175 |
| $C_{11}H_{18}N_2O_3$ | $Y = H$ $R_1 = -CH_2CH_3$<br>$R_2 = -CH(CH_3)CH_2CH_2CH_3$<br>Pentobarbital | Quinine methohydroxide in MeOH | a) $-13.38^o$ (c 2.38, EtOH) (t = $24^o$)<br>b) $+9.3^o$ (c 3.460, EtOH) (t = $20^o$) | A 22 (p.122), B 175<br>B 175 |
| $C_{11}H_{19}N$ | 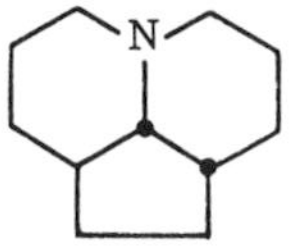 (tricyclic amine structure with N) | Partial quick resolution by chromatography on (+)-lactose | $\alpha_D + 0.08 \pm 0.01^o$ (1 2 dm, Pet E) | B 176 |
| $C_{12}H_{11}NO$ | 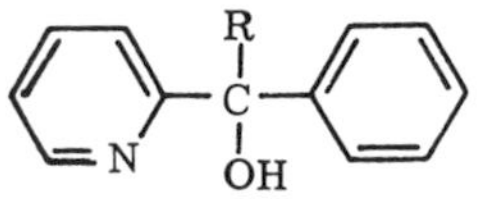 (2-pyridyl)–C(R)(OH)–(phenyl) R = H | (+)-Tartaric acid in EtOH | a) $-86.2^o$ (c 4.682, $CHCl_3$) (t = $18^o$)<br>b) $+84.8^o$ (c 3.352, $CHCl_3$) (t = $18^o$) | B 177 |
| $C_{12}H_{11}NO$ | 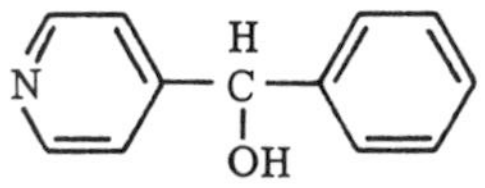 (4-pyridyl)–C(H)(OH)–(phenyl) | (+)-Tartaric acid in EtOH | a) $-55.5^o$ (c 3.656, $CHCl_3$) (t = $18^o$)<br>b) $+52.4^o$ (c 2.083, $CHCl_3$) (t = $18^o$) | B 177 |

TABLE 9d (continued)

| | Substance resolved | Resolving agent and solvent | Rotation $[\alpha]_D$ of resolved compound | Reference |
|---|---|---|---|---|
| $C_{12}H_{14}NOSBr$ | 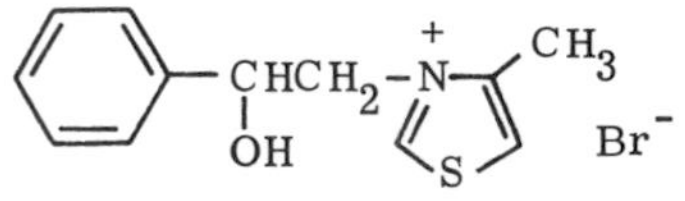 | (+)-Camphor-10-sulfonic acid | + 76.3° (c 1.29, EtCH) (t = 25°) (546 nm) | B 178 |
| $C_{12}H_{16}N_2O$ | 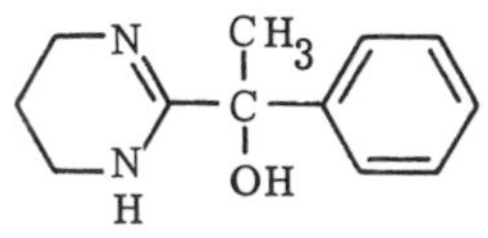 | (-)-Mandelic acid in EtOH-Pet E | + 27.5° (c 1-1.5, MeOH) (546 nm) | A 136 (p.128) |
| $C_{12}H_{16}N_2O_2$ | See $C_{11}H_{13}N_2Br$ above<br>$Y = OCH_3$ | (-)-Mandelic acid in EtOAc-acetone | + 13.8° (MeOH) (546 nm) on the mandelate salt | A 136 |
| $C_{12}H_{16}N_2O_3$ | See $C_{11}H_{14}N_2O_3$ above<br>$Y = CH_3$ $R_1 = -CH_3$ $R_2 =$ 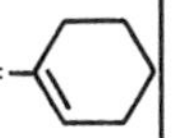<br>Hexobarbital | Quinine methohydroxide in MeOH | a) - 11.94° (c 2.388, EtOH) (t = 20°)<br>b) + 12.17° (c 3.452, EtCH) (t = 20°) | B 179 |
| $C_{12}H_{17}NO_2$ | 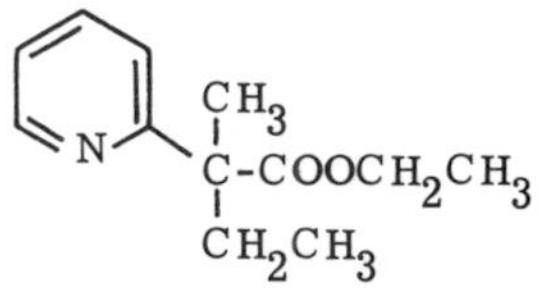 | (-)-O,O'-Dibenzyltartaric acid<br>Recr.: EtOH | a) - 0.50° (neat) (t = 28°)<br>b) + 0.28° (neat) (t = 27°) | B 163 |
| $C_{12}H_{21}N$ | 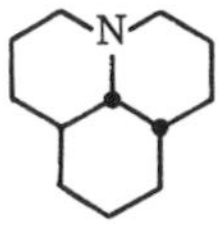 *cis*-*trans*-hexahydro-jujolidine | Partial quick resolution by chromatography on (+)-lactose | a) $\alpha_D$ + 0.13 ± 0.01° (1 2 dm, Pet E)<br>b) $\alpha_D$ - 0.06 ± 0.01° (1 2 dm, Pet E) | B 176 |

TABLE 9d (continued)

| | Substance resolved | Resolving agent and solvent | Rotation $[\alpha]_D$ of resolved compound | Reference |
|---|---|---|---|---|
| $C_{13}H_{13}NO$ | See $C_{12}H_{11}NO$ above<br>$R = CH_3$ | Brucine on the H Phthal ester in acetone | $+38.2^{o}$ (c 7.36, $CHCl_3$) ($t = 22^{o}$) | B 177 |
| $C_{13}H_{13}NO$ | 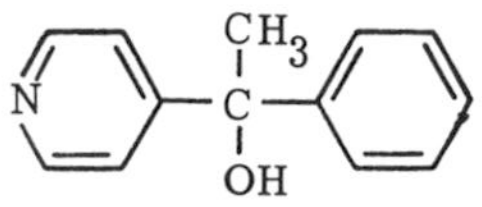 | (+)-Camphor-10-sulfonic acid in EtOH | a) $+3.08^{o}$ (c 3.982, $CHCl_3$) ($t = 31^{o}$)<br>b) $-0.2^{o}$ (c 5.3, $CHCl_3$) ($t = 31^{o}$) | B 180 |
| $C_{13}H_{14}N_2O_3$ | See $C_{11}H_{14}N_2O_3$ above<br>$Y = CH_3$ $R_1 = -CH_2CH_3$ $R_2 = -C_6H_5$ | Quinine methohydroxide in MeOH | a) $+9.3^{o}$ (c 3.118, EtOH) ($t = 20^{o}$)<br>b) $-9.0^{o}$ (c 2.774, EtOH) ($t = 20^{o}$) | B 175 |
| $C_{13}H_{15}N$ | 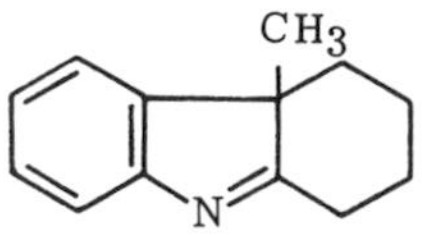 | (+)-Tartaric acid in $H_2O$-acetone | $-198^{o}$ (c 1.89, MeOH) ($t = 20^{o}$) | B 181 |
| $C_{13}H_{16}NOCl$ | See $C_{13}H_{17}NO_2$ below | | | |
| $C_{13}H_{17}NO$ | 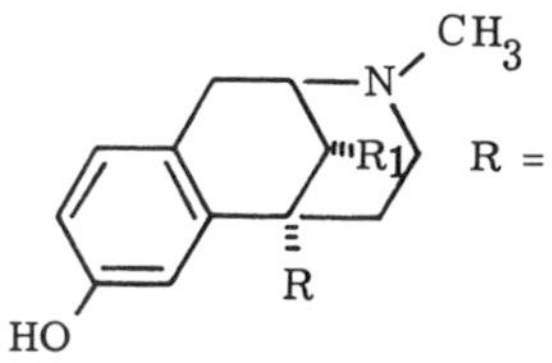 $R = R_1 = H$ | Ammonium (+)-3-bromo-camphor-9-sulfonate on the amine hydrochloride in $H_2O$ | a) $+88.9^{o}$ (MeCH) ($t = 20^{o}$)<br>b) $-89.5^{o}$ (MeOH) ($t = 20^{o}$) | B 182 |
| $C_{13}H_{17}NO_2$ | $Y = OH$ | (+)-Camphor-10-sulfonic acid in EtOAc | a) $+11.4^{o}$ (EtOH) ($t = 25^{o}$) | B 183 |
| | | (-)-Camphor-10-sulfonic acid in EtOAc | b) $-11.6^{o}$ (EtOH) ($t = 25^{o}$) | B 183 |
| $C_{13}H_{16}NOCl$ | $Y = Cl$ | (+)-Camphor-10-sulfonic acid in EtOAc | $-4.2^{o}$ (EtOH) ($t = 27^{o}$) | B 183 |

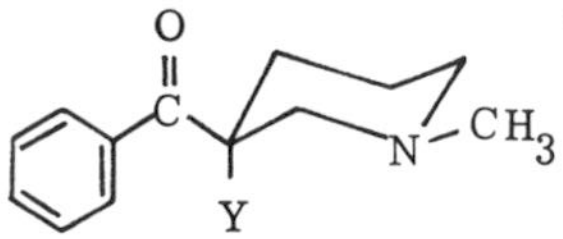

TABLE 9d (continued)

| | Substance resolved | Resolving agent and solvent | Rotation $[\alpha]_D$ of resolved compound | Reference |
|---|---|---|---|---|
| $C_{13}H_{18}N_2$ | $CH_3$-CH-$NR_1R_2$ (indole, N-R); R = $R_1$ = H, $R_2$ = -CH$(CH_3)CH_3$ | (-)-O,O'-Dibenzoyltartaric acid in EtOH | a) - 36 to $-41^o$ ($CHCl_3$) (t = $25^o$)<br>b) + 32 to $+36^o$ ($CHCl_3$) (t = $26^o$) | B 184 |
| | R = $R_1$ = $R_2$ = $CH_3$ | (-)-O,O'-Dibenzoyltartaric acid in EtOH-EtOAc-$Et_2O$ | a) + $19.5^o$ ($CHCl_3$) (t = $26^o$)<br>b) - $7.3^o$ ($CHCl_3$) (t = $26^o$) | B 184 |
| $C_{13}H_{24}N_2O$ | (piperidyl, N-H)-$CH_2$C(=O)$CH_2$-(piperidyl, N-H) | (+)-Mandelic acid<br>Recr.: MeOH-acetone<br>(-)-Mandelic acid | a) $+96.3 \pm 2^o$ (c 0.456, $H_2O$-MeOH 1:1) (t = $22^o$)<br>b) $-96.1 \pm 2^o$ (c 0.489, $H_2O$-MeOH 1:1) (t = $22^o$) | B 185<br>B 185 |
| $C_{14}H_{14}N_4$ | $H_2N$, N = N, $NH_2$, $CH_3$ $CH_3$ (biphenyl) | Ammonium (+)-3-bromo-camphor-9-sulfonate in $H_2O$ on the amine hydrochloride | a) + 55 $\pm$ $2^o$ (c 0.2, MeOH) (t = $22^o$)<br>b) - 40 $\pm$ $3^o$ (c 0.2, MeOH) (t = $23^o$) | B 186 |
| $C_{14}H_{15}NO$ | See $C_{12}H_{11}NO$ above<br>R = $CH_3CH_2$ | (+)-Tartaric acid in EtOH | a) + $65.9^o$ (c 4.52, $CHCl_3$) (t = $22^o$)<br>b) - $66.2^o$ (c 4.825, $CHCl_3$) (t = $22^o$) | B 177 |
| $C_{14}H_{17}NO_2$ | O=, $CH_3$, $CH_3$, N-H, $OCH_3$ | (+)-$\alpha$-Methylbenzyl isocyanate | a) - $54.4^o$ (t = $24^o$)<br>b) + $54.4^o$ (t = $24^o$) | B 187 |
| $C_{14}H_{19}NO$ | See $C_{13}H_{17}NO$ above<br>R = $CH_3$ $R_1$ = H | (-)-Mandelic acid<br>(+)-Mandelic acid<br>both in acetone-MeOH | a) - $72.5^o$ (95% EtOH) (t = $20^o$)<br>b) + $74.5^o$ (95% EtOH) (t = $20^o$) | B 188<br>B 188 |

TABLE 9d (continued)

| | Substance resolved | Resolving agent and solvent | Rotation $[\alpha]_D$ of resolved compound | Reference |
|---|---|---|---|---|
| $C_{14}H_{19}N_3O_2$ | 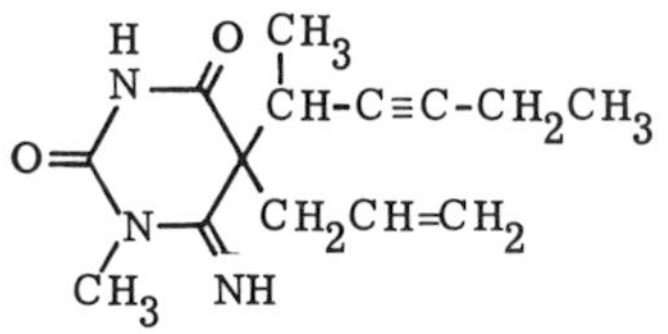 H, O, $CH_3$, N, $CH-C\equiv C-CH_2CH_3$, O, N, $CH_2CH=CH_2$, $CH_3$, NH | (+)-Camphor-10-sulfonic acid in acetone | a) $+113.5^o$ (c 5%, EtOH) (t = $25^o$)<br>b) $-113.6^o$ (c 5%, EtOH) (t = $30^o$) | B 174 |
| $C_{14}H_{21}NO_3$ | 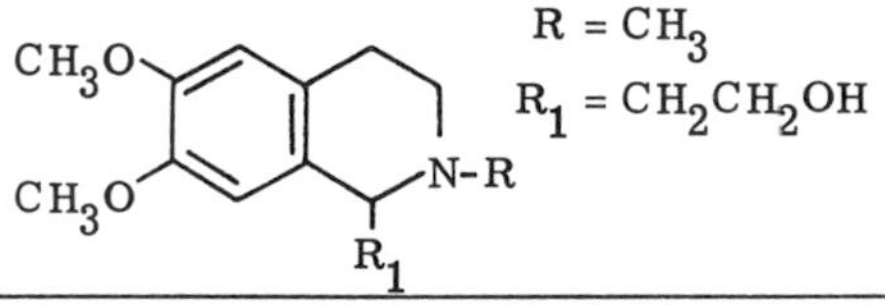 $CH_3O$, $CH_3O$, N-R, $R_1$<br>R = $CH_3$<br>$R_1$ = $CH_2CH_2OH$ | (+)-2-Nitrotartranilic acid in 95 % EtOH | a) $+22.7^o$ (c 2.05, $CHCl_3$) (t = $25^o$)<br>b) $-22.8^o$ (c 2.02, $CHCl_3$) (t = $25^o$) | B 189 |
| $C_{15}H_{13}NO_2$ | 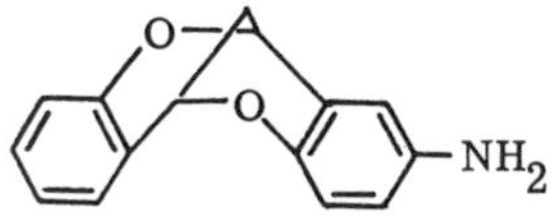 O, O, $NH_2$ | (+)-Tartaric acid in EtOH | a) $+389.0^o$ (MeOH) (t = $25^o$)<br>b) $-393.3^o$ (MeOH) (t = $25^o$) | B 190 |
| $C_{15}H_{15}NO$ | 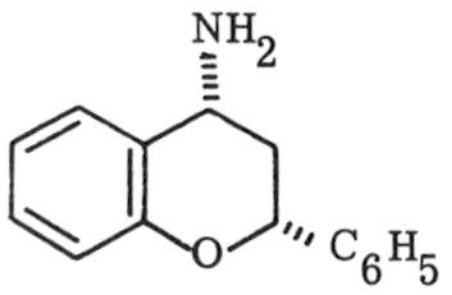 $NH_2$, O, $C_6H_5$<br>2,4-cis-4-Aminoflavan | (+)-Camphor-10-sulfonic acid in py | a) $-13.8^o$ (c 0.5%, EtOH)<br>$-26.5^o$ (c 0.5%, EtOH) on the hydrochloride salt<br>b) $+14^o$ (c 0.5%, EtOH)<br>$+28.6^o$ (c 0.5%, EtOH) on the hydrochloride salt | B 191 |
| | | (-)-O,O'-Dibenzoyltartaric acid in MeOH | $+30.8^o$ (c 0.5%, EtOH) on the hydrochloride salt | B 191 |
| $C_{15}H_{17}NO$ | See $C_{12}H_{11}NO$ above<br>R = $CH(CH_3)_2$ | (+)-Tartaric acid in EtOH | a) $+35.0^o$ (c 0.795, MeOH)<br>b) $-4.4^o$ (c 10.15, MeOH) | B 180 |
| $C_{15}H_{20}NOCl$ | 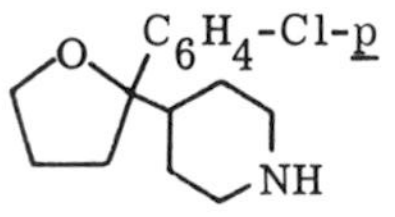 O, $C_6H_4$-Cl-p, NH | (+)-Tartaric acid in EtOH | a) $+15 \pm 0.5^o$ (c 2, $CHCl_3$)<br>b) $-15 \pm 0.05^o$ (c 2, $CHCl_3$) | B 192 |

TABLE 9d (continued)

| | Substance resolved | Resolving agent and solvent | Rotation $[\alpha]_D$ of resolved compound | Reference |
|---|---|---|---|---|
| $C_{15}H_{21}NO$ | See $C_{13}H_{17}NO$ above<br>R = $CH_3CH_2$ $R_1$ = H | (+)-Mandelic acid<br>(-)-Mandelic acid<br>both in acetone-MeOH | a) $- 49.7^o$ (95% EtOH) (t = $20^o$)<br>b) $+ 51^o$ (95% EtOH) (t = $24^o$) | B 188<br>B 188 |
| $C_{15}H_{21}NO_4$ | See $C_{14}H_{21}NO_3$ above<br>R = H $R_1$ = $CH_2COOCH_2CH_3$ | (+)-2-Nitrotartranilic acid in 95% EtOH | a) $+ 50.7^o$ (c 1.84, 95% EtOH) (t = $25^o$)<br>b) $- 51.0^o$ (c 1.88, 95% EtOH) (t = $25^o$) | B 189 |
| | | Partial resolution with (-)-O, O'-dibenzoyltartaric acid | | B 193 |
| $C_{15}H_{23}NO$ | See $C_8H_{17}NO$ above<br>R = $CH_2C_6H_5$ N-Benzylsedridine | (+)-Tartaric acid<br>(-)-Tartaric acid<br>both in EtOH on the O-benzoyl ester | a) $- 42.8 \pm 0.8^o$ (c 4.00%, MeOH) (t = $20^o$)<br>b) $+ 43.5 \pm 0.8^o$ (c 4.01%, MeOH) (t = $21^o$) | B 161<br>B 161 |
| $C_{15}H_{26}N_2$ | 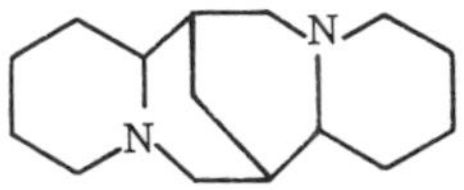 Sparteine | (+)-Camphor-10-sulfonic acid in EtOH<br>(-)-Camphor-10-sulfonic acid in EtOH | a) $+ 24.8 \pm 0.5^o$ (c 1.932, $CHCl_3$) (t = $29^o$)<br>b) $- 24.0 \pm 0.5^o$ (c 2.040, $CHCl_3$) (t = $22^o$)<br>both on the camphorsulfonate salts | B 194<br>B 194 |
| $C_{16}H_{15}NO_5$ | 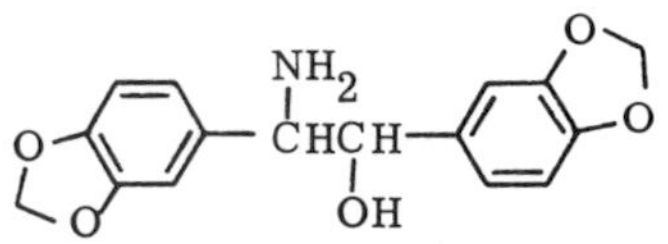 | (+)-Tartaric acid in MeOH | a) $- 200.1^o$ (c 0.48, 95% EtOH) (t = $24^o$)<br>b) $+ 192^o$ (c 0.41, 95% EtOH) (t = $24^o$) | B 195 |
| $C_{16}H_{17}NO$ | 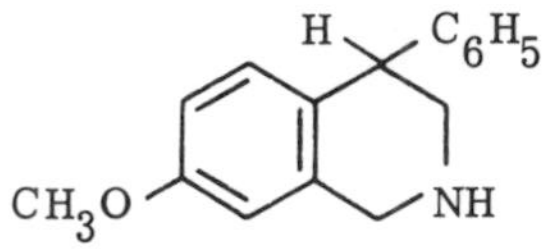 | (-)-Diisopropylidene-2-oxo-L-gulonic acid in EtOH | $- 8.75^o$ (c 1.02, MeOH) (t = $25^o$) | Chapter II, ref. 4 (p. 14) |

TABLE 9d (continued)

| | Substance resolved | Resolving agent and solvent | Rotation $[\alpha]_D$ of resolved compound | Reference |
|---|---|---|---|---|
| $C_{16}H_{18}N_4O$ | 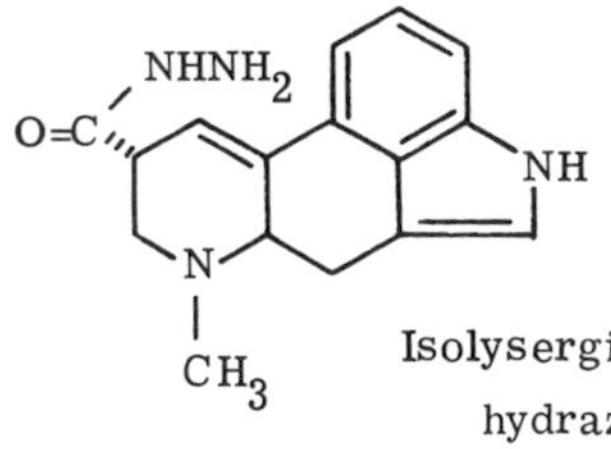 Isolysergic acid hydrazide | (+)-O,O'-Di-p-toluoyltartaric acid in MeOH<br>(-)-O,O'-Di-p-toluoyltartaric acid in MeOH | a) $+452^{\circ}$ (c 0.8, py) (t = $20^{\circ}$)<br>b) $-454^{\circ}$ (c 0.7, py) (t = $20^{\circ}$) | B 196<br>B 196 |
| $C_{16}H_{25}N$ | 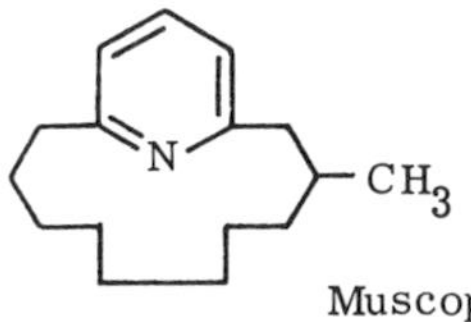 Muscopyridine | (+)-O,O'-Di-p-toluoyltartaric acid in acetone<br>Partial resolution | $+13.31 \pm 0.22^{\circ}$ (c 0.902, $CHCl_3$) (t = $25^{\circ}$) | B 197 |
| $C_{17}H_{17}NO_2$ | 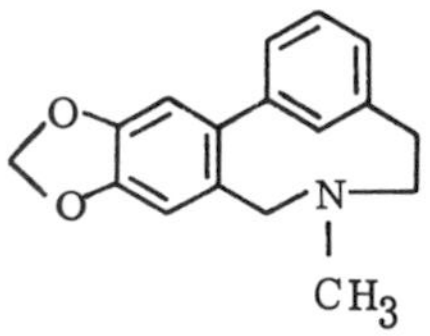 | (+)-O,O'-Di-p-toluoyltartaric acid<br>Partial resolution | $+21^{\circ}$ (c 0.95, $CHCl_3$) | B 198 |
| $C_{17}H_{18}N_2$ | 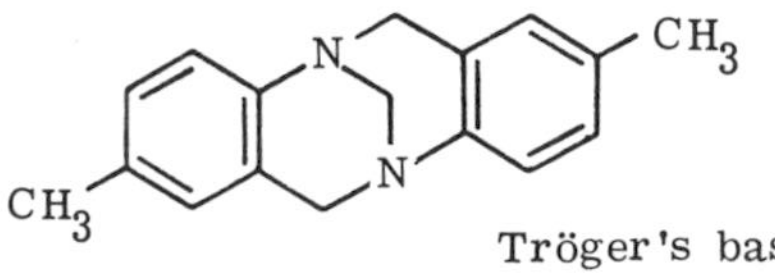 Tröger's base | Chromatography on (+)-lactose hydrate | a) $+287 \pm 7^{\circ}$ (c 0.281, hexane) (t = $17^{\circ}$)<br>b) $-278 \pm 7^{\circ}$ (c 0.292, hexane) (t = $17^{\circ}$) | B 199 |
| | | Also, resolution by chromatography on cellulose acetate (Table 7, ref. 3) (p. 40) | | |
| $C_{17}H_{23}NO$ | 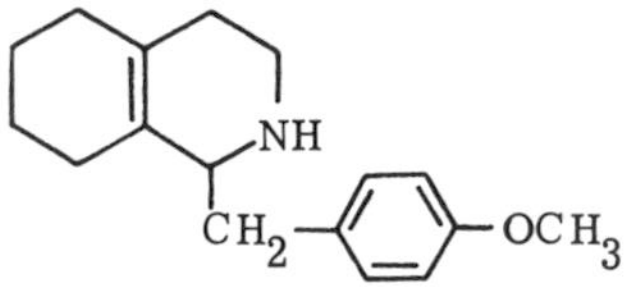 | (-)-Diisopropylidene-2-oxo-L-gulonic acid in $CH_3CN$ | a) $-144^{\circ}$ (c 2.01, MeOH) (t = $25^{\circ}$)<br>b) $+148^{\circ}$ (c 2.02, MeOH) (t = $25^{\circ}$) | Chapter II, ref. 4 (p.14) |

TABLE 9d (continued)

| Formula | Substance resolved | Resolving agent and solvent | Rotation $[\alpha]_D$ of resolved compound | Reference |
|---|---|---|---|---|
| $C_{17}H_{23}NO$ | 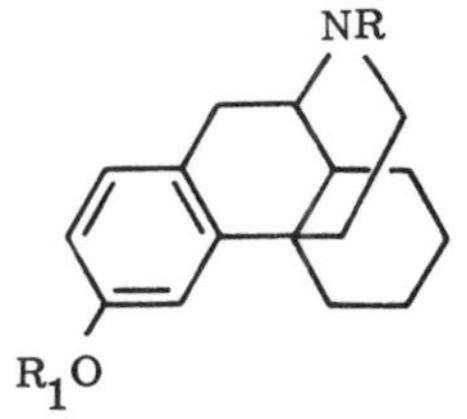 NR; $R_1O$; R = $CH_3$, $R_1$ = H | (+)-Tartaric acid in $H_2O$ | a) - 56.0° (c 3, EtOH) (t = 20°)<br>b) + 56.3° (c 3, EtOH) (t = 20°) | B 200 |
| | | Also, resolutions with (+)- and (-)-O,O'-dibenzoyltartaric acid (B 201), and (-)-diisopropylidene-2-oxo-L-gulonic acid (Chapter II, ref. 4) (p. 14). | | |
| $C_{17}H_{23}NO$ | R = H $R_1$ = $CH_3$<br>3-Methoxymorphinan | (-)-Diisopropylidene-2-oxo-L-gulonic acid in $C_6H_6$ | + 31.2° (c 2.12, MeOH) (t = 25°) | Chapter II, ref. 4 (p. 14) |
| $C_{17}H_{25}NO$ | See $C_{13}H_{17}NO$ above<br>R = $CH_3CH_2CH_2$ $R_1$ = $CH_3$ | (+)-Camphor-10-sulfonic acid in EtOH | a) + 69.8° (95% EtOH) (t = 20°) | B 188 |
| | | (+)-Mandelic acid in acetone | b) - 70.0° (95% EtOH) (t = 20°) | B 188 |
| $C_{18}H_{12}N_2$ | 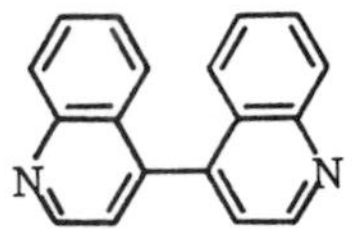 N, N | (+)-Tartaric acid in EtOH | a) + 205° (c 5.44, 1 N HCl) (t = 17.5°)<br>b) - 52.3° (c 4.30, 1 N HCl) (t = 18°)<br>both on the amine hydrochloride salts | B 202 |
| $C_{18}H_{12}N_2$ | 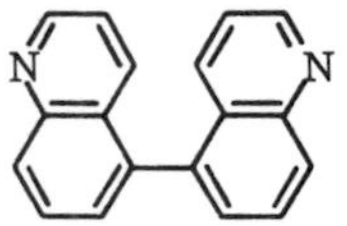 N, N | (+)-Tartaric acid in EtOH | a) + 1.3° (c 5.89, 1 N HCl) (t = 20.5°)<br>b) - 3.24° (c 1.08, 1 N HCl) (t = 20°)<br>both on the amine hydrochloride salts | B 202 |
| $C_{18}H_{20}N_2O$ | 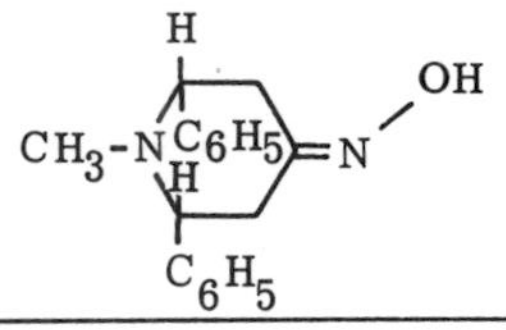 H; OH; $CH_3$-N; $C_6H_5$; H; =N; $C_6H_5$ | (+)-Camphor-10-sulfonic acid in MeOH-$Et_2O$ | + 32.63° (c 2.2, 95% EtOH) (t = 25°) | B 203 |

TABLE 9d (continued)

| Substance resolved | Resolving agent and solvent | Rotation $[\alpha]_D$ of resolved compound | Reference |
|---|---|---|---|
| $C_{18}H_{21}NO_3$ (structure: $CH_3O$, $CH_3O$, H, $C_6H_4$-$OCH_3$-p, NH) | (-)-O,O'-Dibenzoyltartaric acid in EtOH | a) + 38.4° (c 1.22, MeOH) (t = 25°) | B 204 |
| | (-)-Tartaric acid in EtOH | b) - 38.2° (c 1.12, MeOH) (t = 25°) | B 204 |
| $C_{18}H_{23}NO$ (structure: $CH_3O$, R, N-$CH_3$) R = H 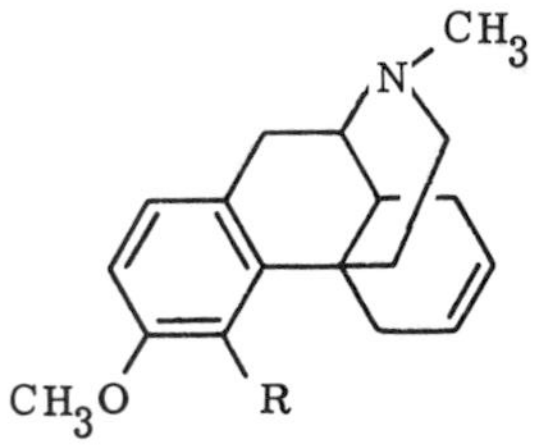 | (-)-O,O'-Dibenzoyltartaric acid in MeOH | a) $\alpha_D^{25}$ + 33.9° (1 1 dm, EtOH) | B 201 |
| | (+)-O,O'-Dibenzoyltartaric acid in MeOH | b) $\alpha_D^{25}$ - 34.3° (1 1 dm, EtOH) | B 201 |
| $C_{19}H_{25}NO_2$ R = $CH_3$ | (-)-O,O'-Dibenzoyltartaric acid in MeOH | a) + 80° (c 1.24, EtOH) (t = 27°) | B 205 |
| | (+)-O,O'-Dibenzoyltartaric acid in MeOH | b) - 79° (c 1.09, EtOH) (t = 27°) | B 205 |
| 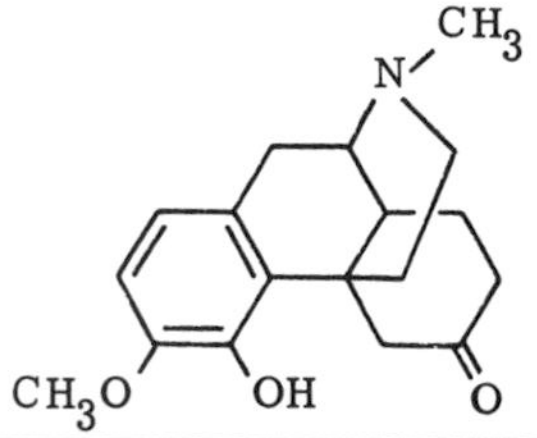 $C_{18}H_{23}NO_3$ Dihydro-thebainone (structure: $CH_3O$, OH, O, N-$CH_3$) | (+)-Tartaric acid in acetone | - 75° (c 0.77, EtOH) (t = 21°) | B 206 |
| 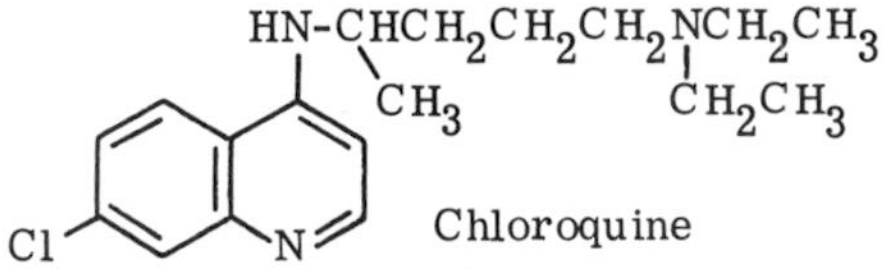 $C_{18}H_{26}N_3Cl$ Chloroquine (structure: Cl, N, HN-$CHCH_2CH_2CH_2NCH_2CH_3$, $CH_3$, $CH_2CH_3$) | (+)-3-Bromocamphor-9-sulfonic acid in EtOH | a) + 12.3° (c 2, 95% EtOH) (t = 25°)<br>b) - 13.2° (c 2, 95% EtOH) (t = 27°) | B 207 |
| $C_{19}H_{21}NO_4$ N-Norromneine (structure: RO, RO, $NR_1$, $R_2$; R = $-CH_2-$, $R_1$ = H, $R_2$ = $-CH_2-$(3,4-di-$OCH_3$-phenyl)) | (-)-N-Acetyl-L-leucine in MeOH-$Et_2O$ | a) - 17° (c 0.154, EtOH) | B 208 |
| | (-)-O,O'-Di-p-toluoyl-tartaric acid in acetone | b) + 16° (c 0.052, EtOH) | B 208 |

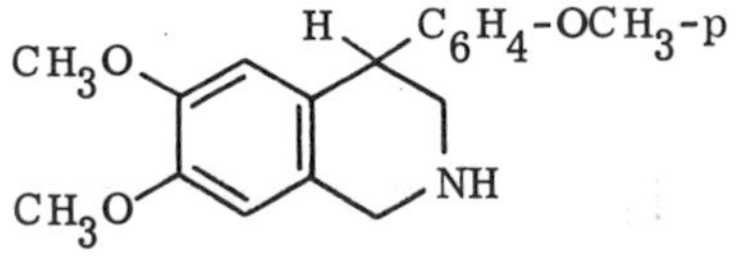

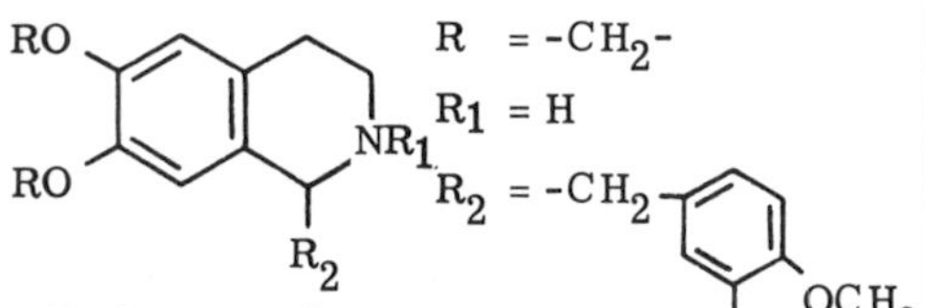

TABLE 9d (continued)

| | Substance resolved | Resolving agent and solvent | Rotation $[\alpha]_D$ of resolved compound | Reference |
|---|---|---|---|---|
| $C_{19}H_{23}NO_3$ | See $C_{19}H_{21}NO_4$ above<br>R = $CH_3$ $R_1$ = $CH_3$ $R_2$ = $-CH_2$-(p-hydroxyphenyl)<br>Armepavine | (+)-Camphor-10-sulfonic acid in $H_2O$ | a) + 118.4° ($CHCl_3$) (t = 20°)<br>b) - 117.6° ($CHCl_3$) (t = 20°) | B 209 |
| | | Also, resolution with (+)- and (-)-O, O'-dibenzoyltartaric acid (B 210) | | |
| $C_{19}H_{23}NO_5$ | See $C_{19}H_{21}NO_4$ above<br>R = H $R_1$ = H $R_2$ = $-CH_2$-(3,4,5-trimethoxyphenyl) | (+)-Tartaric acid<br>in $CHCl_3$-EtOH | a) - 30° (c 1, MeOH) (t = 23°)<br>b) + 29.20° (c 1, MeOH) (t = 22°)<br>both on the hydrochloride salts | B 211 |
| $C_{19}H_{24}N_2OS$ | (phenothiazine, N-$CH_2CH(CH_3)$-$R_3$)<br>$R_1$ = H $R_2$ = $OCH_3$<br>$R_3$ = $-CH_2N(CH_3)_2$ | (+)-Tartaric acid<br>in $H_2O$-EtOH<br>or acetone | a) + 14° ($CHCl_3$)<br>b) - 16° ($CHCl_3$) | B 212 |
| | $R_1$ = $OCH_3$ $R_2$ = H $R_3$ = $-CH_2N(CH_3)_2$ | (-)-O, O'-Dibenzoyltartaric acid in acetone-iso PrOH | - 15° (c 5, $CHCl_3$) (t = 20°) | B 213 |
| $C_{19}H_{24}N_2S$ | See $C_{19}H_{24}N_2OS$ above<br>$R_1$ = $R_2$ = H $R_3$ = $N(CH_2CH_3)_2$ | (+)-Tartaric acid<br>in PrOH | a) + 16.7° (c 4%, EtOH) (t = 25°)<br>b) - 16.7° (c 4%, EtOH) (t = 25°) | B 214 |
| $C_{19}H_{25}NO_2$ | See $C_{18}H_{23}NO$ above | | | |
| $C_{19}H_{26}N_2O_2$ | (6-methoxyquinolin-4-yl)-CH(OH)-(piperidyl, $CH_2CH_2CH_3$) | (-)-Malic acid<br>in 95 % EtOH | a) + 231.5 ± 1.0° (c 0.717, 0.1 N $H_2SO_4$) (t = 18°)<br>b) - 229.0 ± 2° (c 0.58, 0.1 N $H_2SO_4$) (t = 24.5°) | B 215 |

TABLE 9d (continued)

| | Substance resolved | Resolving agent and solvent | Rotation $[\alpha]_D$ of resolved compound | Reference |
|---|---|---|---|---|
| $C_{19}H_{30}N_2O$ | 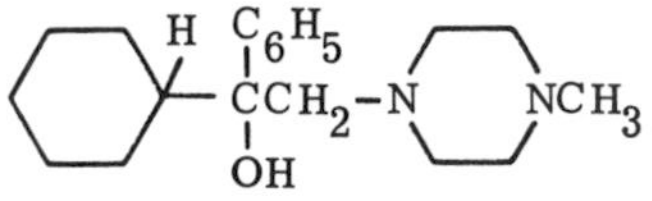 (structure) | (+)-Tartaric acid in MeOH | a) $+6.2^{\circ}$ (c 5, $H_2O$) (t = $25^{\circ}$)<br>b) $-6.2^{\circ}$ (c 5, $H_2O$) (t = $25^{\circ}$)<br>both on the hydrochloride salts | B 216 |
| $C_{20}H_{17}N$ | 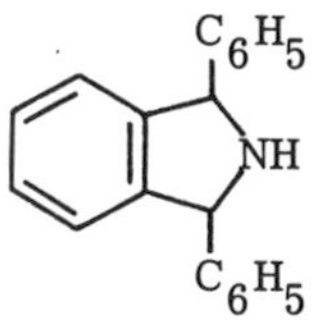 (structure) | (-)-Menthyl chloroformate _via_ formation of diastereomeric carbamates<br>Recr.: EtOH-$CH_3NO_2$ | $+206.2^{\circ}$ (c 1.03, $CHCl_3$) (t = $27^{\circ}$) | B 217 |
| $C_{20}H_{25}NO_4$ | See $C_{19}H_{21}NO_4$ above<br>R = $CH_3$ $R_1$ = $CH_3$ $R_2$ = $-CH_2$-(3-OCH₃, 4-OH-phenyl) 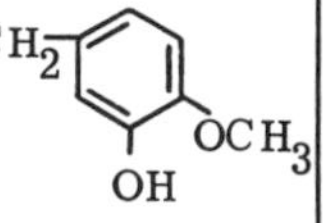<br>Laudanine | (+)-Tartaric acid<br>(-)-Tartaric acid<br>both in EtOH on the benzoate ester | a) $+94.8 \pm 1.4^{\circ}$ (c 0.4, $CHCl_3$) (t = $22^{\circ}$)<br>b) $-94.7 \pm 1.3^{\circ}$ ($CHCl_3$) | B 218<br>B 218 |
| | R = $CH_3$ $R_1$ = $CH_3$ $R_2$ = $-CH_2$-(3-OH, 4-$CH_3O$-phenyl) 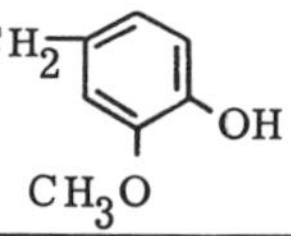<br>Pseudocodamine | (+)-Tartaric acid | a) $+24.7 \pm 2^{\circ}$ (c 0.46, $CHCl_3$) (t = $22^{\circ}$)<br>b) $-22.4 \pm 2^{\circ}$ ($CHCl_3$) | B 218 |
| | R = $CH_3$ / H $R_1$ = $CH_3$ $R_2$ = $-CH_2$-(3-OCH₃, 4-$CH_3O$-phenyl) 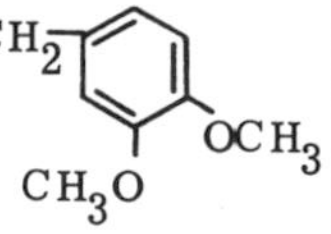<br>Codamine | a) (-)-O,O'-Dibenzoyltartaric acid in MeOH<br>b) (+)-O,O'-Dibenzoyltartaric acid in MeOH | a) $-68.8 \pm 1.9^{\circ}$ (c 1.0, EtOH) (t = $17^{\circ}$)<br>b) $+66.1 \pm 1.5^{\circ}$ (c 1.0, EtOH) (t = $15^{\circ}$) | B 219<br>B 219 |
| $C_{20}H_{26}N_2O_2$ | 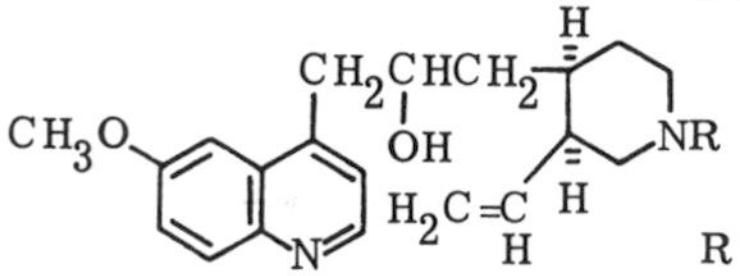 (structure) R = H | (-)-O,O'-Dibenzoyltartaric acid | $+39.0^{\circ}$ (c 1.0, $CHCl_3$) (t = $25^{\circ}$) | B 220 |

TABLE 9d (continued)

| Formula | Substance resolved | Resolving agent and solvent | Rotation $[\alpha]_D$ of resolved compound | Reference |
|---|---|---|---|---|
| $C_{20}H_{27}NO$ | 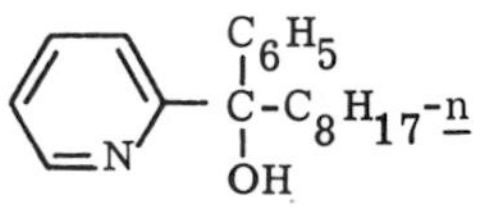 2-pyridyl-$C(C_6H_5)(OH)$-$C_8H_{17}$-n | (+)-Camphor-10-sulfonic acid in EtOH | a) $+3.96^{o}$ (c 3.22, EtOH) (t = $35^{o}$)<br>b) $-2.64^{o}$ (c 3.4, EtOH) (t = $35^{o}$) | B 221 |
| $C_{20}H_{27}NO$ | 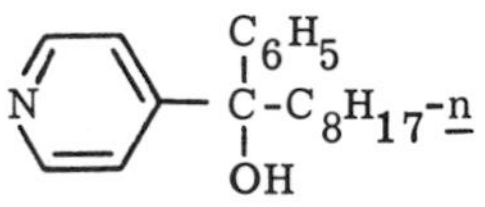 4-pyridyl-$C(C_6H_5)(OH)$-$C_8H_{17}$-n | (+)-Tartaric acid in EtOH | a) $+4.2^{o}$ (c 5.8, EtOH) (t = $35^{o}$)<br>b) $-3.50^{o}$ (c 6.73, EtOH) (t = $35^{o}$) | B 221 |
| $C_{21}H_{21}NO_6$ | See $C_{19}H_{21}NO_4$ above<br>R = $CH_3$ $R_1$ = $CH_3$ $R_2$ 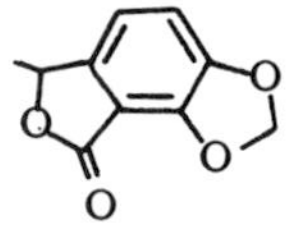 = (structure)<br>Corlumine | (+)-Tartaric acid in MeOH | a) $-73.0^{o}$ (c 1.32, $CHCl_3$) (t = $26^{o}$)<br>b) $+75.0^{o}$ (c 1.18, $CHCl_3$) (t = $26^{o}$) | B 222 |
| $C_{21}H_{26}NO_2Br$ | 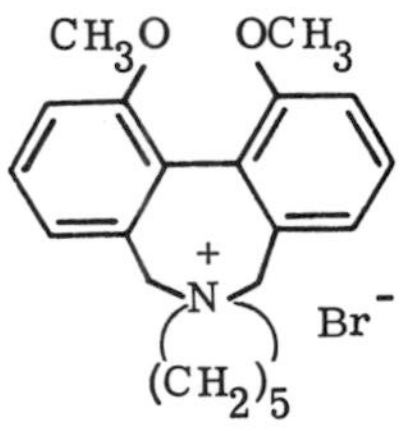 (structure: $CH_3O$, $OCH_3$, $N^+$, $(CH_2)_5$, $Br^-$) | Silver (+)-camphor-10-sulfonate in $H_2O$ | a) $-3.4 \pm 0.2^{o}$ (c 2.35, $CH_3CN$) (t = $22^{o}$) (579 nm)<br>b) $+3.6 \pm 0.4^{o}$ (c 1.11, $CH_3CN$) (t = $22^{o}$) (579 nm)<br>both on the iodide salts | B 223 |
| $C_{21}H_{26}N_2O_3$ | Yohimbine | (-)-N-Acetyl-L-leucine in MeOH-$Et_2O$<br>Also, resolution with (-)-camphor-10-sulfonic acid | | B 224<br>B 225 |
| | β-Yohimbine | Resolution with L-camphor-10-sulfonic acid in MeOH | | B 224 |
| $C_{22}H_{29}NO_5$ | 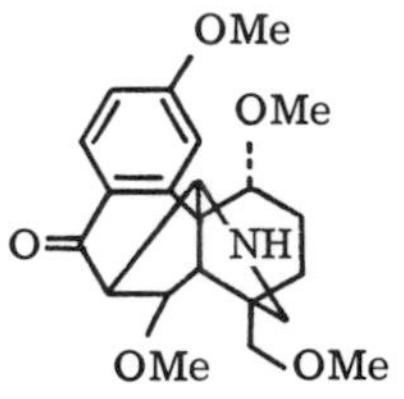 (structure: OMe, OMe, NH, O=, OMe, OMe) | Kinetic resolution with (+)- and (-)-camphor-10-sulfonyl chloride. Unreacted amine isolated and enantiomer obtained from amide formed. | | B 226 |

TABLE 9d (continued)

| | Substance resolved | Resolving agent and solvent | Rotation $[\alpha]_D$ of resolved compound | Reference |
|---|---|---|---|---|
| $C_{25}H_{27}N$ | 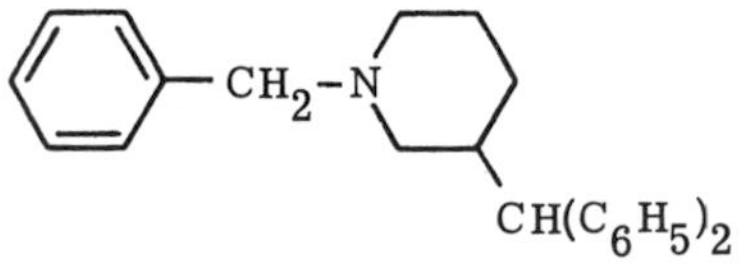 $C_6H_5-CH_2-N$ (piperidine), 3-$CH(C_6H_5)_2$ | (+)-Tartaric acid<br>(-)-Tartaric acid<br>both in acetone | a) $-6.33^o$ (c 17.5, EtOH) (t = $25^o$)<br>b) $+7.80^o$ (c 21.5, EtOH) (t = $25^o$) | B 227<br>B 227 |
| $C_{25}H_{32}N_2O_6$ | 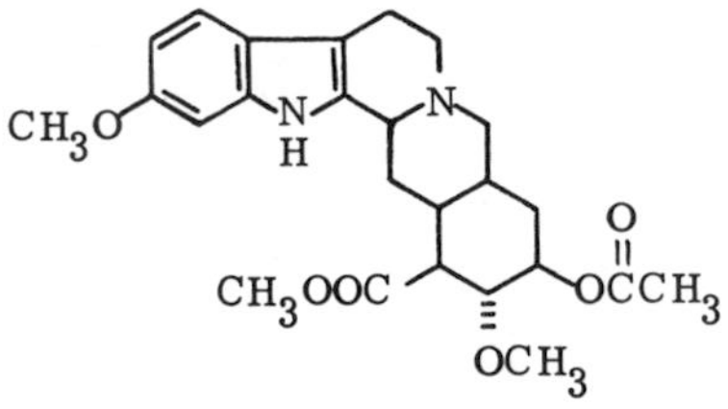 $CH_3O$, N, H, N, $CH_3OOC$, $OCH_3$, $OCCH_3$ (O) | (+)-O,O'-Di-p-toluoyl-tartaric acid<br>(-)-O,O'-Di-p-toluoyl-tartaric acid<br>both in acetone | a) $-134^o$ (c 1.04, $CHCl_3$) (t = $24^o$)<br>b) $+130^o$ (t = $24^o$) | B 228<br>B 228 |
| $C_{27}H_{30}N_2O_3$ | See $C_{20}H_{26}N_2O_2$ above<br>R = $C_6H_5CO$- | (-)-O,O'-Dibenzoyltartaric acid in MeOH | $+39.6^o$ (c 1.425, $CHCl_3$) (t = $25^o$) | B 229 |
| $C_{29}H_{40}N_2O_4$ | 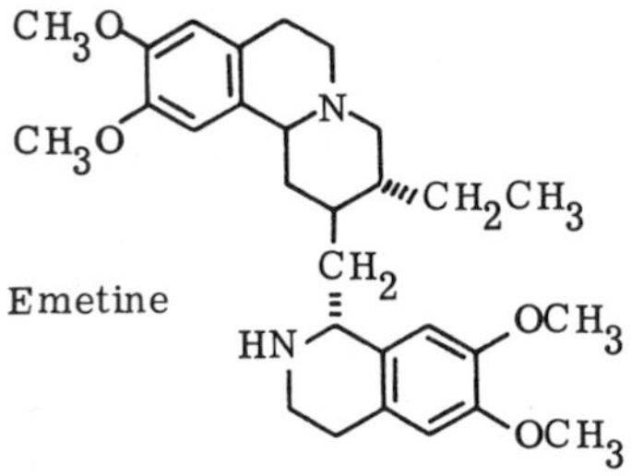 $CH_3O$, $CH_3O$, N, $CH_2CH_3$, $CH_2$, HN, $OCH_3$, $OCH_3$<br>Emetine | (+)-O,O'-Dibenzoyltartaric acid in MeOH-EtOAc on the dihydrochloride salt<br>Also, resolution with (-)-N-acetyl-L-leucine | $+26.65 \pm 1.75^o$ (c 0.571, $CHCl_3$) (t = $19^o$) on the dihydroiodide salt | B 230<br>B 231 |
| $C_{33}H_{40}N_2O_9$ | Reserpine | See $C_{25}H_{32}N_2O_6$ above | | B 228 |

## References to Tables 9a-d
## Resolutions of Amines and other Nitrogen Bases


(B 1) H.E. Smith, S.L. Cook and M.E. Warren, Jr., J. Org. Chem., 29, 2265 (1964).

(B 2) P. Bruck, I.N. Denton and A.H. Lamberton, J. Chem. Soc., 1956, 921.

(B 3) R.A. Moss and D.W. Reger, J. Amer. Chem. Soc., 91, 7539 (1969); D.W. Reger, Ph.D. Thesis, Rutgers University, 1970.

(B 4) R.H. Hohn, A. Chakravorty and G.O. Dudek, J. Amer. Chem. Soc., 86, 379 (1964).

(B 5) R.H. Mazur, J. Org. Chem., 35, 2050 (1970).

(B 6) P. Karrer and P. Dinkel, Helv. Chim. Acta, 36, 122 (1953).

(B 7) D.C. Iffland and N.T. Buu, J. Org. Chem., 32, 1230 (1967).

(B 8) R.J. Dearborn and J.A. Stekol, U.S. Patent 2,528,267 (Oct. 31, 1950) [Chem. Abstr., 45, 2984c (1951)].

(B 9) F. Nerdel and H. Liebig, Chem. Ber., 87, 221 (1954).

(B 10) A.C. Cope, W.R. Moore, R.D. Bach and H.J.S. Winkler, J. Amer. Chem. Soc., 92, 1243 (1970).

(B 11) F. Nerdel and H. Liebig, Justus Liebigs Ann. Chem., 621, 42 (1959).

(B 12) A.P. Terent'ev and V.M. Potapov, Zh. Obshch. Khim., 27, 1092 (1957) [Chem. Abstr., 52, 3709 (1958)].

(B 13) W. Theilacker and H.-G. Winkler, Chem. Ber., 87, 690 (1954).

(B 14) A. Ault, Org. Syn., 49, 93 (1969).

(B 15) E.H. White and C.A. Aufdermarsh, Jr., J. Amer. Chem. Soc., 83, 1179 (1961).

(B 16) A.W. Ingersoll, Organic Syntheses, Coll. Vol. II, 1943, p. 506.

(B 17) A.P. Terent'ev, G.V. Panova, G.N. Koval' and O.V. Toptygina, Zh. Obshch. Khim., 40, 1409 (1970) [Engl. transl., p. 1393].

(B 18) H.D. De Witt and A.W. Ingersoll, J. Amer. Chem. Soc., 73, 5782 (1951).

(B 19) A.P. Terent'ev and V.M. Potapov, Zh. Obshch. Khim., 26, 1225 (1956) [Chem. Abstr., 50, 16709a (1956)].

(B 20) E. Felder, D. Pitrè and S. Boveri, Helv. Chim. Acta, 52, 329 (1969).

(B 21) B. Helferich and W. Portz, Chem. Ber., 86, 1034 (1953).

(B 22) J.F. Tocanne and C.A. Asselineau, Bull. Soc. Chim. Fr., 1965, 3348.

(B 23) (a) F.G. Mann and J.W.G. Porter, J. Chem. Soc., 1944, 456; (b) F.G. Mann and J. Reid, ibid., 1950, 3384.

(B 24) J.J. Lafferty and B. Loev, U.S. Patent 3,147,305 (Sept. 1, 1964) [Chem. Abstr., 61, 13243e (1964)].

(B 25) (a) M.E. Warren, Jr., and H.E. Smith, J. Amer. Chem. Soc., 87, 1757 (1965); (b) A.J. Little, J.M'Lean and F.J. Wilson, J. Chem. Soc., 1940, 336.

(B 26) V.M. Potapov and A.P. Terent'ev, Zh. Obshch. Khim., 30, 666 (1960) [Engl. transl., p. 688].

(B 27) V. Sánchez del Olmo, Mutarrotación y Prototropía de N-Bencil-Bencilideniminas Dialquil Sustituídas, Monografias de Ciencia Moderna, No. 69, C S IC., Madrid, 1964, p. 22; R. Pérez Ossorio and V. Sánchez del Olmo, An. Real Soc. Españ. Fis. Quim., 56B, 915 (1960).

(B 28) V.M. Potapov and A.P. Terent'ev, Zh. Obshch. Khim., 28, 3323 (1958) [Chem. Abstr., 53, 14028h (1959)].

(B 29) O. Červinka, E. Kroupová and O. Bělovský, Collect. Czech. Chem. Commun., 33, 3551 (1968).

(B 30) P. Pratesi and A. La Manna, Farmaco, Ed. Sci., 11, 33 (1956) [Chem. Abstr., 50, 13800b (1956)].

(B 31) A.A. Bothner-By, R.S. Schutz, R.F. Dawson and M.L. Solt, J. Amer. Chem. Soc., 84, 52 (1962).

(B 32) W.R. Vaughan and R. Perry, Jr., J. Amer. Chem. Soc., 75, 3168 (1953).

(B 33) O. Červinka, V. Dudek and L. Hub, Collect. Czech. Chem. Commun., 35, 724 (1970); idem., Z. Chem., 9, 267 (1969).

(B 34) R. Pérez Ossorio and F. Gómez Herrera, An. Real Soc. Españ. Fis. Quim., 50B, 875 (1954).

(B 35) D. J. Severn and E. M. Kosower, J. Amer. Chem. Soc., 91, 1710 (1969).

(B 36) E. H. White and J. E. Stuber, J. Amer. Chem. Soc., 85, 2168 (1963).

(B 37) A. W. Ingersoll and H. D. De Witt, J. Amer. Chem. Soc., 73, 3360 (1951).

(B 38) A. C. Cope, C. F. Howell and A. Knowles, J. Amer. Chem. Soc., 84, 3190 (1962).

(B 39) V. V. Young, U.S. Patent 3,085,110 (Apr. 9, 1963) [Chem. Abstr., 59, 9837h (1963)].

(B 40) S. Senoh and I. Mita, J. Pharm. Soc. Jap., 72, 1096 (1952) [Chem. Abstr., 47, 6364g (1952)].

(B 41) A. Fredga, B. Sjöberg and R. Sandberg, Acta Chem. Scand., 11, 1609 (1957).

(B 42) Brit. Patent 1,078,186 (Aug. 2, 1967) [Chem. Abstr., 68, 68644x (1968)].

(B 43) A. G. Mohan and R. T. Conley, J. Org. Chem., 34, 3259 (1969).

(B 44) G. R. Clemo, C. Gardner and R. Raper, J. Chem. Soc., 1939, 1958.

(B 45) C. J. Collins, Z. K. Cheema, R. G. Werth and B. M. Benjamin, J. Amer. Chem. Soc., 86, 4913 (1964).

(B 46) Y. Ogata and K. Takagi, J. Org. Chem., 35, 1642 (1970).

(B 47) D. Pitrè and L. Fumagalli, Farmaco, Ed. Sci., 17, 130 (1962) [Chem. Abstr., 57, 16548h (1962)].

(B 48) V. Ghislandi and U. Conte, Farmaco, Ed. Sci., 23, 1022 (1968) [Chem. Abstr., 70, 28245c (1969)].

(B 49) B. M. Benjamin, P. Wilder, Jr., and C. J. Collins, J. Amer. Chem. Soc., 83, 3654 (1961).

(B 50) A. C. Cope, W. R. Funke and F. N. Jones, J. Amer. Chem. Soc., 88, 4693 (1966).

(B 51) M. Oki, Bull. Chem. Soc. Jap., 26, 161 (1953) [Chem. Abstr., 48, 13665 g (1954)].

(B 52) C. J. Collins, W. A. Bonner and C. T. Lester, J. Amer. Chem. Soc., 81, 466 (1959).

(B 53) H. E. Zimmerman, K. G. Hancock and G. C. Licke, J. Amer. Chem. Soc., 90, 4892 (1968).

(B 54) R. H. Sullivan, U.S. Patent 3,116,332 (Dec. 31, 1963) [Chem. Abstr., 60, 6746f (1964)].

(B 55) R. L. Clark, W. H. Jones, W. J. Raich and K. Folkers, J. Amer. Chem. Soc., 76, 3995 (1954); R. L. Clark and W. H. Jones, U.S. Patent 2,646,445 (July 21, 1953) [Chem. Abstr., 48, 7628b (1954)].

(B 56) J.R. Piper and T. P. Johnston, J. Org. Chem., 29, 1657 (1964).

(B 57) D. Pitrè and E.B. Grabitz, Chimia, 23, 399 (1969).

(B 58) F. H. Radke, R. B. Fearing and S.W. Fox, J. Amer. Chem. Soc., 76, 2801 (1954).

(B 59) G. Zoja, French Patent 1,439,850 (May 27, 1966) [Chem. Abstr., 66, 55027g (1967)].

(B 60) Span. Patent 357,033 (Mar. 1, 1970) [Chem. Abstr., 74, 87369b (1971)].

(B 61) I. A. Halmos and T. E. Ricketts, Brit. Patent 1,188,185 (Apr. 15, 1970) [Chem. Abstr., 73, 24902 d (1970)].

(B 62) J. Seres, Mrs. I. Doroczi and J. Lengyel, Hung. Patent 157,225 (Apr. 8, 1970) [Chem. Abstr., 73, 34782 d (1970)].

(B 63) G. Toth, Hung. Teljes 667 (July 24, 1970) [Chem. Abstr., 74, 76008g (1971)].

(B 64) F. H. Dickey, W. Fickett and H. J. Lucas, J. Amer. Chem. Soc., 74, 944 (1952).

(B 65) T. Taguchi, T. Takatori and M. Kojima, Chem. Pharm. Bull. (Tokyo), 10, 245 (1962); M. Takatori, M. Kojima and T. Taguchi, Yakugaku Zasshi, 88, 366 (1968).

(B 66) U. Folli and D. Iarossi, Gazz. Chim. Ital., 99, 1297 (1969).

(B 67) D.G. Farnum and G.R. Carlson, J. Amer. Chem. Soc., 92, 6700 (1970).

(B 68) J.K. Coward and T.C. Bruice, J. Amer. Chem. Soc., 91, 5339 (1969).

(B 69) A. D'Amico, L. Bertolini and C. Monreale, Chim. Ind. (Milan), 38, 93 (1956).

(B 70) T. Kametani, H. Sugi, H. Yagi, K. Fukumoto and S. Shibuya, J. Chem. Soc. C, 1970, 2213.

(B 71) T. Kappe and M.D. Armstrong, J. Med. Chem., 7, 569 (1964).

(B 72) B.F. Tullar, J. Amer. Chem. Soc., 70, 2067 (1948).

(B 73) K. Wismayr, R. Kilches, O. Schmid and G. Zoelss, Ger. Patent 1,247,331 (Aug. 17, 1967) [Chem. Abstr., 68, 68696r (1968)].

(B 74) B.M. Benjamin and C.J. Collins, J. Amer. Chem. Soc., 83, 3662 (1961).

(B 75) P. Pratesi, Farmaco, Ed. Sci., 8, 41 (1953) [Chem. Abstr., 48, 635g (1954)].

(B 76) W. Krauss and H. Ohle, Chem. Tech. (Berlin), 5, 79 (1953).

(B 77) J. Controulis, M.C. Rebstock and H.M. Crooks, Jr., J. Amer. Chem. Soc., 71, 2463 (1949).

(B 78) H.M. Crooks, Jr., M.C. Rebstock, J. Controulis and Q.R. Bartz, U.S. Patent 2,483,884 (Oct. 4, 1949) [Chem. Abstr., 45, 179h (1951)].

(B 79) J. Kučera, Collect. Czech. Chem. Commun., 20, 968 (1955).

(B 80) H. Ikeda and H. Ikeda, J. Sci. Res. Inst. Tokyo, 45, No. 1237, 8 (1951).

(B 81) S. Tatsuoka, J. Ueyanagi, M. Miyamoto and T. Kinoshita, J. Pharm. Soc. Japan, 71, 612 (1951) [Chem. Abstr., 46, 465 (1952)].

(B 82) R. Huisgen and C. Rüchardt, Justus Liebigs Ann. Chem., 601, 21 (1956).

(B 83) G.P. Men'shikov and G.M. Borodina, Zh. Obshch. Khim., 17, 1569 (1947) [Chem. Abstr., 42, 2245a (1948)].

(B 84) M. Nagawa and Y. Murase, Takamine Kenkyujo Nempo, 8, 15 (1956) [Chem. Abstr., 52, 308d (1958)].

(B 85) V.D'Amato and R. Pagani, Brit. Patent 738,064 (Oct. 5, 1955) [Chem. Abstr., 50, 10773g (1956)].

(B 86) Fr. Patent 1,498,674 (Oct. 20, 1967) [Chem. Abstr., 69, 77572v (1968)]. Also, M. Tsuruga, S. Murakami, K. Kondo, S. Akabane, K. Washikita and T. Koshino, U.S. Patent 3,478,101 (Nov. 11, 1969).

(B 87) L.R. Overby and A.W. Ingersoll, J. Amer. Chem. Soc., 82, 2067 (1960).

(B 88) R.H. Manske and T.B. Johnson, J. Amer. Chem. Soc., 51, 1906 (1929).

(B 89) E. Späth and R. Göhring, Monatsh. Chem., 41, 319 (1920).

(B 90) R. Howe, E.H.P. Young and A.D. Ainley, J. Med. Chem., 12, 998 (1969).

(B 91) R.A. Cutler, R.J. Stenger and C.M. Suter, J. Amer. Chem. Soc., 74, 5475 (1952).

(B 92) M.C. Rebstock and L.L. Bambas, J. Amer. Chem. Soc., 77, 186 (1955).

(B 93) R.K. Hill, J. Amer. Chem. Soc., 80, 1611 (1958).

(B 94) R.V. Heinzelman, J. Amer. Chem. Soc., 75, 921 (1953).

(B 95) E. Kerschbaum and K. Benedikt, Monatsh. Chem., 83, 1090 (1952).

(B 96) J.-C. Jallageas and E. Casadevall, C.R. Acad. Sci., Paris, Ser. C, 269, 1141 (1969).

(B 97) A. Pohland, L.R. Peters and H.R. Sullivan, J. Org. Chem., 28, 2483 (1963).

(B 98) A.F. Casy and J.L. Myers, J. Pharm. Pharmacol., 16, 455 (1964).

(B 99) A. Casadevall, E. Casadevall and M. Mion, Bull. Soc. Chim. Fr., 1968, 4498.

(B 100) S.I. Goldberg and F.-L. Lam, J. Amer. Chem. Soc., 91, 5113 (1969).

(B 101) (a) J. Weijlard, K. Pfister, 3 rd., E.F. Swanezy, C.A. Robinson and M. Tishler, J. Amer. Chem. Soc., 73, 1216 (1951);

(b) P. Pratesi, A. La Manna and L. Fontanella, Farmaco, Ed. Sci., 10, 673 (1955) [Chem. Abstr., 50, 10057a (1956)].

(B 102) G. Berti and G. Bellucci, Tetrahedron Lett., 1964, 3853.

(B 103) J.H. Hunt and D. Mc Hale, J. Chem. Soc., 1957, 2073.

(B 104) H. Gerlach, Helv. Chim. Acta, 49, 2481 (1966).

(B 105) V.V. Young, J. Amer. Pharm. Assoc., 40, 261 (1951).

(B 106) B.M. Benjamin, H.J. Schaeffer and C.J. Collins, J. Amer. Chem. Soc., 79, 6160 (1957).

(B 107) W. Stühmer and H.-H. Frey, Arch. Pharm. (Weinheim), 286, 22 (1953).

(B 108) M.C. Rebstock, C.D. Stratton and L.L. Bambar, J. Amer. Chem. Soc., 77, 24 (1955).

(B 109) R. Howe and B.S. Rao, J. Med. Chem., 11, 1118 (1968).

(B 110) D. Shapiro, H. Segal and H.M. Flowers, J. Amer. Chem. Soc., 80, 1194 (1958).

(B 111) C.A. Grob and E.F. Jenny, Helv. Chim. Acta, 35, 2106 (1952).

(B 112) H.E. Carter and D. Shapiro, J. Amer. Chem. Soc., 75, 5132 (1953).

(B 113) H. Corrodi and E. Hardegger, Helv. Chim. Acta, 40, 193 (1957).

(B 114) A. Pohland and H.R. Sullivan, J. Amer. Chem. Soc., 77, 3400 (1955).

(B 115) E.J. Corey, I. Vlattas and K. Harding, J. Amer. Chem. Soc., 91, 535 (1969).

(B 116) A.A. Larsen, B.F. Tullar, B. Elpern and J.S. Buck, J. Amer. Chem. Soc., 70, 4194 (1948).

(B 117) E.E. Howe and M. Sletzinger, J. Amer. Chem. Soc., 71, 2935 (1949).

(B 118) W.R. Brode and M.W. Hill, J. Org. Chem., 13, 191 (1948).

(B 119) E.E. Howe and M. Tishler, U.S. Patent 2,644,010 (June 30, 1953) [Chem. Abstr., 48, 5222h (1954)].

(B 120) L.H. Smith, Brit. Patent 1,136,918 (Dec. 18, 1968) [Chem. Abstr., 70, 87382t (1969)].

(B 121) D.G. Neilson and D.A.V. Peters, J. Chem. Soc., 1962, 1309.

(B 122) F.P. Dwyer, F.L. Garvan and A. Shulman, J. Amer. Chem. Soc., 81, 290 (1959).

(B 123) E. Balieu, P.M. Boll and E. Larsen, Acta Chem. Scand., 23, 2191 (1969).

(B 124) R.G. Asperger and C.F. Liu, Inorg. Chem., 4, 1492 (1965).

(B 125) R.S. Downing and F.L. Urbach, J. Amer. Chem. Soc., 91, 5977 (1969).

(B 126) F.P. Dwyer and T.E. MacDermott, J. Amer. Chem. Soc., 85, 2916 (1963).

(B 127) W.E. Pereira, Jr., M. Solomon and B. Halpern, Aust J. Chem., 24, 1103 (1971).

(B 128) D.G. Neilson and L.H. Roach, J. Chem. Soc., 1965, 1658.

(B 129) D.G. Neilson and D.F. Ewing, J. Chem. Soc. C, 1966, 393.

(B 130) R. Ghirardelli and H.J. Lucas, J. Amer. Chem. Soc., 79, 734 (1957).

(B 131) R. Roger and D.G. Neilson, J. Chem. Soc., 1959, 688.

(B 132) R. Roger and D.G. Neilson, J. Chem. Soc., 1961, 3181.

(B 133) D.J. Cram and J.S. Bradshaw, J. Amer. Chem. Soc., 85, 1108 (1963).

(B 134) P.D. Bartlett and N.A. Porter, J. Amer. Chem. Soc., 90, 5317 (1968).

(B 135) S. Wawzonek, J. Chua, E.L. Yeakey and W. McKillip, J. Org. Chem., 28, 2376 (1963).

(B 136) I. Lifschitz and J.G. Bos, Rec. Trav. Chim. Pays-Bas, 59, 173 (1940); see B. Bosnich and S.B. Wild, J. Amer. Chem. Soc., 92, 459 (1970) for the rotation.

(B 137) F.A. McGinn, A.K. Lazarus, M. Siegel, J.E. Ricci and K. Mislow, J. Amer. Chem. Soc., 80, 476 (1958).

(B 138) D.F. Ewing and D.G. Neilson, J. Chem. Soc. C, 1966, 390.

(B 139) R.E. Carter and L. Dahlgren, Acta Chem. Scand., 24, 633 (1970).

(B 140) C.G. Overberger, N.P. Marullo and R.G. Hiskey, J. Amer. Chem. Soc., 83, 1374 (1961).

(B 141) W. Theilacker and R. Hopp, Chem. Ber., 92, 2293 (1959).

(B 142) A.D. Thomsen and H. Lund, Acta Chem. Scand., 23, 3582 (1969).

(B 143) M.J. O'Connor, R.E. Ernst and R.H. Holm, J. Amer. Chem. Soc., 90, 4561 (1968).

(B 144) A.S. Cooke and M.M. Harris, J. Chem. Soc., 1963, 2365.

(B 145) W. Stühmer and G. Messwarb, Arch. Pharm. (Weinheim), 286, 221 (1953).

(B 146) C.H. Stammer, A.N. Wilson, F.W. Holly and K. Folkers, J. Amer. Chem. Soc., 77, 2346 (1955).

(B 147) A. Mishra and R.J. Crawford, Can. J. Chem., 47, 1515 (1969).

(B 148) O. Cervinka, O. Belovsky and L. Koralova, Z. Chem., 9, 448 (1969).

(B 149) French Patent 1,559,885 (Mar. 14, 1969) [Chem. Abstr., 72, 55890g (1970)].

(B 150) M. Brenner and H.R. Rickenbacker, Helv. Chim. Acta, 41, 181 (1958). See also, Neth. Appl. 6,604,321 (Oct. 2, 1967) [Chem. Abstr., 68, 95676r (1968)].

(B 151) (a) L.H. Sternbach and S. Kaiser, J. Amer. Chem. Soc., 74, 2215 (1952); idem, ibid., 74, 2219 (1952).

(B 152) M.S. Toy and C.C. Price, J. Amer. Chem. Soc., 82, 2613 (1960).

(B 153) G. Bulteau, S. African Patent 68 02,593 (Oct. 18, 1968) [Chem. Abstr., 71, 30354b (1969)].

(B 154) R. Heymes and G. Amiard, Fr. Addn. 91,765 (Aug. 9, 1968) [Chem. Abstr., 71, 91498y (1969)] to Fr. Patent 1,492,854 (Aug. 25, 1967).

(B 155) N.B. Chapman and J.F.A. Williams, J. Chem. Soc., 1953, 2797.

(B 156) H. Singh and B. Razdan, Tetrahedron Lett., 1966, 3243.

(B 157) N.J. Leonard and D.L. Felley, J. Amer. Chem. Soc., 72, 2537 (1950).

(B 158) N.J. Leonard and W.J. Middleton, J. Amer. Chem. Soc., 74, 5776 (1952).

(B 159) A.J. Aasen and C.C.J. Culvenor, J. Org. Chem., 34, 4143 (1969).

(B 160) F. Galinovsky and H. Mulley, Montash. Chem., 79, 426 (1948).

(B 161) C. Schöpf, E. Gams, F. Koppernock, R. Rausch and R. Walbe, Justus Liebigs Ann. Chem., 732, 181 (1970).

(B 162) E.J. Corey, R.J. Mc Caully and H.S. Sachdev, J. Amer. Chem. Soc., 92, 2476 (1970).

(B 163) W. von E. Doering and V.Z. Pasternack, J. Amer. Chem. Soc., 72, 143 (1950).

(B 164) E.J. Corey and J.I. Schulman, Tetrahedron Lett., 1968, 3655.

(B 165) N.J. Leonard, R.C. Sentz and W.J. Middleton, J. Amer. Chem. Soc., 75, 1674 (1953).

(B 166) Ya. L. Gol'dfarb and L.M. Smorgonskiĭ, Izv. Akad. Nauk SSSR, Otdel. Khim. Nauk, 1946, 557 [Chem. Abstr., 42, 6364d (1948)].

(B 167) B. Witkop, J. Amer. Chem. Soc., 71, 2559 (1949).

(B 168) K. Winterfeld and C. Heinen, Justus Liebigs Ann. Chem., 578, 171 (1952).

(B 169) A. García-Alvarez and I. Ribas-Marquês, An. Real Soc. Espan. Fis. Quim., Ser. B., 61, 573 (1965).

(B 170) M. Bullock, Brit. Patent 1,127,852 (Sept. 18, 1968) [Chem. Abstr., 70, 11698e (1969)].

(B 171) Brit. Patent 1,169,310 (Nov. 5, 1969) [Chem. Abstr., 72, 55453s (1970)].

(B 172) F. Dewilde and G.G. Frot, Ger. Offen., 2,027,030 (Dec. 10, 1970) [Chem. Abstr., 74, 53786v (1971)].

(B 173) E.E. van Tamelen and J.S. Baran, J. Amer. Chem. Soc., 80, 4659 (1958).

(B 174) W.J. Doran, J. Org. Chem., 25, 1737 (1960).

(B 175) J. Knabe and K. Philipson, Arch. Pharm. (Weinheim), 299, 231 (1966).

(B 176) N.J. Leonard and W.J. Middleton, J. Amer. Chem. Soc., 74, 5117 (1952).

(B 177) A.G. Davies, J. Kenyon and K. Thaker, J. Chem. Soc., 1956, 3394.

(B 178) J.C. Sheehan and D.H. Hunneman, J. Amer. Chem. Soc., 88, 3666 (1966).

(B 179) J. Knabe and R. Kräuter, Arch. Pharm. (Weinheim), 298, 1 (1965).

(B 180) K.A. Thakar and U.S. Pathak, J. Indian Chem. Soc., 41, 555 (1964).

(B 181) H. Fritz and E. Stock, Justus Liebigs Ann. Chem., 721, 82 (1969).

(B 182) E.L. May and M. Takeda, J. Med. Chem., 13, 805 (1970).

(B 183) E.E. Smissman and G. Hite, J. Amer. Chem. Soc., 82, 3375 (1960).

(B 184) J.D. Albright and H.R. Snyder, J. Amer. Chem. Soc., 81, 2239 (1959).

(B 185) M.M. El-Olemy and A.E. Schwarting, J. Org. Chem., 34, 1352 (1969).

(B 186) W. Theilacker and F. Baxmann, Justus Liebigs Ann. Chem., 581, 117 (1953).

(B 187) R.B. Woodward, Pure Appl. Chem., 17, 519 (1968).

(B 188) J.H. Ager, A.E. Jacobson and E.L. May, J. Med. Chem., 12, 288 (1969).

(B 189) T.R. Montzka, T.L. Pindell and J.D. Matiskella, J. Org. Chem., 33, 3993 (1968).

(B 190) H.E. Hennis and C.-S. Wang, J. Org. Chem., 34, 1907 (1969).

(B 191) R. Bognár, J.W. Clark-Lewis, A. Liptákné-Tökés and M. Rákosi, Aust. J. Chem., 23, 2015 (1970); R. Bognár, A.L.-Tökés and M. Rákosi, Magy. Kem. Foly., 76, 271 (1970) [Chem. Abstr., 73, 76565r (1970)].

(B 192) Brit. Patent 1,133,302 (Nov. 13, 1968) [Chem. Abstr., 70, 68187s (1969)].

(B 193) A.R. Battersby, R. Binks and T.P. Edwards, J. Chem. Soc., 1960, 3474.

(B 194) N. J. Leonard and R.E. Beyler, J. Amer. Chem. Soc., 71, 757 (1949).

(B 195) G.G. Lyle and M.L. Durand, J. Org. Chem., 32, 3295 (1967).

(B 196) A. Stoll and A. Hoffmann, Helv. Chim. Acta, 26, 922 (1943).

(B 197) K. Biemann, G. Büchi and B.H. Walker, J. Amer. Chem. Soc., 79, 5558 (1957).

(B 198) E.W. Warnhoff and S. Valverde Lopez, Tetrahedron Lett., 1967, 2723.

(B 199) V. Prelog and P. Wieland, Helv. Chim. Acta, 27, 1127 (1944).

(B 200) O. Schnider and A. Grüssner, Helv. Chim. Acta, 34, 2211 (1951).

(B 201) M. Gates and W.G. Webb, J. Amer. Chem. Soc., 80, 1186 (1958).

(B 202) M. Crawford and I.F.B. Smyth, J. Chem. Soc., 1952, 4133.

(B 203) R.E. Lyle and G.G. Lyle, J. Org. Chem., 24, 1679 (1959).

(B 204) A. Brossi, G. Grethe, S. Teitel, W.C. Wildman and D.T. Bailey, J. Org. Chem., 35, 1100 (1970).

(B 205) M. Gates and G. Tschudi, J. Amer. Chem. Soc., 78, 1380 (1956).

(B 206) D. Elad and D. Ginsburg, J. Amer. Chem. Soc., 76, 312 (1954).

(B 207) B. Riegel and L.T. Sherwood, Jr., J. Amer. Chem. Soc., 71, 1129 (1949).

(B 208) J.-I. Kunitomo, E. Yuge, Y. Nagai and K. Fujitani, Chem. Pharm. Bull. (Tokyo), 16, 364 (1968).

(B 209) J. Knabe and P. Horn, Arch. Pharm. (Weinheim), 300, 547 (1967).

(B 210) D.G. Farber and A. Giacomazzi, An. Asoc. Quim. Argent., 58, 133 (1970); idem., Chem. Ind. (London), 1968, 57.

(B 211) K. Mashimo, E. Yamato, A. Kiyomoto and H. Nakajima, S. African Patent 68 02,416 (Sept. 11, 1968) [Chem. Abstr., 70, 68198w (1969)].

(B 212) S. Pataki and G. Gelegonya, Hung. Patent 157,158 (Mar. 5, 1970) [Chem. Abstr., 72, 132762w (1970)].

(B 213) G. Tokar, Mrs. I. Krasznai and J. Somlo, Hung. Patent 155,314 (Nov. 22, 1968) [Chem. Abstr., 70, 87834s (1969)].

(B 214) W. Drell, J. Amer. Chem. Soc., 75, 2506 (1953).

(B 215) H.S. Mosher, R. Forker, H.R. Williams and T.S. Oakwood, J. Amer. Chem. Soc., 74, 4627 (1952).

(B 216) H.E. Zaugg, R.J. Michaels, H.J. Glenn, L.R. Swett, M. Freifelder, G.R. Stone and A.W. Weston, J. Amer. Chem. Soc., 80, 2763 (1958).

(B 217) L.A. Carpino, Chem. Commun., 1966, 858.

(B 218) B. Frydman, R. Bendisch and V. Deulofeu, Tetrahedron, 4, 342 (1958).

(B 219) B.K. Cassels and V. Deulofeu, Tetrahedron, Suppl., 8, 845 (1966).

(B 220) J. Gutzwiller and M. Uskoković, J. Amer. Chem. Soc., 92, 204 (1970).

(B 221) A.A. Shaikh and K.A. Tahker, J. Indian Chem. Soc., 45, 378 (1968).

(B 222) W.M. Whaley and M. Meadow, J. Chem. Soc., 1953, 1067.

(B 223) G.H. Blaven, D.M. Hall, M.S. Lesslie and E.E. Turner, J. Chem. Soc., 1952, 854.

(B 224) L. Töke, K. Honty and C. Szántay, Chem. Ber., 102, 3248 (1969).

(B 225) E.E. van Tamelen, M. Shamma, A.W. Burgstahler, J. Wolinsky, R. Tamm and P.E. Aldrich, J. Amer. Chem. Soc., 80, 5006 (1958).

(B 226) K. Wiesner, E.W.K. Jay and L. Poon-Jay, Experientia, 27, 363 (1971).

(B 227) R.P. Mull and R.H. Mizzoni, Brit. Patent 1,142,030 (Feb. 5, 1969) [Chem. Abstr., 70, 68186r (1969)].

(B 228) R.B. Woodward, F.E. Bader, H. Bickel, A.J. Frey and R.W. Kierstad, Tetrahedron, 2, 1 (1958).

(B 229) J. A. W. Gutzwiller and M. R. Uskokovic, Ger. Offen. 1,933,600 (Jan. 8, 1970) [Chem. Abstr., 72, 90698x (1970)].

(B 230) M. Barasch, J. M. Osborn and J. C. Wickens, J. Chem. Soc., 1959, 3530.

(B 231) A. Brossi, M. Baumann and O. Schnider, Helv. Chim. Acta, 42, 1515 (1959).

RESOLUTIONS OF ALCOHOLS AND ENOLS

TABLE 10a Acyclic Alcohols*)

| | Substance resolved | Derivative used+ | Resolving agent and solvent | Rotation $[\alpha]_D$ of resolved compound | Reference |
|---|---|---|---|---|---|
| $C_3H_5OF_3$ | $CF_3$-CH(OH)-$CH_3$ | β-Alkoxypropionic acid | Quinine in PrOH-$H_2O$ (20% v/v) | - 5.65° (neat) (t = 25°) | Al 1 |
| | | Preparative gas chromatographic resolution | | | Al 2 |
| $C_4H_8O$ | $CH_2$=CH-CH(OH)-$CH_3$ | H Phthal | Brucine in acetone | a) +33.2° (neat) (t = 30°)<br>b) $\alpha_D^{25}$ - 8.45° (1 1 dm, neat) | Al 3<br>Al 4 |
| $C_4H_9OI$ | $CH_3$-CH(I)-CH(OH)-$CH_3$ (threo) | — | Brucine in $CHCl_3$ via asymmetric destruction | $\alpha_D^{25}$ +3.06° | Al 5 |
| $C_4H_{10}O$ | $CH_3CH_2$-CH(OH)-$CH_3$ | H Phthal | Brucine in acetone | a) +13.89 ± 0.01° (neat) (t = 20°) | Al 6 |
| | | H Phthal | (-)-Menthylamine via amide formation Recr.: EtOH (85 %) then Pet E | b) - 13.5° (neat) (t = 20°) | Al 7 |
| $C_5H_{10}O$ | $CH_3$-CH=CH-CH(OH)-$CH_3$ (H, H trans) | H Phthal | Brucine in acetone-$CHCl_3$ (1:2) | $\alpha_D^{25}$ +0.413° (1 1 dm, neat) | Al 8, 9 |
| $C_5H_{12}O$ | $CH_3$-CH($CH_3$)-CH(OH)-$CH_3$ | — | L-Valine via the N-Tosyl ester | $\alpha_D$ + 3.87°<br>+ 5.0° (neat) (t = 27°) | Al 10<br>Al 16 |
| $C_5H_{12}O$ | $CH_3CH_2CH_2$CH(OH)$CH_3$ | H Phthal | Brucine in acetone | $\alpha_D^{20}$ + 10.3° (1 1 dm, neat)<br>apparent optical purity: 92 % | Al 8, 11 |

* References to Tables 10a-d begin on p. 217.

\+ For abbreviations used in this table, see p. 42.

TABLE 10a (continued)

| | Substance resolved | Derivative used | Resolving agent and solvent | Rotation $[\alpha]_D$ of resolved compound | Reference |
|---|---|---|---|---|---|
| $C_6H_{11}OF_3$ | $CH_3\text{-}C(CH_3)_2\text{—}CH(OH)\text{-}CF_3$ | — | 3β-Acetoxy-$\Delta^5$-etienic acid<br>Ester Recr.: isoPrOH | $+5.55^o$ (neat) ($t = 23^o$) | Al 12 |
| $C_6H_{14}O$ | $CH_3\text{-}C(CH_3)_2\text{—}CH(OH)\text{-}CH_3$ | H Phthal | Brucine<br>in acetone | a) $+8.10^o$ (neat) ($t = 25^o$)<br>b) $-4.41^o$ (neat) ($t = 25^o$) | Al 13 |
| | | Preparative gas chromatographic resolution | | | Al 14 |
| $C_6H_{14}O_2$ | $CH_3OCH_2CH_2CH_2CH(OH)CH_3$ | H Phthal | Brucine<br>in acetone-$C_6H_6$ (4:1) | a) $\alpha_D^{25}$ $+12.38 \pm 0.1^o$ (l 1 dm, neat)<br>b) $\alpha_D^{25}$ $-5.11^o$ (l 1 dm, neat) | Al 15 |
| $C_7H_{16}O$ | $CH_3\text{-}C(CH_3)_2\text{-}CH_2CH(OH)CH_3$ | H Phthal | Brucine<br>in acetone-$H_2O$ | $+24.8^o$ (neat) ($t = 27^o$) | Al 16 |
| $C_7H_{16}O$ | $CH_3\text{-}C(CH_3)_2\text{—}CH(OH)\text{-}CH_2CH_3$ | H Phthal | Brucine<br>in acetone | $\alpha_D^{23}$ $+27.40 \pm 0.03^o$ (l 1 dm, neat) | Al 17 |
| $C_7H_{16}O$ | $CH_3CH_2CH_2CH(OH)\text{-}CH(CH_3)\text{-}CH_3$ | H Phthal | Brucine<br>in EtOAc | a) $\alpha_D^{18}$ $+0.10^o$ (l 0.5 dm, neat)<br>b) $\alpha_D^{18}$ $-0.19^o$ (l 0.5 dm, neat) | Al 18 |
| $C_8H_{18}O$ | $CH_3\text{-}C(CH_3)_2\text{—}CH(OH)\text{-}CH(CH_3)\text{-}CH_3$ | H Phthal | Brucine and Strychnine<br>Recr.: acetone<br>Brucine and Cinchonine<br>Recr.: acetone | a) $\alpha_D^{28}$ $-8.94^o$ (l 1 dm, neat)<br>b) $\alpha_D^{28}$ $+9.06^o$ (l 1 dm, neat) | Al 17 |
| $C_8H_{18}O$ | $CH_3\text{-}CH(CH_3)\text{-}CH_2\text{-}C(CH_3)(OH)\text{-}CH_2CH_3$ | H Phthal | Brucine<br>in acetone | a) $-2.72^o$ (neat) ($t = 21^o$)<br>b) $+1.55^o$ (neat) ($t = 24^o$) | Al 19 |

TABLE 10a (continued)

| | Substance resolved | Derivative used | Resolving agent and solvent | Rotation $[\alpha]_D$ of resolved compound | Reference |
|---|---|---|---|---|---|
| $C_8H_{17}OD$ | $CH_3-(CH_2)_5-CD(OH)-CH_3$ | H Phthal | Brucine in acetone | $-9.23^o$ (neat) ($t = 25^o$) | Al 20 |
| $C_8H_{18}O$ | $CH_3-(CH_2)_5-CH(OH)-CH_3$ | H Phthal | Brucine in acetone | a) $+9.6^o$ (neat) ($t = 26^o$)<br>b) $-9.7^o$ (neat) ($t = 24^o$) | Al 16, 21, 22 |
| $C_9H_{18}O$ | n-$C_3H_7-CH=C(CH_2CH_3)-CH(OH)-CH_3$ | H Phthal | Brucine in acetone | a) $-12.8^o$ (c 3.05, $CS_2$) ($t = 20^o$)<br>b) $+12.4^o$ (c 3.05, $CS_2$) ($t = 20^o$) | Al 23 |
| $C_9H_{20}O$ | $CH_3-C(CH_3)_2-CH(OH)-CH_2-CH(CH_3)-CH_3$ | H Phthal | Strychnine<br>Recr.: acetone | $+57.5^o$ (c 20.4, MeOH) ($t = 26^o$) | Al 17 |
| $C_9H_{20}O$ | $CH_3-C(CH_3)_2-CH_2-CH(CH_3)-CH_2CH_2OH$ | H Phthal | Brucine in acetone | a) $+4.5 \pm 0.3^o$ (c 4.0, $CHCl_3$) ($t = 20^o$)<br>b) $-4.5 \pm 0.3^o$ (c 4.0, $CHCl_3$) ($t = 20^o$) | Al 24 |
| $C_9H_{20}O$ | $CH_3-C(CH_3)_2-CH(OH)-CH_2CH_2CH_2CH_3$ | H Phthal | Strychnine<br>Recr.: acetone-MeOH | a) $\alpha_D^{24}$ $+17.12^o$ (l 0.5 dm, neat)<br>b) $\alpha_D^{24}$ $-16.39^o$ (l 0.5 dm, neat) | Al 17 |
| $C_{10}H_{22}O$ | n-$C_8H_{17}-CH(OH)-CH_3$ | H Phthal | Brucine<br>Recr.: acetone<br>(+)-α-Methylbenzylamine in acetone | a) $+8.50^o$ (neat) ($t = 25^o$)<br>b) $-8.56^o$ (neat) ($t = 25^o$) | Al 25 |
| $C_{11}H_{20}O$ | $CH_3-C(CH_3)_2-C \equiv C-CH(OH)-C(CH_3)_2-CH_3$ | H Phthal | Brucine | $+4.8^o$ (c 1.00, $CHCl_3$) ($t = 22^o$) (578 nm) | Al 26 |

## TABLE 10b Alicyclic Alcohols

| | Substance resolved | Derivative used | Resolving agent and solvent | Rotation $[\alpha]_D$ of resolved compound | Reference |
|---|---|---|---|---|---|
| $C_5H_{10}O$ | 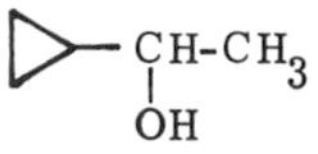 | H Phthal | Brucine in acetone | $\alpha_D^{28}$ + 20.1° (1 1 dm, neat) | Al 27 |
| $C_6H_{10}O$ | 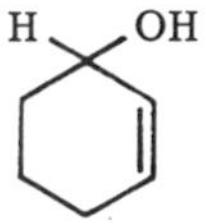 | H Phthal | Brucine in MeOH | a) + 110.6° (c 0.94, $CHCl_3$) (t = 21°) | Al 28a |
| | | H Phthal | Dehydroabietylamine in acetone | b) - 15.2° (c 5.32, $CHCl_3$) (t = 25°) | Al 28b |
| $C_7H_{10}O$ | 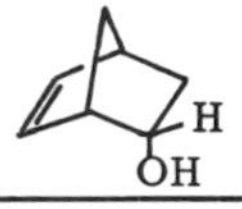 | H Phthal | Brucine in MeOH | - 11.3° (c 1.5, $CHCl_3$) (t = 27°) [sample contains 8% exo-isomer] | Al 29 |
| $C_7H_{12}O$ | 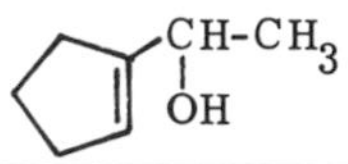 | H Phthal | Brucine in acetone | - 10.21° (neat) (t = 25°) | Al 30 |
| $C_7H_{12}O$ | 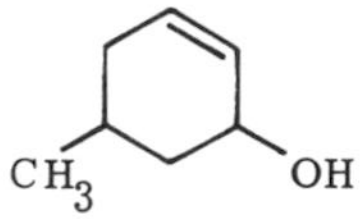 cis | H Phthal | Cinchonidine in acetone | a) - 7.0° (neat) (t = 25°) | Al 31 |
| | | | Brucine in acetone | b) + 6.95° (neat) (t = 30°) | |
| | trans | H Phthal | Brucine in acetone | a) - 265° (neat) (t = 23°) optically pure | Al 31, 32 |
| | | | | b) + 127.0° (c 19.4, acetone) (t = 27°) | |
| $C_7H_{12}O$ | 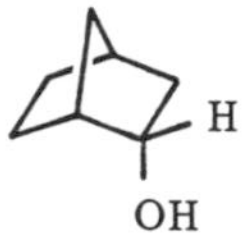 | H Phthal | Brucine in acetone Partial resolution | a) - 3.21° (c 4.897, $CHCl_3$) (t = 24°) | Al 33 |
| | | | | b) + 2.55° (c 5.433, $CHCl_3$) (t = 30°) | |
| $C_7H_{12}O$ | 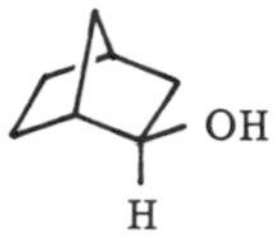 | H Phthal | Cinchonidine in acetone | a) - 2.41° (c 10.04, $CHCl_3$) (t = 24°) | Al 34 |
| | | | | b) + 2.44° (c 9.99, $CHCl_3$) (t = 24°) | |

TABLE 10b (continued)

| | Substance resolved | Derivative used | Resolving agent and solvent | Rotation $[\alpha]_D$ of resolved compound | Reference |
|---|---|---|---|---|---|
| $C_7H_{14}O$ | CH3, OH — cis | H Phthal | Brucine in acetone | a) $-71.9^o$ ($t = 25^o$) on the H Phthal derivative<br>b) $+21.2^o$ (neat) ($t = 27^o$) | Al 35 |
| | trans | H Phthal | (+)-α-(2-Naphthyl)-ethylamine in $Et_2O$<br>(-)-α-(2-Naphthyl)-ethylamine in $Et_2O$ | a) $+43.1^o$ (neat) ($t = 20^o$)<br>b) $-72.3^o$ ($t = 25^o$) on the H Phthal derivative | Al 35 |
| | | Partial resolution of trans-isomer with 3β-acetoxy-$\Delta^5$-etienic acid | | | Al 36 |
| $C_7H_{14}O$ | CH3, OH | H Phthal | (-)-Menthylamine via amide formation<br>Recr.: aqueous EtOH (80-90%) | $\alpha_D^{14}$ $+3.70^o$ (1 1 dm, neat) | Al 7 |
| $C_8H_{12}O$ | H, OH | H Phthal | (-)-Ephedrine in EtOH | $+68.2^o$ (c 1.1, $CHCl_3$) ($t = 23^o$) | Al 37 |
| | | Partial resolution with cinchonidine | | | Al 38 |
| $C_8H_{12}O$ | H, OH | H Phthal | Cinchonidine in acetone | $+219^o$ (c 0.6, $CHCl_3$) ($t = 25^o$) | Al 38 |
| $C_8H_{12}O$ | OH | H Phthal | Brucine in acetone<br>Ephedrine in $Et_2O$ | a) $-21.9^o$ (c 1.6, $CHCl_3$) ($t = 24^o$) (546 nm)<br>b) $+3.08$ (c 10.4, MeOH) | Al 39<br>Al 40 |
| $C_8H_{14}O$ | CH-CH3, OH | H Phthal | Brucine in acetone | $-1.48^o$ (neat) ($t = 25^o$) | Al 30 |

TABLE 10b (continued)

| | Substance resolved | Derivative used | Resolving agent and solvent | Rotation $[\alpha]_D$ of resolved compound | Reference |
|---|---|---|---|---|---|
| $C_8H_{14}O$ | 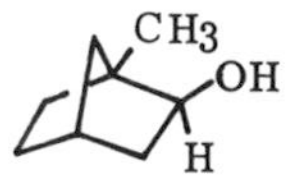 | H Phthal | Cinchonine in $CHCl_3$-EtOAc Partial resolution | $+0.48^o$ (c 5.0, $CHCl_3$) (t = $30^o$) | Al 41 |
| $C_8H_{14}O$ | 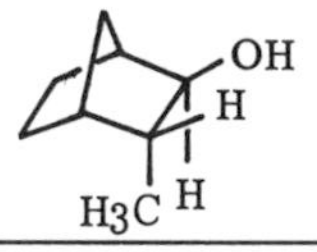 | H Phthal | (-)-Ephedrine in acetone Partial resolution | $-1.85^o$ (EtOH) (t = $26^o$) on the acetate ester | Al 42 |
| $C_8H_{14}O$ | 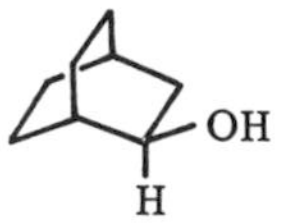 | H Phthal | Brucine in EtOAc Partial resolution | a) $-7.45^o$ (c 0.750, $CHCl_3$) (t = $25^o$) optical purity ca. 19 % <br> b) $+6.80^o$ (c 0.615, $CHCl_3$) (t = $25^o$) | Al 43, 44 |
| $C_8H_{14}O$ |  | H Phthal | Brucine in acetone Partial resolution | $-19.2^o$ (c 1.5, $CHCl_3$) (t = $25^o$) | Al 43, 45 |
| $C_8H_{16}O$ | 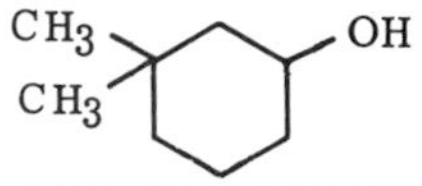 | H Phthal | Brucine Recr.: acetone-MeOH | $+1.13^o$ (neat) (t = $25^o$) | Al 21 |
| $C_9H_{14}O$ | 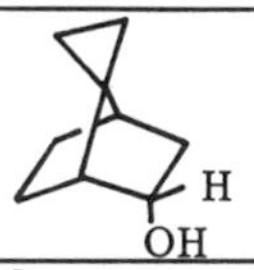 | H Phthal | Brucine in EtOH | $-7.6^o$ (c 1.0, $CH_2Cl_2$) (t = $27^o$) | Al 29 |
| $C_9H_{14}O_2$ | 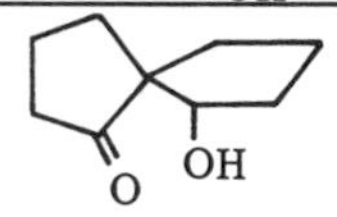 | p-Aminobenzoate ester | (+)-Camphor-10-sulfonic acid in MeOH | $+120^o$ (c 1, $CHCl_3$) on the p-aminobenzoate ester | Al 46 |
| $C_9H_{16}O$ | 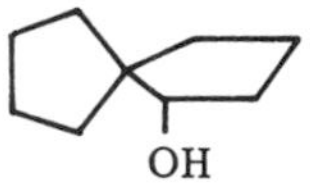 | H Phthal | Cinchonidine in EtOAc | a) $-32.2^o$ (c 0.560, $C_6H_6$) (t = $25^o$) <br> b) $+31.6^o$ (c 0.507, $C_6H_6$) (t = $25^o$) | Al 47 |
| $C_9H_{16}O$ | 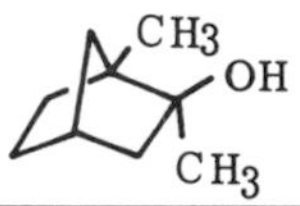 | H Phthal | Brucine in acetone-$H_2O$ | $-22.59^o$ (c 7.33, $CHCl_3$) (t = $30^o$) | Al 41 |

TABLE 10b (continued)

| | Substance resolved | Derivative used | Resolving agent and solvent | Rotation $[\alpha]_D$ of resolved compound | Reference |
|---|---|---|---|---|---|
| $C_9H_{18}O$ | HO, $(CH_3)_2CH$ | — | 3β-Acetoxy-$\Delta^5$-etienic acid<br>Ester Recr.: acetone | + 46° (c 0.53, $CHCl_3$) | Al 48 |
| $C_{10}H_{12}O$ | H, OH | — | 3β-Acetoxy-$\Delta^5$-etienic acid<br>Ester Recr.: acetone | - 87° (c 2.91, $CHCl_3$) (t = 24°) | Al 49 |
| $C_{10}H_{14}O_2$ | $CH_3$, H, HO, $CH_2$-CH=CH-$CH_3$, O<br>cis-Cinerolone | — | (+)-trans-Chrysanthemic acid via ester formation<br>Recr. of ester-semicarbazone ex $Et_2O$-Pet E | a) + 10.5° (c 11.7, EtOH) (t = 25°)<br>b) - 10.7° (c 8.9, EtOH) (t = 25°) | Al 50 |
| $C_{10}H_{18}O$ | H OH | H Phthal | Cinchonidine | + 21.1° (c 1.6, $CHCl_3$) (t = 30°)<br>(365 nm) | Al 51 |
| $C_{10}H_{18}O$ | H, H, H, HO | — | 3β-Acetoxy-$\Delta^5$-etienic acid<br>Ester Recr.: hexane | - 22° ($CHCl_3$) | Al 52 |
| $C_{10}H_{18}O$ | $CH_3$, $CH_3$, $CH_3$, OH, H | H Phthal | Cinchonine | - 26.6° (c 0.6, EtOH) (t = 20°) | Al 53 |
| $C_{10}H_{20}O$ | HO, $(CH_3)_3C$ | — | 3β-Acetoxy-$\Delta^5$-etienic acid<br>Esters separated by chromatography | a) - 44.4° (c 0.46, $CHCl_3$) (t = 28°)<br>b) + 44.2° (c 0.79, $CHCl_3$) (t = 28°) | Al 54 |
| $C_{10}H_{20}O$ | $(CH_3)_3C$, OH<br>cis | H Phthal | Brucine<br>in acetone | + 7.9° (c 3.1, $CHCl_3$) (t = 30°) | Al 55 |
| | trans | — | 3β-Acetoxy-$\Delta^5$-etienic acid<br>Esters separated by chromatography | + 19.8° (c 0.85, $CHCl_3$) (t = 28°) | Al 55 |

TABLE 10b (continued)

| | Substance resolved | Derivative used | Resolving agent and solvent | Rotation $[\alpha]_D$ of resolved compound | Reference |
|---|---|---|---|---|---|
| $C_{10}H_{20}O$ | 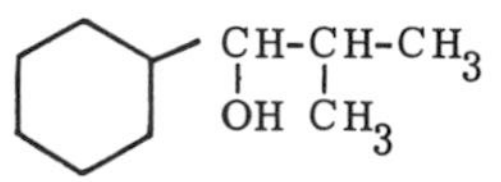 cyclohexyl–CH(OH)–CH($CH_3$)–$CH_3$ | H Phthal | Strychnine followed by brucine Recr.: MeOH-acetone | a) + 9.90° (neat) (t = 25°)<br>b) - 8.41° (neat) (t = 26°) | Al 56 |
| $C_{10}H_{20}O$ | 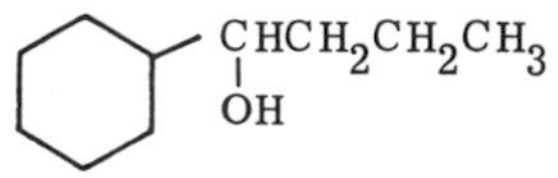 cyclohexyl–CH(OH)$CH_2CH_2CH_3$ | H Phthal | Strychnine Recr.: MeOH-acetone | + 15.6° (neat) (t = 29°) | Al 56 |
| $C_{10}H_{20}O$ | Menthol see Table 2 (p. 19) | — | (-)-Menthoxyacetic acid Ester Recr.: MeOH<br>N-(-)-Menthyl-p-sulfamyl-benzoic acid Ester Recr.: Pet E | + 49.5° (c 2, 95% EtOH) (t = 20°)<br>a) + 48.6° (t = 20°)<br>b) - 48.4° (t = 20°) | Table 2, ref 12a (p. 23)<br>Table 2, ref 12b<br>Table 2, ref 12b |
| $C_{13}H_{22}O$ | 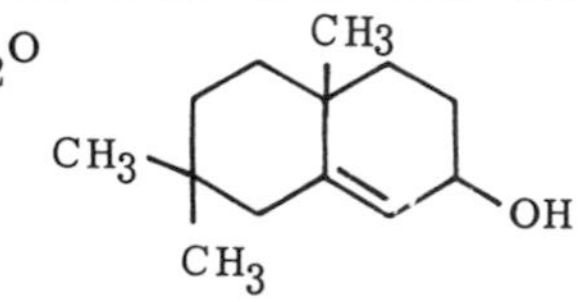 $CH_3$, $CH_3$, $CH_3$, OH | — | 3β-Acetoxy-$\Delta^5$-etienic acid Esters separated by chromatography | - 132° (c 0.59, $CHCl_3$) on the corresponding ketone | Al 57 |
| $C_{13}H_{26}O$ | 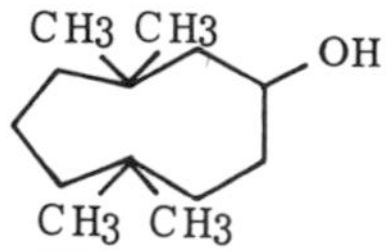 $CH_3$ $CH_3$, OH, $CH_3$ $CH_3$ | H(3-$NO_2$)Phthal | Cinchonidine in acetone | a) - 13.0 ± 1° (c 5, $CHCl_3$) (t = 25°)<br>b) + 12.6 ± 1° (c 5, $CHCl_3$) (t = 25°) | Al 58 |
| $C_{16}H_{26}O_4$ | See (A) below | H Phthal | Strychnine | a) + 59.4° (c 1.13, $CHCl_3$) (t = 24°)<br>b) - 57.8° (c 1.11, $CHCl_3$) (t = 24°) | Al 59 |
| $C_{17}H_{30}O$ | See (B) below | — | 3α-Acetoxy-11-ketoetianic acid Esters separated by chromatography | - 29° (c 0.84, $CHCl_3$) | Al 60 |

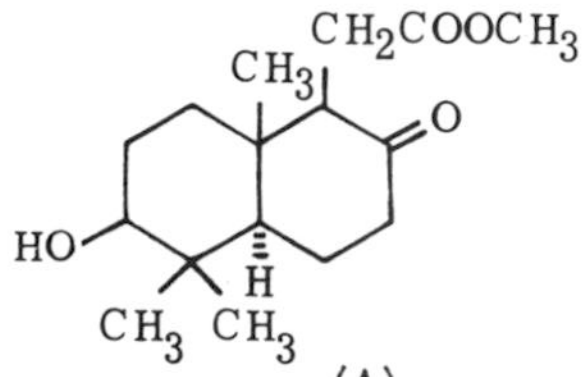

$CH_2COOCH_3$, $CH_3$, O, HO, H, $CH_3$ $CH_3$

(A)

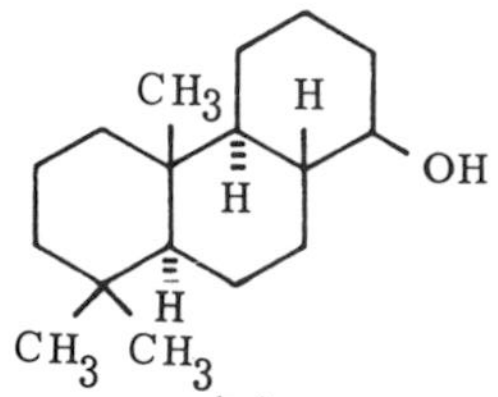

$CH_3$, H, H, OH, H, $CH_3$ $CH_3$

(B)

TABLE 10c Aryl Alcohols and Phenols

| | Substance resolved | Derivative used | Resolving agent and solvent | Rotation $[\alpha]_D$ of resolved compound | Reference |
|---|---|---|---|---|---|
| $C_8H_7OF_3$ | 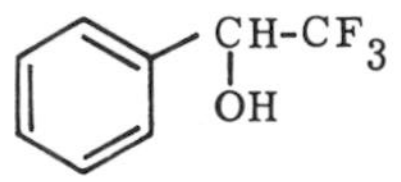 CH-CF$_3$<br>OH | Glycolic acid | (-)-Amphetamine in EtOAc | a) - 31.85$^o$ (neat) (t = 26$^o$) | Al 61, 62 |
| | | Glycolic acid | (+)-Amphetamine in EtOAc | b) + 31.82$^o$ (neat) (t = 26$^o$) | Al 61, 62 |
| | | Resolution *via* ester formation with 3$\beta$-acetoxy-$\Delta^5$-etienic acid | | | Al 62 |
| $C_8H_9NO_3$ | 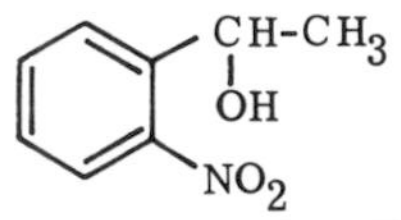 CH-CH$_3$<br>OH<br>NO$_2$ | H Phthal | Brucine in acetone | + 318$^o$ (c 0.54, MeOH) (t = 20$^o$) on the H phthalate ester | Al 63 |
| $C_8H_{10}O$ | 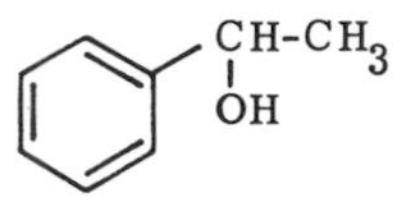 CH-CH$_3$<br>OH | H Phthal | Brucine in acetone | a) $\alpha^{17}_D$ - 11.04$^o$ (l 0.25 dm, neat) | Al 64a, 65, 66 |
| | | | | b) + 38.69$^o$ (neat) (t = 22$^o$) | Al 64a, 64b |
| | | For simplified or rapid procedures see | | | Al 64a, 67 |
| $C_9H_{10}O$ | CH-CH=CH$_2$<br>OH | H Phthal | Quinidine in acetone | a) - 5.19$^o$ (neat) (t = 22$^o$)<br>b) $\alpha^{17}_{5893}$ + 5.10$^o$ (l 0.5 dm) | Al 68, 69<br>Al 69 |
| $C_9H_{12}O$ | 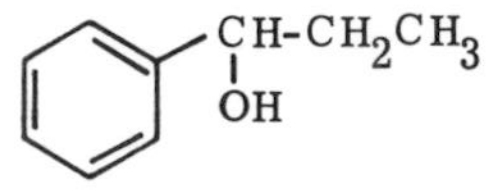 CH-CH$_2$CH$_3$<br>OH | H Succ | Cinchonidine in CHCl$_3$-acetone (1:10) | - 47.03$^o$ (acetone) (t = 20$^o$) | Al 70 |
| | | H Phthal | Quinidine | a) $\alpha^{22}_D$ + 29.16$^o$ (l 1, neat)<br>b) $\alpha^{22}_D$ - 28.50$^o$ (l 1, neat) | Al 71 |
| $C_9H_{12}O$ | CH$_2$-CH-CH$_3$<br>OH | H Phthal | Brucine in acetone | + 27.05$^o$ (neat) (t = 23$^o$)<br>+ 38.69 ± 0.01$^o$ (c 13.35, CHCl$_3$) (t = 24$^o$) | Al 72<br>Al 73 |
| $C_9H_{12}O_2$ | 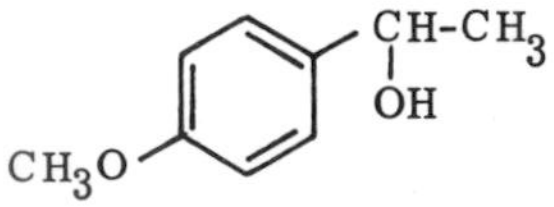 CH$_3$O- ... CH-CH$_3$<br>OH | H Phthal | Cinchonidine | - 44.4$^o$ (CHCl$_3$) (t = 25$^o$) | Al 74 |

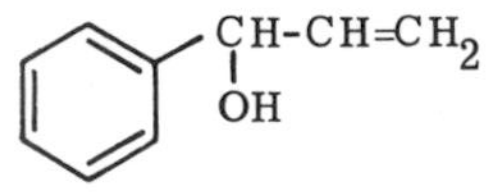

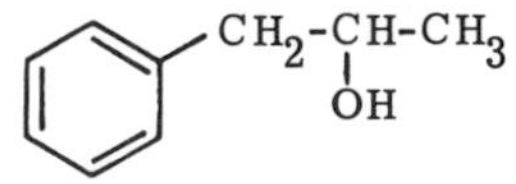

TABLE 10c (continued)

| | Substance resolved | Derivative used | Resolving agent and solvent | Rotation $[\alpha]_D$ of resolved compound | Reference |
|---|---|---|---|---|---|
| $C_{10}H_{12}O$ | $C_6H_5$-CH(OH)-CH=CH-$CH_3$ | H Phthal | Quinidine in MeOAc | $\alpha^{30}_{589}$ - 50.415° (neat)<br>optical purity 63 ± 4 % | Al 75, 76 |
| $C_{10}H_{12}O$ | $C_6H_5$-CH=CH-CH(OH)-$CH_3$ | H Phthal | Cinchonidine in EtOAc | + 34.7° (c 5.04, $CHCl_3$) (t = 30°)<br>optical purity 44 ± 4 % | Al 75 |
| $C_{10}H_{13}NO$ | 2-(1-hydroxyethyl)indoline: CH(OH)-$CH_3$, H, N-H | H Phthal | Brucine in EtOAc | - 21.1° (c 0.556, MeOH) (t = 28°) | Al 77 |
| $C_{10}H_{14}O$ | 4-$CH_3CH_2$-$C_6H_4$-CH(OH)-$CH_3$ | H Phthal | Brucine in acetone<br>Cinchonidine in acetone | a) - 46.5° (neat) (t = 25°)<br>b) + 46.5° (neat) (t = 25°) | Al 78 |
| $C_{10}H_{14}O$ | $C_6H_5$-CH(OH)-CH($CH_3$)-$CH_3$ | H Phthal | Brucine<br>Recr.: acetone | + 20.6° (neat) (t = 23°) | Al 79, 80 |
| $C_{10}H_{14}O$ | $C_6H_5$-C($CH_3$)(OH)-$CH_2CH_3$ | H Phthal | Brucine in acetone | a) - 18.4° (neat) (t = 22°) *<br>a) + 17.45° (neat) (t = 22°) *<br>b) - 14.56° (neat) (t = 26°)<br>* note reversal of sign | Al 18<br>Al 81<br>Al 81 |
| $C_{10}H_{14}O$ | $C_6H_5$-CH($CH_3$)-CH(OH)-$CH_3$ | | | | |
| | erythro | H(3-$NO_2$)Phthal | Cinchonine in acetone | a) + 0.68° (neat) (t = 25°) | Al 82 |
| | | H(3-$NO_2$)Phthal | Cinchonidine in acetone | b) - 0.69° (neat) (t = 25°) | Al 82 |
| | threo | H Phthal | Brucine in acetone | a) + 30.9° (neat) (t = 25°)<br>b) - 30.2° (neat) (t = 25°) | Al 82, 83 |

TABLE 10c (continued)

| Formula | Substance resolved | Derivative used | Resolving agent and solvent | Rotation $[\alpha]_D$ of resolved compound | Reference |
|---|---|---|---|---|---|
| $C_{10}H_{14}O$ | $C_6H_5$-CH(OH)-$CH_2CH_2CH_3$ | H Phthal | Strychnine in $H_2O$-MeOH (1:3) | - 45.93° (c 6.1, $C_6H_6$) (t = 27°) | Al 84 |
| $C_{10}H_{14}O$ | 2-HO-$C_6H_4$-CH($CH_3$)-$CH_2CH_3$ | — | (+)-Camphor-10-sulfonic acid *via* ester formation<br>Recr.: EtOH | a) + 18.1° (neat) (t = 23°) | Al 85, 86 |
| | | — | (-)-Camphor-10-sulfonic acid *via* ester formation | b) - 16.7° (neat) (t = 23°) | Al 85 |
| $C_{11}H_{16}O$ | 2,4,6-($CH_3$)$_3C_6H_2$-CH(OH)-$CH_3$ | H Phthal | Brucine in acetone | + 52° (c 1.53, EtOH) (t = 20°) | Al 87 |
| $C_{11}H_{16}O$ | $C_6H_5$-CH(OH)-C($CH_3$)$_2$-$CH_3$ | H Phthal | Brucine | a) + 25.9° (c 2.241, $C_6H_6$) | Al 80, 88 |
| | | H Phthal | Cinchonidine in acetone | b) - 30.3 ± 0.4° (c 3.640, acetone) (t = 23°) | Al 88 |
| $C_{11}H_{16}O$ | $C_6H_5$-CH(OH)-$CH_2$-CH($CH_3$)-$CH_3$ | H Phthal | Strychnine in MeOH | - 32.3° (c 16.70, *n*-heptane) (t = 26°) | Al 80 |
| $C_{11}H_{16}O$ | $C_6H_5$-CH($CH_2CH_3$)-CH(OH)-$CH_3$<br>*erythro* | H(3-$NO_2$)Phthal | Cinchonine in acetone | a) - 20.37° (neat) (t = 25°)<br>b) + 20.98° (neat) (t = 25°) | Al 89 |
| | *threo* | H Phthal | Brucine in acetone | - 16.89° (neat) (t = 25°) | Al 89 |

TABLE 10c (continued)

| Formula | Substance resolved | Derivative used | Resolving agent and solvent | Rotation $[\alpha]_D$ of resolved compound | Reference |
|---|---|---|---|---|---|
| $C_{11}H_{16}O$ | $C_6H_5$-CH($CH_3$)-CH(OH)-$CH_2CH_3$ erythro | H(3-$NO_2$)Phthal | Cinchonidine in 95% EtOH | a) + 7.34° (neat) (t = 25°)<br>b) - 7.32° (neat) (t = 25°) | Al 89 |
| | threo | H Phthal | Strychnine and brucine in acetone | + 12.40° (neat) (t = 25°) | Al 89 |
| $C_{11}H_{16}O_2$ | $CH_3O$-$C_6H_4$-CH($CH_3$)-CH(OH)-$CH_3$ erythro | H Phthal | Brucine in 2-butanone | - 8.25° (c 1.96, $CHCl_3$) (t = 25°) on the tosylate ester | Al 90 |
| | threo | H Phthal | Brucine in 2-butanone | - 11.3° (c 2.0, $CHCl_3$) (t = 26°) on the tosylate ester | Al 90 |
| $C_{12}H_{12}O$ | 1-naphthyl-$CH_3$-CH-OH | H Phthal | Brucine | - 74.7° (c 4.0, EtOH) (t = 20°) | Al 53 |
| $C_{12}H_{15}OBr$ | 2-(2-bromophenyl)cyclohexanol (HO, Br) | — | (-)-Menthoxyacetic acid<br>Ester Recr.: hexane | - 68° (c 5, MeOH) | Al 91 |
| $C_{12}H_{18}O$ | $C_6H_5$-CH($CH_2CH_3$)-CH(OH)-$CH_2CH_3$ erythro | H Phthal | Strychnine in $CHCl_3$-MeOH (1:6) | a) - 28.5° (c 6%, $CHCl_3$) (t = 26°)<br>b) + 28.3° (c 6%, $CHCl_3$) (t = 25°) | Al 92 |
| | threo | H Phthal | Strychnine in $CHCl_3$-acetone (1:4) | a) - 21.2° (c 5%, $CHCl_3$) (t = 28-29°)<br>b) + 20.9° (c 5%, $CHCl_3$) (t =28-29°) | Al 92 |
| $C_{13}H_7OD_5$ | $C_6H_5$-CH(OH)-$C_6D_5$ | H Phthal | Brucine in EtOAc | + 0.8° ($C_6H_6$) (t = 20°) | Al 93 |

TABLE 10c (continued)

| Formula | Substance resolved | Derivative used | Resolving agent and solvent | Rotation $[\alpha]_D$ of resolved compound | Reference |
|---|---|---|---|---|---|
| $C_{13}H_{11}OCl$ | $C_6H_5$-CH(OH)-$C_6H_4$-Cl | H Phthal | Brucine in EtOAc | a) - 32.1$^o$ (c 6.7, $CHCl_3$) (t = 20$^o$)<br>b) + 13.7$^o$ (c 20.0, $CHCl_3$) (t = 20$^o$) | Al 74, 94, 95<br>Al 94 |
| $C_{13}H_{18}O$ | 2-(2-$CH_3$-phenyl)cyclohexanol (HO) *cis* | — | (-)-Menthoxyacetic acid | - 102$^o$ (c 0.128, MeOH) (t = 25$^o$) | Al 96 |
| | | | Esters separated by column chromatography | | |
| | *trans* | — | (-)-Menthoxyacetic acid<br>Ester Recr.: $CH_3CN$ and column chromatography | a) + 70.6$^o$ (c 10, $CHCl_3$) (t = 25$^o$)<br>b) - 70.6$^o$ (c 10, $CHCl_3$) (t = 25$^o$) | Al 96 |
| $C_{14}H_{12}O_2$ | $C_6H_5$-C(=O)-CH(OH)-$C_6H_5$ | H Phthal | Quinidine in MeOH | a) + 118.4$^o$ (c 2.50, acetone) (t = 25$^o$)<br>b) - 118.3$^o$ (c 2.40, acetone) (t = 25$^o$) | Al 97 |
| $C_{14}H_{14}O$ | $C_6H_5$-CH(OH)-$C_6H_4$-$CH_3$<br>For phenyl-o-tolyl carbinol, see Table 8c, $C_{16}H_{16}O_3$ (p. 89) | H Phthal | Quinidine in acetone | - 13.13$^o$ (c 0.63, $CHCl_3$) (t = 24$^o$) (435 nm) | Al 98 |
| $C_{14}H_{14}O_2$ | $C_6H_5$-CH(OH)-$C_6H_4$-$OCH_3$ | H Phthal | Cinchonidine in EtOAc | - 20.02$^o$ (c 1.38, $CHCl_3$) (t = 24$^o$) (435 nm) | Al 98 |
| $C_{14}H_{16}O$ | 1-naphthyl-C($CH_3$)(OH)-$CH_2CH_3$ | H Phthal | Brucine in acetone | a) - 7.6$^o$ (c 4.6, EtOH) (t = 16.5$^o$)<br>b) + 7.6$^o$ (EtOH) | Al 99 |

TABLE 10c (continued)

| | Substance resolved | Derivative used | Resolving agent and solvent | Rotation $[\alpha]_D$ of resolved compound | Reference |
|---|---|---|---|---|---|
| $C_{14}H_{22}O$ | erythro | H Phthal | Brucine in MeOH | a) + 19.08° (c 5-7%, $CHCl_3$) (t = 24-27°)<br>b) - 19.11° (c 5-7%, $CHCl_3$) (t = 24-27°) | Al 100 |
| | threo | H Phthal | Strychnine in acetone | a) + 14.56° (neat) (t = 25°)<br>b) - 14.50° (neat) (t = 25°) | Al 100 |
| $C_{15}H_{14}O_3$ | | H Phthal | Quinidine in MeOH | a) + 72° (c 1.00, acetone) (t = 25-26°)<br>b) - 71.8° (c 1.00, acetone) (t = 25°) | Al 101 |
| $C_{15}H_{16}O$ | | H Phthal | Quinine in $EtOH-H_2O$ | a) - 25.2° (c 0.600, $CS_2$) (t = 21°) | Al 102 |
| | | H Phthal | Strychnine in $EtOH-H_2O$ | b) + 24.1° (c 1.100, $CS_2$) (t = 20°) | Al 102 |
| $C_{15}H_{16}O$ | | H Phthal | Brucine in acetone-$H_2O$ | a) - 8.2° (c 9%, $CHCl_3$) (t = 27°)<br>b) + 8.05° (c 9%, $CHCl_3$) (t = 25°) | Al 103 |
| $C_{15}H_{16}O$ | erythro | H(3-$NO_2$)Phthal | Brucine in $CHCl_3$-MeOH | a) - 68.8° (c 3.0, $CHCl_3$) (t = 21°)<br>b) + 68.7° (c 4.9, $CHCl_3$) (t = 23°) | Al 104 |
| | threo | H(3-$NO_2$)Phthal | Cinchonidine in $CHCl_3$-MeOH | a) + 47.2° (c 5, $CHCl_3$) (t = 21°)<br>b) - 47.2° (c 5.1, $CHCl_3$) (t = 21°) | Al 104 |

TABLE 10c (continued)

| Substance resolved | | Derivative used | Resolving agent and solvent | Rotation $[\alpha]_D$ of resolved compound | Reference |
|---|---|---|---|---|---|
| $C_{15}H_{16}O$ | $C_6H_5-CH_2-C(CH_3)(OH)-C_6H_5$ | H Phthal | Brucine | - 8.3° (c 2, MeOH) (t = 18°) | Al 105 |
| $C_{15}H_{16}O_2$ | $C_6H_5-CH(OH)-C_6H_4-OCH_2CH_3$ | H Phthal | Quinidine in acetone | - 13.7° (c 1.525, $C_6H_6$) (t = 22°) | Al 106 |
| $C_{15}H_{18}O$ | $C_6H_5-C(OH)(CH_2CH_3)-C_6H_4-CH_3$ | H Phthal | Brucine in EtOAc | a) - 10.783° (c 4.3, acetone) (t = 31°)<br>b) + 10.81° (c 4.3, acetone) (t = 31°) | Al 107 |
| $C_{15}H_{22}O$ | [structure: benzene ring bridged by a carbon chain bearing OH] | H Phthal | Brucine in acetone | a) - 8.6° (c 13.7, $CHCl_3$) (t = 27°) (546 nm)<br>b) + 8.5° (c 14, $CHCl_3$) (t = 27°) (546 nm) | Al 108 |
| $C_{16}H_{16}O$ | [structure: tetrahydronaphthalene ring with H, bonded to a C₆H₄–OH ring] | — | (+)-Camphor-10-sulfonic acid<br>Ester Recr.: MeOH | a) - 26.9° (c 2.69, MeOH) (t = 25°)<br>b) + 27.0° (c 2.37, MeOH) (t = 25°) | Al 109 |
| $C_{16}H_{16}O$ | [structure: bridged aromatic ring system with H and OH] | H Phthal | Brucine | a) + 76.9° (c 1.0, $CHCl_3$) (t = 25°) (546 nm)<br>b) - 71.7° (c 1.0, $CHCl_3$) (t = 25°) (546 nm) | Al 110 |

TABLE 10c (continued)

| | Substance resolved | Derivative used | Resolving agent and solvent | Rotation $[\alpha]_D$ of resolved compound | Reference |
|---|---|---|---|---|---|
| $C_{16}H_{17}NO_2$ | 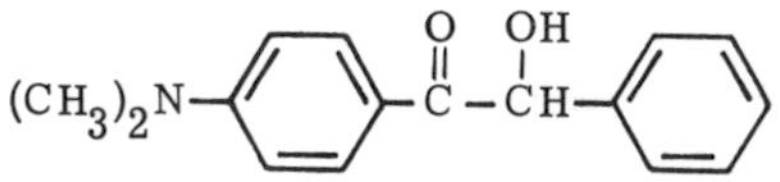 $(CH_3)_2N$–C₆H₄–CO–CH(OH)–C₆H₅ | H Phthal | Quinidine apparently in MeOH (see Ref. Al 97) | a) $-22.6^o$ (c 1.04, acetone) (t = $25^o$)<br>b) $+22.5^o$ (c 1.06, acetone) (t = $25^o$) | Al 111 |
| $C_{16}H_{24}O$ | 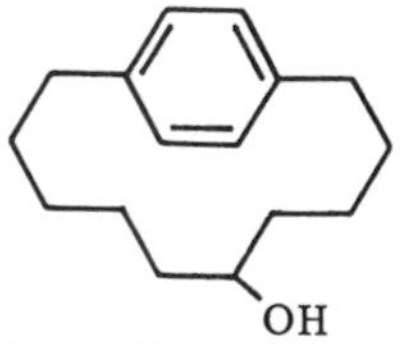 OH | $H(3\text{-}NO_2)$Phthal | Strychnine in $CHCl_3$-MeOH (1:5) | a) $-14.1^o$ (c 3.3, $CHCl_3$) (t = $29^o$)<br>b) $+14.3^o$ (c 3.3, $CHCl_3$) (t = $24^o$)<br>both on the $H(3\text{-}NO_2)$Phthal ester | Al 112 |
| $C_{19}H_{16}O$ | 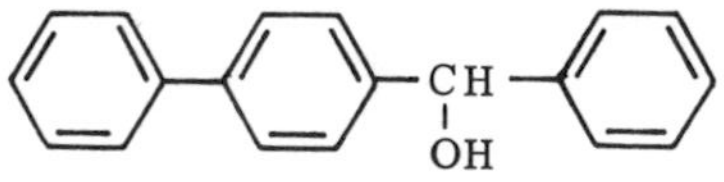 CH–OH | H Phthal | Quinine in MeOH | $+7.5^o$ (c 2.5, $CHCl_3$)(t=$25^o$) (546 nm) | Al 113 |
| $C_{19}H_{16}O_4$ | Warfarin see (A) below | — | Quinidine in $CHCl_3$-acetone (2:3) | a) $-148.0 \pm 0.5^o$ (c 1.2, 0.5 N NaOH) (t = $25^o$)<br>b) $+149.0 \pm 0.5^o$ (c 2, 0.5 N NaOH) (t = $25^o$) | Al 114 |
| $C_{19}H_{18}O_3$ | see (B) below | — | Quinidine in EtOH | $-142 \pm 1^o$ (c 1.2, 1 N NaOH) (t = $25^o$) | Al 114 |
| $C_{19}H_{22}O_2$ | see (C) below | H Succ | Dehydroabietylamine acetate + $Et_3N$ in EtOAc | a) $-132.8^o$ (c 0.5-1%, $CHCl_3$) (t = $25^o$) | Al 115 |
| | | H Succ | (-)-Ephedrine in EtOAc | b) $+135^o$ (c 0.5-1%, $CHCl_3$) (t = $25^o$) | Al 115 |
| | | Resolution via ester formation with (-)-menthoxyacetic acid | | | Al 116 |

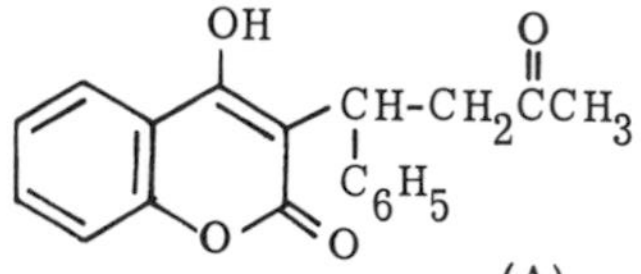

OH; $CH\text{-}CH_2CCH_3$ (C=O); $C_6H_5$; O; O

(A)

OH; $CH\text{-}CH_2CH_2CH_3$; $C_6H_5$; O; O

(B)

$CH_3$ OH; $CH_3O$

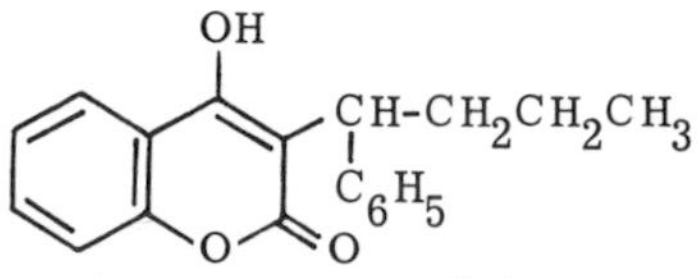

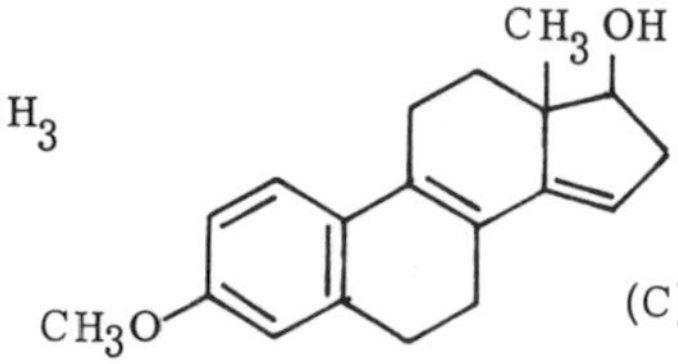

(C)

TABLE 10c (continued)

| | Substance resolved | Derivative used | Resolving agent and solvent | Rotation $[\alpha]_D$ of resolved compound | Reference |
|---|---|---|---|---|---|
| $C_{19}H_{26}O_2$ | Estradiol 3-methyl ether see (A) below | — | (+)-O,O'-Diacetyltartaric acid monomethyl ester chloride *via* ester formation | a) $-76^{o}$ (c 1.25, EtOH) ($t = 22^{o}$)<br>b) $+76^{o}$ (c 1.41, EtOH) ($t = 22^{o}$) | Al 117 |
| | | For alternative resolutions, see | | | Al 115, 118 |
| $C_{20}H_{14}O$ | see (B) below | H Succ | Quinine in EtOAc | a) $-21.73^{o}$ (c 3.50, MeOH) ($t = 30^{o}$)<br>b) $+21.70^{o}$ (MeOH) ($t = 30^{o}$) | Al 119 |
| $C_{20}H_{18}O$ | see (C) below | H Succ | Cinchonidine in $C_6H_6$ | a) $-123^{o}$ (EtOH) ($t = 25^{o}$) | Al 120 |
| | | H Succ | (-)-2-Amino-1,1-diphenyl-propanol-1 in EtOH | b) $+120^{o}$ (EtOH) ($t = 25^{o}$) | Al 120 |
| $C_{20}H_{28}O_2$ | see (D) below | H Sulfate | (-)-2-Amino-1-butanol in $CH_2Cl_2$-$Et_2O$ (1:1) | a) $+53 \pm 3^{o}$ ($CHCl_3$) ($t = 23^{o}$)<br>b) $-41 \pm 3^{o}$ ($CHCl_3$) ($t = 23^{o}$) | Al 121 |
| $C_{21}H_{20}O$ | see (E) below | H Phthal | Brucine Recr.: acetone | $-16.4^{o}$ (c 4.1, MeOH) ($t = 17^{o}$) | Al 105 |
| $C_{21}H_{20}O$ | $(C_6H_5)_3C$-CH($OH$)-$CH_3$ | H Phthal | Brucine in acetone | $-29.1^{o}$ (c 2.4, $CHCl_3$) ($t = 22^{o}$) | Al 87 |
| $C_{21}H_{20}O_2$ | see (F) below | H Phthal | Brucine | $-21.8^{o}$ (c 3.6, MeOH) ($t = 17^{o}$) | Al 105 |
| $C_{21}H_{28}O_3$ | see (G) below | — | Digitonin in 90% EtOH | ca. $+83^{o}$ ($CHCl_3$) ($t = 24^{o}$) | Al 122 |

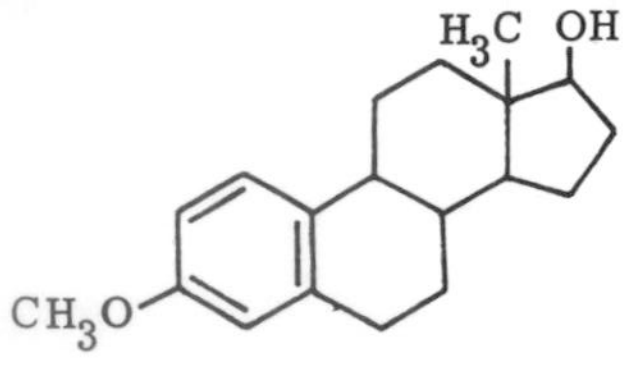

(A) 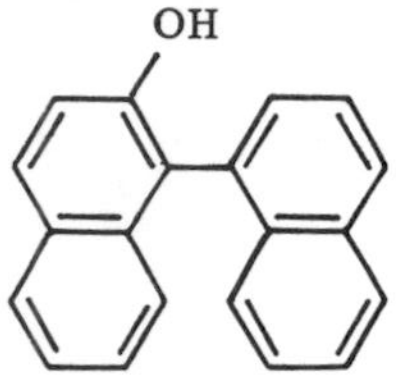 (B) 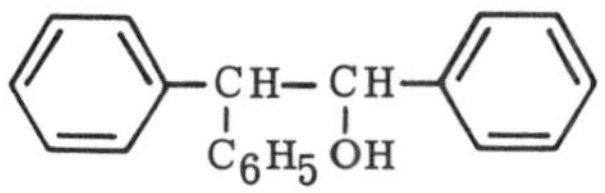 (C) 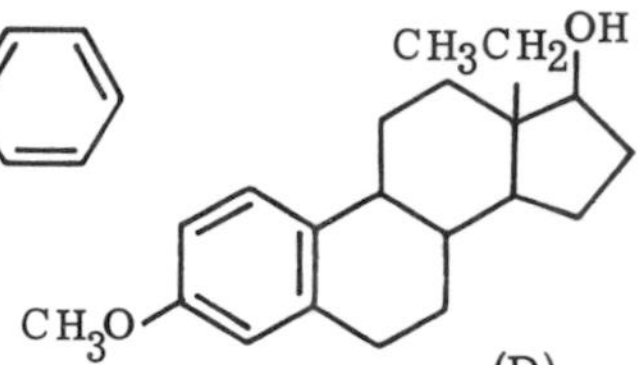 (D)

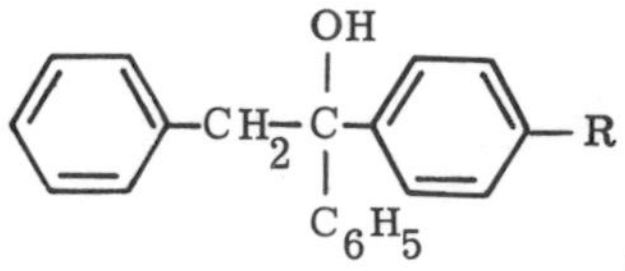

(E) R = $CH_3$

(F) R = $OCH_3$

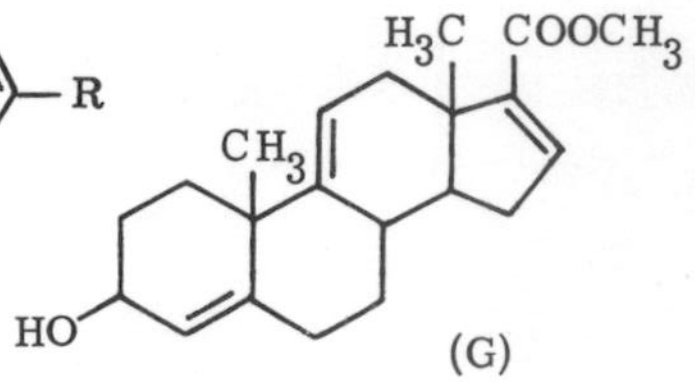

(G)

TABLE 10c (continued)

| | Substance resolved | Derivative used | Resolving agent and solvent | Rotation $[\alpha]_D$ of resolved compound | Reference |
|---|---|---|---|---|---|
| $C_{24}H_{20}O$ | 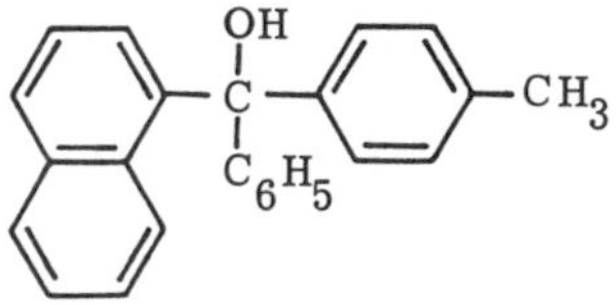 OH, $C_6H_5$, $CH_3$ | β-Alkoxypropionic acid | Brucine in acetone | $+3.12^{\circ}$ (c 1.0092, $CHCl_3$) ($t=25^{\circ}$) (436 nm) | Al 123, 107 |
| $C_{26}H_{22}O$ | $(C_6H_5)_3C\text{-}CH(OH)\text{-}C_6H_5$ | — | (-)-Menthoxyacetic acid<br>Esters separated by column chromatography | $+43^{\circ}$ (c 1.73, $CHCl_3$) ($t = 20^{\circ}$) | Al 87 |
| $C_{26}H_{32}O_2$ | 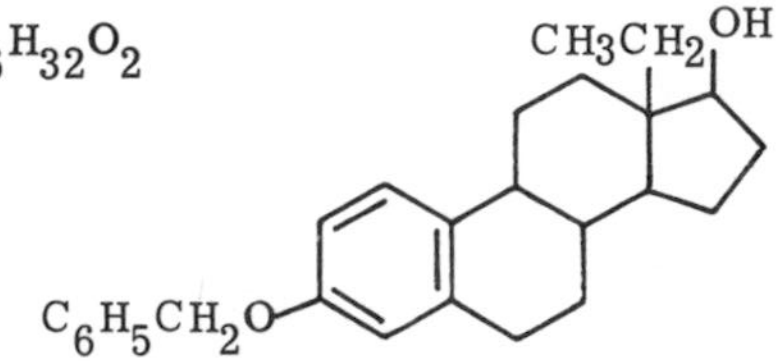 $CH_3CH_2$, OH, $C_6H_5CH_2O$ | — | (-)-Menthoxyacetic acid<br>Ester Recr.: Pet E | a) $-42.2^{\circ}$ (c 0.5-1%, $CHCl_3$) ($t = 25^{\circ}$)<br>b) $+42.6^{\circ}$ (c 0.5-1%, $CHCl_3$) ($t = 25^{\circ}$) | Al 115 |
| $C_{26}H_{40}O$ | 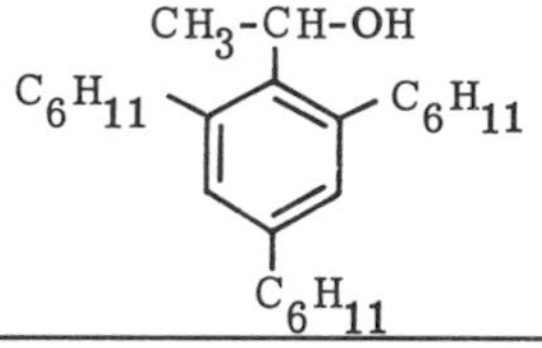 $CH_3\text{-}CH\text{-}OH$, $C_6H_{11}$, $C_6H_{11}$, $C_6H_{11}$ | H Phthal | Quinine in acetone | $+33^{\circ}$ (c 1.09, EtOH) ($t = 18^{\circ}$) | Al 87 |
| $C_{29}H_{22}O$ | 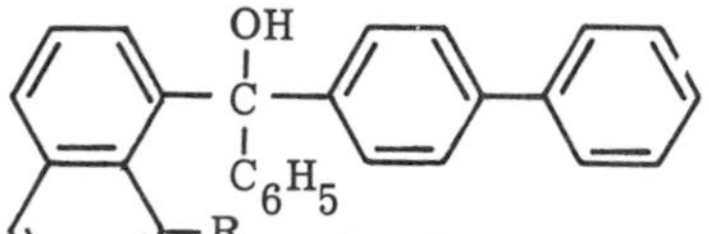 OH, $C_6H_5$, R, R = H (A) | Glycolic acid | Brucine in acetone<br>Cinchonidine in EtOAc | a) $+17.40^{\circ}$ (c 0.98, $CHCl_3$) ($t = 23^{\circ}$)<br>b) $-16.5^{\circ}$ (c 1.2, $CHCl_3$) ($t = 23^{\circ}$)<br>both on the glycolic acid derivative | Al 124<br>Al 124 |
| $C_{30}H_{24}O$ | Formula (A) in $C_{29}H_{22}O$ above R = $CH_3$ | — | Spontaneous resolution: from $Et_2O$<br>from cyclohexane | a) $+76^{\circ}$ (c 2.0, $CHCl_3$)<br>b) $-76^{\circ}$ (c 0.3, $CHCl_3$) | Al 125 |

TABLE 10d Diols and Polyols

| | Substance resolved | Derivative used | Resolving agent and solvent | Rotation $[\alpha]_D$ of resolved compound | Reference |
|---|---|---|---|---|---|
| $C_3H_8O_2$ | $CH_3$-CH(OH)-$CH_2$(OH) | — | (-)-Menthone <u>via</u> fractional distillation of ketal | a) + 16.35° (neat) (t = 25°)<br>b) - 15.02° (neat) (t = 25°) | Al 126 |
| $C_6H_{12}O_2$ | | — | (-)-Menthoxyacetic acid<br>Ester Recr.: $C_6H_6$ | - 46.5° ($H_2O$) (t = 25°) | Al 127 |
| $C_6H_{12}O_2$ | | H Phthal | Brucine in acetone | a) - 15.2° (c 10, MeOH) (t = 20°) | Al 128 |
| | | H Phthal | Strychnine in acetone-$CHCl_3$ | b) + 16.2° (c 10, MeOH) (t = 20°) | Al 128 |
| $C_6H_{14}O_2$ | $CH_3$-CH(OH)-$CH_2CH_2$-CH(OH)-$CH_3$ | H Phthal | Brucine in acetone-MeOH<br>Partial resolution | a) - 26.6° ($CHCl_3$) (t = 29°)<br>b) + 21.2° (c 9.74, $CHCl_3$) (t = 29°) | Al 129 |
| $C_8H_{16}O_2$ | | — | (-)-Menthyl isocyanate in $C_6H_6$ | + 21.4° (c 4, $C_6H_6$) (t = 24°) | Al 130 |
| $C_8H_{16}O_2$ | | H Phthal | Strychnine<br>Recr.: $CHCl_3$-Pet E then trichloroethylene | + 16.9 (c 1.33, EtOH) (t = 22°) | Al 131 |
| $C_9H_{16}O_2$ | | — | (-)-ω-Camphanic acid<br>Esters separated by chromatography | - 53° (c 0.3, EtOH) | Al 132 |
| $C_{10}H_{12}O_2$ | | — | (-)-Menthoxyacetic acid<br>Ester Recr.: Pet E | a) - 134° (c 1.0, $CHCl_3$) (t = 21°)<br>b) + 132° (c 1.0, $CHCl_3$) (t = 21°) | Al 133 |

TABLE 10d (continued)

| Substance resolved | Derivative used | Resolving agent and solvent | Rotation $[\alpha]_D$ of resolved compound | Reference |
|---|---|---|---|---|
| $C_{10}H_{14}O_3$ | H Phthal | Quinidine and Brucine in acetone | a) $+23.4^o$ (c1.96, MeOH) (t = $28^o$)<br>b) $-11.6^o$ (c 2.1, MeOH) (t = $28^o$) | Al 134 |
| $C_{11}H_{12}N_2O_5Cl$ Chloramphenicol | H Succ | Strychnine in acetone-$CHCl_3$ | $+17.1^o$ (c 4.9%, EtOH) (t = $18^o$) | Al 135 |
| $C_{14}H_{14}O_2$ Hydrobenzoin | — | Spontaneous resolution from $Et_2O$-Pet E | $\alpha_D$ $-97.0^o$ ($CHCl_3$)<br>$\alpha_D$ $+97.6^o$ ($CHCl_3$) | Al 136 |
| $C_{14}H_{14}O_2$ | — | (-)-Menthoxyacetic acid<br>Esters Recr.: Pet E | a) $-132^o$ (c 0.5, $CHCl_3$) (t = $21^o$)<br>b) $+122^o$ (c 0.5, $CHCl_3$) (t = $21^o$) | Al 137 |
| $C_{15}H_{22}O_3$ | H Succ | Brucine in acetone | $-225 \pm 2^o$ (c 1, $CHCl_3$) (t = $25^o$) | Al 138 |
| $C_{15}H_{22}O_4$ trans-Humulinic acid | —<br>— | (+)-$\alpha$-Methylbenzylamine in acetone<br>(-)-$\alpha$-Methylbenzylamine in acetone | a) $+28.03^o$ (MeOH)<br>b) $-28.36^o$ (MeOH) | Al 139 |
| $C_{16}H_{18}O_2$ | — | Resolved by chromatography on cellulose-2.5 acetate | | Al 140 |

TABLE 10d (continued)

| Substance resolved | Derivative used | Resolving agent and solvent | Rotation $[\alpha]_D$ of resolved compound | Reference |
|---|---|---|---|---|
| $C_{16}H_{18}O_2$ 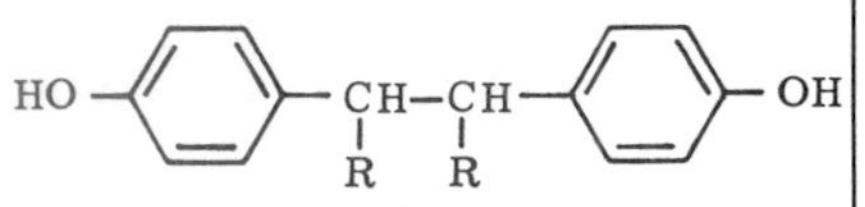 HO–C₆H₄–CH(R)–CH(R)–C₆H₄–OH; R = $CH_3$- | — | 3-Bromocamphor-9-sulfonic acid<br>Esters Recr.: $C_6H_6$ | a) $-103^o$ (c 1.07, dioxane)<br>b) $+65.5^o$ (c 0.93, dioxane) | Al 141 |
| $C_{18}H_{16}O_7$ 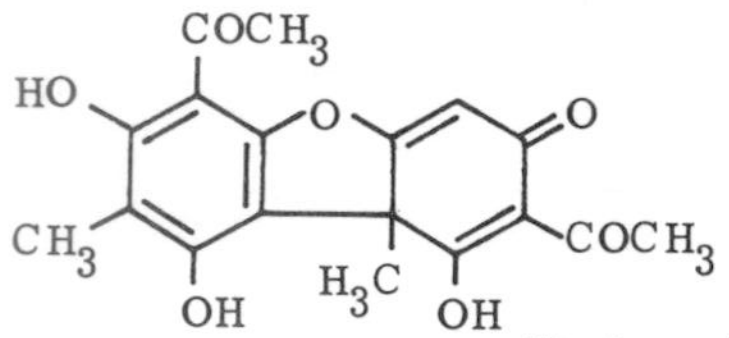 Usnic acid | — | Brucine in MeOH-acetone | a) $+467^o$ (c 5.0, $CHCl_3$) (t = $20^o$)<br>b) $-367^o$ (c 5.0, $CHCl_3$) (t = $20^o$) | Al 142 |
| $C_{18}H_{22}O_2$ Hexestrol<br>See $C_{16}H_{18}O_2$ above<br>R = $CH_3CH_2$- | — | 3-Bromocamphor-9-sulfonic acid<br>Esters Recr.: acetone-EtOH | a) $+39.8^o$ (c 1.13, $CHCl_3$)<br>b) $-39.5^o$ (c 1.14, $CHCl_3$) | Al 141, 143 |
| $C_{19}H_{22}O_5$ 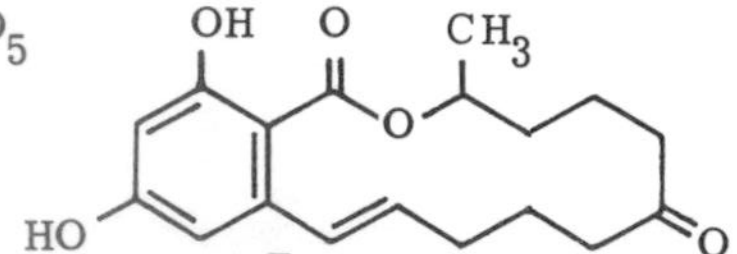 Zearalenone | 4-Methyl ether | (-)-Menthoxyacetic acid *via* ester formation<br>Recr.: acetone-Pet E | $-190^o$ ($CHCl_3$) | Al 144 |
| $C_{22}H_{13}NO_4$ 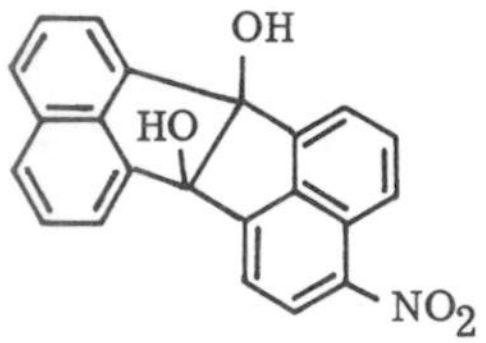 | p-Carboxy-boronate ester | Quinine<br>Recr.: EtOAc | $+57.5^o$ (c 2.76, acetone) (t = $25^o$) (5461 Å) | Al 145 |
| $C_{28}H_{50}O_2$ 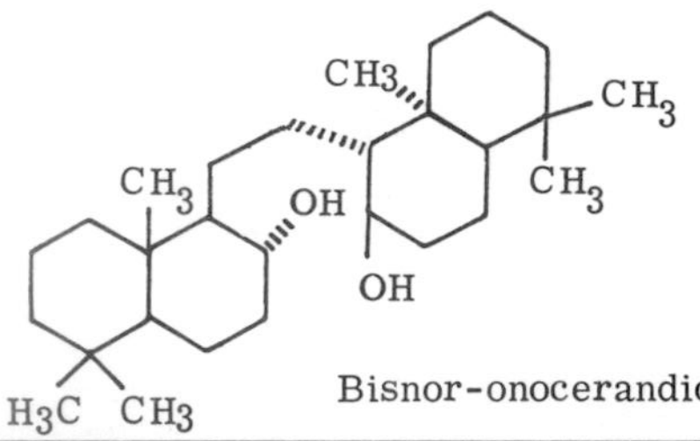 Bisnor-onocerandiol | — | (-)-Menthoxyacetic acid<br>Esters separated by chromatography | a) $+10^o$ ($CHCl_3$) (t = $22^o$)<br>b) $-10^o$ ($CHCl_3$) (t = $24^o$) | Al 146 |

## References to Tables 10a-d. Resolutions of Alcohols and Enols


(Al1) J.W.C. Crawford, J. Chem. Soc. C, 1967, 2332.

(Al2) J.A. Dale, D.L. Dull and H.S. Mosher, J. Org. Chem., 34, 2543 (1969).

(Al3) K.L. Olivier and W.G. Young, J. Amer. Chem. Soc., 81, 5811 (1959). See also K.B. Wiberg, ibid., 74, 3891 (1952).

(Al4) W.G. Young, F.F. Caserio, Jr., and D.D. Brandon, Jr., J. Amer. Chem. Soc., 82, 6163 (1960).

(Al5) H.J. Lucas and H.K. Garner, J. Amer. Chem. Soc., 72, 2145 (1950).

(Al6) S.W. Kantor and C.R. Hauser, J. Amer. Chem. Soc., 75, 1744 (1953); A.J. Speziale and R.C. Freeman, ibid., 82, 909 (1960); M. Henchman and R. Wolfgang, ibid., 83, 2991 (1961).

(Al7) J.P. Egerton Human and J.A. Mills, J. Chem. Soc., 1949, Suppl. Issue No. 1, S 77.

(Al8) H.L. Goering and W.I. Kimoto, J. Amer. Chem. Soc., 87, 1748 (1965). See also H.L. Goering and R.W. Greiner, ibid., 79, 3464 (1957).

(Al9) E.R. Alexander and R.W. Kluiber, J. Amer. Chem. Soc., 73, 4304 (1951).

(Al10) B. Halpern and J.W. Westley, Aust. J. Chem., 19, 1533 (1966).

(Al11) J. Cason and R.A. Coad, J. Amer. Chem. Soc., 72, 4695 (1950).

(Al12) D.M. Feigl and H.S. Mosher, J. Org. Chem., 33, 4242 (1968).

(Al13) N.C. Deno and M.S. Newman, J. Amer. Chem. Soc., 73, 1920 (1951); P. Newman, P. Rutkin and K. Mislow, ibid., 80, 465 (1958).

(Al14) G.S. Ayers, J.H. Mossholder and R.E. Morse, J. Chromatogr., 51, 407 (1970).

(Al15) E.R. Novak and D.S. Tarbell, J. Amer. Chem. Soc., 89, 73 (1967).

(Al16) K. Mislow, R.E. O'Brien and H. Schaefer, J. Amer. Chem. Soc., 84, 1940 (1962).

(A117) W.M. Foley, F.J. Welch, E.M. La Combe and H.S. Mosher, J. Amer. Chem. Soc., 81, 2779 (1959).

(A118) A.G. Davies, J. Kenyon and W.F. Salamê, J. Chem. Soc., 1957, 3148.

(A119) W. von E. Doering and H.H. Zeiss, J. Amer. Chem. Soc., 72, 147 (1950).

(A120) D.J. Cram, D.A. Scott and W.D. Nielsen, J. Amer. Chem. Soc., 83, 3696 (1961).

(A121) H.L. Goering and F.H. McCarron, J. Amer. Chem. Soc., 78, 2270 (1956). See also A.W. Ingersoll, in Organic Reactions, Vol. 2, R. Adams, Ed., Wiley, New York, 1944, p. 376.

(A122) R.L. Letsinger and J.G. Traynham, J. Amer. Chem. Soc., 72, 849 (1950).

(A123) C.L. Arcus and D.G. Smyth, J. Chem. Soc., 1955, 34.

(A124) D.G. Jones and G.H. Whitfield, J. Chem. Soc., 1953, 2795.

(A125) F.S. Prout, J. Cason and A.W. Ingersoll, J. Amer. Chem. Soc., 70, 298 (1948).

(A126) W.T. Borden and E.J. Corey, Tetrahedron Lett., 1969, 313.

(A127) M. Vogel and J.D. Roberts, J. Amer. Chem. Soc., 88, 2262 (1966).

(A128) (a) L. Otzet, J. Pascual and J. Sistarê, An. Real Soc. Espan. Fis. Quim., Ser. B, 62, 965 (1967); (b) R.K. Hill and J.W. Morgan, J. Org. Chem., 33, 927 (1968).

(A129) D.A. Lightner and W.A. Beavers, J. Amer. Chem. Soc., 93, 2677 (1971).

(A130) J. English, Jr. and V. Lamberti, J. Amer. Chem. Soc., 74, 1909 (1952).

(A131) H.L. Goering and J.P. Blanchard, J. Amer. Chem. Soc., 76, 5405 (1954); H.L. Goering, T.D. Nevitt and E.F. Silversmith, ibid., 77, 4042 (1955).

(A132) H.L. Goering and J.T. Doi, J. Amer. Chem. Soc., 82, 5850 (1960).

(A133) S. Winstein and D. Trifan, J. Amer. Chem. Soc., 74, 1147 (1952); J.A. Berson, J.S. Walia, A. Remanick, S. Suzuki, P. Reynolds-Warnhoff and D. Willner, ibid., 83, 3986 (1961).

(A134) S. Winstein and D. Trifan, J. Amer. Chem. Soc., 74, 1154 (1952); J.A. Berson and S. Suzuki, ibid., 81, 4088 (1959).

(A135) R. Bäckström and B. Sjöberg, Ark. Kemi, 26, 549 (1967).

(A136) B. Sjöberg, D.J. Cram, L. Wolf and C. Djerassi, Acta. Chem. Scand., 16, 1079 (1962).

(A137) K. Mislow and J.G. Berger, J. Amer. Chem. Soc., 84, 1956 (1962).

(A138) H.L. Goering and D.L. Towns, J. Amer. Chem. Soc., 85, 2295 (1963).

(A139) G.N. Fickes, J. Org. Chem., 34, 1513 (1969).

(A140) J.A. Berson, D. Wege, G.M. Clarke and R.G. Bergman, J. Amer. Chem. Soc., 91, 5594 (1969).

(A141) H.L. Goering, C. Brown, S. Chang, J.V. Clevenger and K. Humski, J. Org. Chem., 34, 624 (1969).

(A142) J.A. Berson, R.G. Bergman, J.H. Hammons and A.W. Mc Rowe, J. Amer. Chem. Soc., 89, 2581 (1967).

(A143) H.M. Walborsky, M.E. Baum and A.A. Youssef, J. Amer. Chem. Soc., 83, 988 (1961).

(A144) J.A. Berson and D. Willner, J. Amer. Chem. Soc., 86, 609 (1964).

(A145) H.L. Goering and G.N. Fickes, J. Amer. Chem. Soc., 90, 2848 (1968).

(A146) E. Hardegger, E. Maeder, H.M. Semarne and D.J. Cram, J. Amer. Chem. Soc., 81, 2729 (1959).

(A147) H. Christol, D. Duval and G. Solladié, Bull. Soc. Chim. Fr., 1968, 4151.

(A148) C. Djerassi, P.A. Hart and C. Beard, J. Amer. Chem. Soc., 86, 85 (1964).

(A149) R.B. Woodward and T.J. Katz, Tetrahedron, 5, 70 (1959).

(A150) F.B. LaForge and N. Green, J. Org. Chem., 17, 1635 (1952).

(A151) H.L. Goering and R.F. Myers, J. Amer. Chem. Soc., 91, 3386 (1969).

(A152) C. Djerassi and J. Staunton, J. Amer. Chem. Soc., 83, 736 (1961).

(A153) G. Barbieri, V. Davoli, I. Moretti, F. Montanari and G. Torre, J. Chem. Soc. C, 1969, 731.

(A154) C. Djerassi, P.A. Hart and E.J. Warawa, J. Amer. Chem. Soc., 86, 78 (1964); C. Djerassi, E.J. Warawa, J.M. Berdahl and E.J. Eisenbraun, ibid., 83, 3334 (1961).

(A155) C. Djerassi, E.J. Warawa, R.E. Wolff and E.J. Eisenbraun, J. Org. Chem., 25, 917 (1960).

(A156) E.P. Burrows, F.J. Welch and H.S. Mosher, J. Amer. Chem. Soc., 82, 880 (1960).

(A157) C. Djerassi, J. Burakevitch, J.W. Chamberlin, D. Elad, T. Toda and G. Stork, J. Amer. Chem. Soc., 86, 465 (1964).

(A158) A.T. Blomquist and G.A. Miller, J. Amer. Chem. Soc., 83, 243 (1961).

(A159) G. Stork, A. Meisels and J.E. Davies, J. Amer. Chem. Soc., 85, 3419 (1963).

(A160) R.B. Turner, K.H. Gänshirt, P.E. Shaw and J.D. Tauber, J. Amer. Chem. Soc., 88, 1776 (1966).

(A161) W.H. Pirkle, S.D. Beare and T.G. Burlingame, J. Org. Chem., 34, 470 (1969).

(A162) D.M. Feigl and H.S. Mosher, J. Org. Chem., 33, 4242 (1968).

(A163) U. Nagai, T. Shishido, R. Chiba and H. Mitsuhashi, Tetrahedron, 21, 1709 (1965).

(A164) (a) P.E. Verkade, K.S. de Vries, and B.M. Wepster, Rec. Trav. Chim. Pays-Bas, 83, 1149 (1964); (b) J.D. Drumheller and L.J. Andrews, J. Amer. Chem. Soc., 77, 3290 (1955).

(A165) S. Mitsui and Y. Kudo, Tetrahedron, 23, 4271 (1967).

(A166) A.D. Shields, Ph.D. Dissertation, Northwestern University, 1954.

(A167) A. Streitwieser, Jr. and L. Reif, J. Amer. Chem. Soc., 86, 1993 (1964).

(A168) H.L. Goering and R.E. Dilgren, J. Amer. Chem. Soc., 81, 2556 (1959).

(A169) C.L. Arcus and H.E. Strauss, J. Chem. Soc., 1952, 2669.

(A170) H. Kwart and D.P. Hoster, J. Org. Chem., 32, 1896 (1967).

(A171) J.L. Norula and K.L. Dhawan, J. Inst. Chem., 41 (Pt. 2), 53 (1969).

(A172) S. Winstein, M. Brown, K.C. Schreiber and A.H. Schlesinger, J. Amer. Chem. Soc., 74, 1140 (1952).

(A173) J.E. Nordlander and W.J. Kelly, J. Amer. Chem. Soc., 91, 996 (1969).

(A174) H.L. Goering, R.G. Briody and G. Sandrock, J. Amer. Chem. Soc., 92, 7401 (1970).

(A175) H.L. Goering, G.S. Koermer and E.C. Lindsay, J. Amer. Chem. Soc., 93, 1230 (1971).

(A176) Y. Pocker and M.J. Hill, J. Amer. Chem. Soc., 91, 3243 (1969).

(A177) E.J. Corey, H.S. Sachdev, J.Z. Gougoutas and W. Saenger, J. Amer. Chem. Soc., 92, 2488 (1970).

(A178) E.L. Eliel and J.T. Kofron, J. Amer. Chem. Soc., 75, 4585 (1953).

(A179) D.J. Cram and J.E. McCarty, J. Amer. Chem. Soc., 79, 2866 (1957).

(A180) R. MacLeod, F.J. Welch and H.S. Mosher, J. Amer. Chem. Soc., 82, 876 (1960).

(A181) H.H. Zeiss, J. Amer. Chem. Soc., 73, 2391 (1951).

(A182) D.J. Cram, J. Amer. Chem. Soc., 71, 3863 (1949).

(A183) R.S. Bly and R.L. Veazey, J. Amer. Chem. Soc., 91, 4221 (1969).

(A184) K. Mislow and C.L. Hamermesh, J. Amer. Chem. Soc., 77, 1590 (1955).

(A185) F. Hawthorne and D.J. Cram, J. Amer. Chem. Soc., 74, 5859 (1952).

(A186) R.E. Ernst, M.J. O'Connor and R.H. Holm, J. Amer. Chem. Soc., 90, 5735 (1968).

(A187) V. Prelog, E. Philbin, E. Watanabe and M. Wilhelm, Helv. Chim. Acta, 39, 1086 (1956).

(A188) S. Winstein and B.K. Morse, J. Amer. Chem. Soc., 74, 1133 (1952).

(A189) D.J. Cram and R. Davis, J. Amer. Chem. Soc., 71, 3871 (1949).

(A190) S. Winstein and G.C. Robinson, J. Amer. Chem. Soc., 80, 169 (1960).

(A191) T.G. Cochran, D.V. Wareham and A.C. Huitric, J. Pharm. Sci., 60, 180 (1971).

(A192) D.J. Cram, F.A. Abd Elhafez and H. Weingartner, J. Amer. Chem. Soc., 75, 2293 (1953).

(A193) Y. Pocker, Proc. Chem. Soc., 1961, 140.

(A194) G.H. Green and J. Kenyon, J. Chem. Soc., 1950, 751.

(A195) J.L. Kice, R.L. Scriven, E. Koubek and M. Barnes, J. Amer. Chem. Soc., 92, 5608 (1970).

(A196) D.R. Galpin and A.C. Huitric, J. Org. Chem., 33, 921 (1968); idem, J. Pharm. Sci., 57, 447 (1968).

(A197) J. Kenyon and R.L. Patel, J. Chem. Soc., 1965, 435.

(A198) H.L. Goering and H. Hopf, J. Amer. Chem. Soc., 93, 1224 (1971).

(A199) A.G. Davies, J. Kenyon and K. Thaker, J. Chem. Soc., 1957, 3151.

(A1100) D.J. Cram, F.A. Abd Elhafez and H.L. Nyquist, J. Amer. Chem. Soc., 76, 22 (1954).

(A1101) J. Kenyon and R.L. Patel, J. Chem. Soc. C, 1966, 97.

(A1102) M.P. Balfe, M.K. Hargreaves and J. Kenyon, J. Chem. Soc., 1951, 375.

(A1103) D.J. Cram and F.A. Abd Elhafez, J. Amer. Chem. Soc., 76, 28 (1954).

(A1104) F.A. Abd Elhafez and D.J. Cram, J. Amer. Chem. Soc., 74, 5846 (1952).

(A1105) B. Bielawski and A. Chrzaszczewska, Lodz Tow. Nauk. Wydz. III Acta Chim., 11, 105 (1966) [Chem. Abstr., 67, 2837 k (1967)].

(Al106) J. Kenyon and P.R. Sharan, J. Chem. Soc., 1963, 4084; K.A. Thaker, J. Indian Chem. Soc., 40, 293 (1963).

(Al107) K.A. Thaker and N.S. Dave, J. Sci. Ind. Res. (India), 21 B, 374 (1962).

(Al108) D.J. Cram and M. Goldstein, J. Amer. Chem. Soc., 85, 1063 (1963).

(Al109) W.L. Bencze, B. Kisis, R.T. Puckett and N. Finch, Tetrahedron, 26, 5407 (1970).

(Al110) R.E. Singler, R.C. Helgeson and D.J. Cram, J. Amer. Chem. Soc., 92, 7625 (1970).

(Al111) R.L. Patel, J. Chem. Soc. C, 1966, 801.

(Al112) D.J. Cram and M. Cordon, J. Amer. Chem. Soc., 77, 4090 (1955).

(Al113) W.D. Kollmeyer and D.J. Cram, J. Amer. Chem. Soc., 90, 1779 (1968).

(Al114) B.D. West, S. Preis, C.H. Schroeder and K.P. Link, J. Amer. Chem. Soc., 83, 2676 (1961).

(Al115) G.C. Buzby, Jr., D. Hartley, G.A. Hughes, H. Smith, B.W. Gadsby and A.B.A. Jansen, J. Med. Chem., 10, 199 (1967).

(Al116) T. Miki, K. Haraga and T. Asako, Chem. Pharm. Bull. (Tokyo), 13, 1285 (1965).

(Al117) (a) M. Hübner, K. Ponsold, H.-J. Sieman and S. Schwarz, Z. Chem., 8, 380 (1968); (b) Ger. (East) Patent, 62,052 (Jun. 5, 1968) [Chem. Abstr., 70, 58126g (1969)]; (c) Brit. Patent 1,139,019 (Jan. 8, 1969) [Chem. Abstr., 70, 97062c (1969)].

(Al118) F. Reihl, Brit. Patent 1,159,649 (July 30, 1969) [Chem. Abstr., 71, 91768m (1969)].

(Al119) J.A. Berson and M.A. Greenbaum, J. Amer. Chem. Soc., 80, 653 (1958).

(Al120) C.J. Collins, J.B. Christie and V.F. Raaen, J. Amer. Chem. Soc., 83, 4267 (1961).

(Al121) E.W. Cantrall, C. Krieger and R.B. Brownfield, Ger. Offen. 1,942,453 (Mar. 5, 1970) [Chem. Abstr., 72, 111702 m (1970)]; and personal communication from E.W. Cantrall.

(Al122) R.B. Woodward, F. Sondheimer, D. Taub, K. Heusler and W.M. Mc Lamore, J. Amer. Chem. Soc., 74, 4223 (1952).

(Al123) K.G. Rutherford, J.L.H. Batiste and J.M. Prokipcak, Can. J. Chem., 47, 4073 (1969).

(Al124) B.L. Murr, C. Santiago and S. Wang, J. Amer. Chem. Soc., 91, 3827 (1969).

(Al125) B.L. Murr and L.W. Feller, J. Amer. Chem. Soc., 90, 2966 (1968).

(Al126) W.L. Howard and J.D. Burger, U.S. Patent 3,491,152 (Jan. 20, 1970) [Chem. Abstr., 72, 110779e (1970)].

(Al127) D.M. Jerina, H. Ziffer and J.W. Daly, J. Amer. Chem. Soc., 92, 1056 (1970); see also S. Winstein and R. Heck, ibid., 74, 5584 (1952).

(Al128) W. Rigby, J. Chem. Soc., 1949, 1588.

(Al129) R.M. Dodson and V.C. Nelson, J. Org. Chem., 33, 3966 (1968).

(Al130) G.A. Haggis and L.N. Owen, J. Chem. Soc., 1953, 389.

(Al131) E.J. Corey and J.I. Shulman, Tetrahedron Lett., 1968, 3655.

(Al132) H. Gerlach, Helv. Chim. Acta, 51, 1587 (1968).

(Al133) J.W. Cook, J.D. Loudon and W.F. Williamson, J. Chem. Soc., 1950, 911.

(Al134) K.A. Thaker and S.H. Patel, J. Sci. Ind. Research (India), 20B, 327 (1961) [Chem. Abstr., 56, 2319a (1961)].

(Al135) British Patent 687,749 (Feb. 18, 1953) [Chem. Abstr., 48, 2101i (1954)].

(Al136) L.F. Fieser, Organic Experiments, D.C. Heath & Co., Boston, 1964, p. 229.

(Al137) J. Booth, E. Boyland and E.E. Turner, J. Chem. Soc. 1950, 2808.

(Al138) R.M. Lukes and L.H. Sarett, J. Amer. Chem. Soc., 76, 1178 (1954).

(Al139) J. Dierckens and M. Verzele, J. Inst. Brew., London, 75, 391 (1969).

(Al140) D. Seebach and H. Daum, J. Amer. Chem. Soc., 93, 2795 (1971).

(Al141) D.J. Collins and J.J. Hobbs, Aust. J. Chem., 23, 1605 (1970).

(Al142) F.M. Dean, P. Halewood, S. Mongkolsuk, A. Robertson and W.B. Whalley, J. Chem. Soc., 1953, 1250.

(Al143) H.H. Inhoffen, D. Kopp, S. Maric, J. Bekurdts and R. Selimogu, Tetrahedron Lett., 1970, 999.

(Al144) D. Taub, N.N. Girotra, R.D. Hoffsommer, C.-H. Kuo, H.L. Slates, S. Weber and N.L. Wendler, Tetrahedron, 24, 2443 (1968); D. Taub and C.-H. Kuo, French Patent 1,572,929 (June 27, 1969) [Chem. Abstr., 72, 111085n (1970)].

(Al145) W.C. Agosta, J. Amer. Chem. Soc., 89, 3926 (1967).

(Al146) E. Romann, A.J. Frey, P.A. Stadler and A. Eschenmoser, Helv. Chim. Acta, 40, 1900 (1957).

## TABLE 11 RESOLUTIONS OF ALDEHYDES AND KETONES*)

| Substance resolved | | Resolving agent and solvent | Rotation $[\alpha]_D$ of resolved compound | Reference |
|---|---|---|---|---|
| $C_3H_3OBrClF$ | 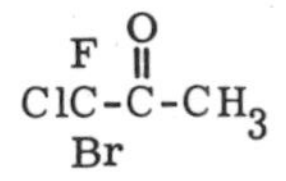 $ClC(F)(Br)-C(=O)-CH_3$ | (-)-Menthydrazide<br>Recr.: $C_6H_6$ | a) + 0.39° (neat) (t = 21°)<br>b) - 0.34° (neat) (t = 21°) | K 1 |
| $C_7H_{12}O$ | 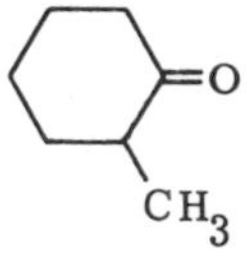 | (-)-Menthydrazide in EtOH-HOAc-NaOAc | a) - 12.64° (neat) (t = 20°)* | K 2 |
| | | (-)-α-Methylbenzylamine bisulfite | + 20.8° (c 1.68, EtOH) (t = 20°) on the semicarbazone derivative | K 3 |
| | | (+)-α-Methylbenzylamine bisulfite | b) - 21.0° (c 1.48, EtOH) (t = 20°) on the semicarbazone derivative | K 3 |
| | | both in $H_2O$-$Et_2O$ | - 11.8° (neat) (t = 24°)* | K 4 |
| | | Also, resolutions with (-)-5-(α-methylbenzyl)-semioxamazide (K 5) and with brucine on the 4-(4-carboxyphenyl)-semicarbazide derivative (K 6).<br>* Note apparent inconsistency in sign of rotation (refs. K3 and K4) | | |
| $C_9H_{14}O_2$ | 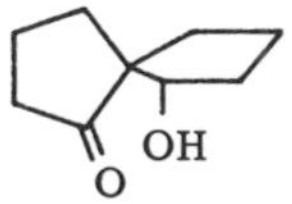 OH, O | See Table 10 b.<br>For the diketone $C_9H_{12}O_2$, see Table 10d, $C_9H_{16}O_2$. ref. Al 132 (p. 224). | | |
| $C_9H_{14}O_3$ | 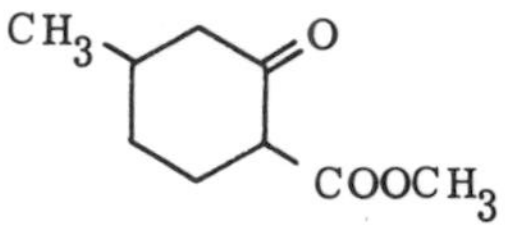 $CH_3$, O, $COOCH_3$ | (+)-Menthol<br>via ester interchange<br>Recr.: hexane | - 104° (c 2.2, $CHCl_3$) (t = 25°) | K 7 |
| $C_9H_{16}O$ | 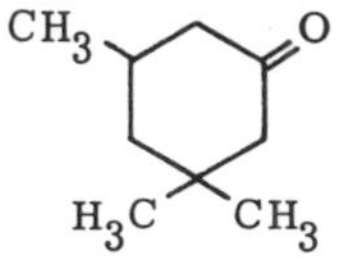 $CH_3$, O, $H_3C$, $CH_3$ | Diethyl (+)-tartrate<br>via gas chromatography<br>on the ketal derivatives | a) - 9.45° (c 0.9, hexane) (t = 20°)<br>b) + 10.3° (c 0.8, hexane) (t = 20°) | K 8 |

*) References to Table 11 begin on p. 234.

TABLE 11 (continued)

| | Substance resolved | Resolving agent and solvent | Rotation $[\alpha]_D$ of resolved compound | Reference |
|---|---|---|---|---|
| $C_{10}H_6OBrCl$ | 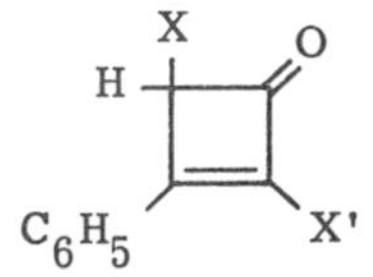 (structure: H, X / O / $C_6H_5$, X') X = Br, X' = Cl | Brucine in $CHCl_3$ *via* asymmetric destruction Partial resolution | $+2.24^o$ (acetone) | K 9 |
| $C_{10}H_6OCl_2$ | 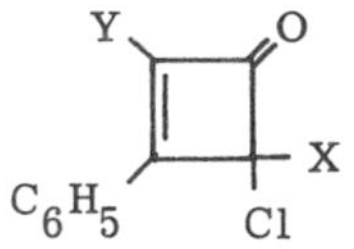 (structure: Y / O / $C_6H_5$, Cl, X) Y = Cl, X = H | Brucine in $CHCl_3$ *via* asymmetric destruction Partial resolution | $+5.4^o$ (c 0.2, $CHCl_3$) (t = $25^o$) | K 10 |
| $C_{10}H_6OClF$ | Y = H, X = F | Brucine in $CHCl_3$ *via* asymmetric destruction Partial resolution | $+12.8^o$ ($CHCl_3$) | K 9 |
| $C_{10}H_6OClI$ | See $C_{10}H_6OBrCl$ above X = I X' = Cl | Brucine in acetone *via* asymmetric destruction Partial resolution | $+2.0^o$ (acetone) | K 9 |
| $C_{10}H_6OFI$ | X = I X' = F | Brucine in $CHCl_3$ *via* asymmetric destruction Partial resolution | $+4.5^o$ ($CHCl_3$) | K 9 |
| $C_{10}H_{11}OCl_3$ | (structure: $CH_3$, $Cl_3C$, $H_3C$, O, $CH_3$) | (+)-2-(Isopropylideneaminooxy)-propionic acid | a) $+26.4 \pm 0.2^o$ (c 8.11, $C_6H_6$) (t = $25^o$) | K 11 |
| | | (-)-2-(Isopropylideneaminooxy)-propionic acid | b) $-25.5 \pm 0.2^o$ (c 8.00, $C_6H_6$) (t = $20^o$) | K 11 |
| | | *via* $\alpha$-aminooxypropionic acid derivatives. Recr.: $C_6H_6$-hexane | | |

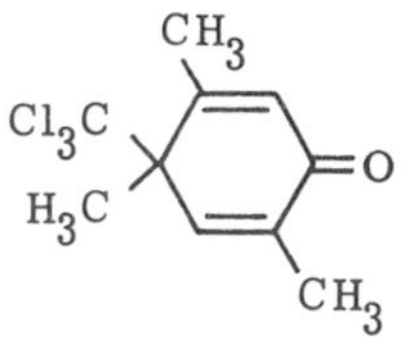

TABLE 11 (continued)

| | Substance resolved | Resolving agent and solvent | Rotation $[\alpha]_D$ of resolved compound | Reference |
|---|---|---|---|---|
| $C_{10}H_{14}O$ | 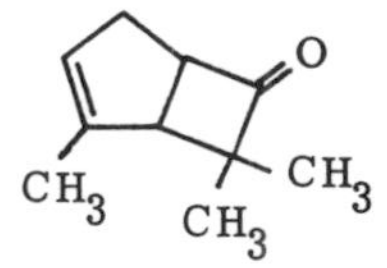 $CH_3$ $CH_3$ $CH_3$ O | (+)-Dehydroabietylamine<br>*via* Schiff's base formation<br>Kinetic resolution | a) $+3.24^{\circ}$ (neat) ($t = 24^{\circ}$)<br>unreacted ketone<br>b) $-4.35^{\circ}$ (neat) ($t = 24^{\circ}$)<br>recovered from Schiff's base | K 12 |
| | | Also, similar kinetic resolutions with (+)-*threo*-2-amino-1-phenyl-propane-1,3-diol acetonide and with (-)-$\alpha$-methylbenzylamine (K 12) | | |
| $C_{10}H_{14}O$ | 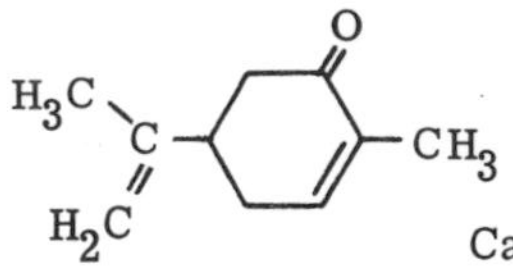 Carvone | (-)-Menthydrazide<br>Recr.: 95% EtOH<br>Partial resolution | a) $+94.9^{\circ}$ (c 1.0, MeOH) ($t = 25^{\circ}$)<br>b) $-196^{\circ}$ (c 1.0, MeOH) ($t = 25^{\circ}$)<br>both on the diastereomeric menthydrazone derivatives | K 13 |
| $C_{10}H_{14}O_2$ | *cis*-Cinerolone | See Table 10b | | |
| $C_{10}H_{16}O$ | 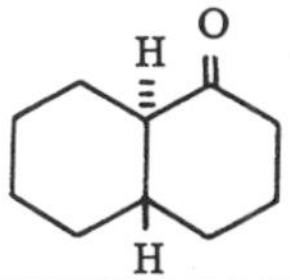 H O H | (-)-Menthydrazide<br>in EtOH-HOAc-NaOAc | $+9.5^{\circ}$ (c 1.26, EtOH) | K 14 |
| $C_{10}H_{16}O$ | 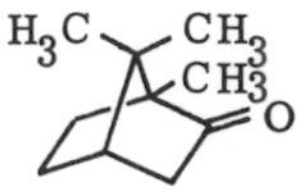 Camphor | (-)-Menthydrazide<br>in EtOH-HOAc-NaOAc | a) $\alpha_D^{25}$ $-1.01^{\circ}$ (1 2 dm, c 1.2, EtOH)<br>$[M]_D^{25}$ $-64^{\circ}$<br>b) $\alpha_D^{25}$ $+0.98^{\circ}$ (1 2 dm, c 1.2, EtOH)<br>$[M]_D^{25}$ $+62^{\circ}$ | K 15 |
| | | "Unsuccessful" resolution with (-)-menthydrazide (K 16), i.e. only partial resolution obtained. Also, gas chromatographic resolution on ketals formed with (-)-2,3-butanediol (K 17):<br>$[\alpha]_D^{25}$ $+33.7 \pm 0.7$ (c 1.69, 95% EtOH) and $[\alpha]_D^{25}$ $-34.1 \pm 1.1^{\circ}$ (c 1.30, 95% EtOH) | | |

TABLE 11 (continued)

| | Substance resolved | Resolving agent and solvent | Rotation $[\alpha]_D$ of resolved compound | Reference |
|---|---|---|---|---|
| $C_{10}H_{16}O$ | Fenchone | Resolution with (-)-menthydrazide failed. | | K 18 |
| $C_{10}H_{18}O$ | 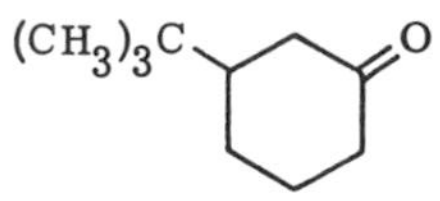 | (+)-$\alpha$-Methylbenzylamine bisulfite in $EtOH\text{-}Et_2O\text{-}H_2O$<br>(-)-$\alpha$-Methylbenzylamine bisulfite in $EtOH\text{-}Et_2O\text{-}H_2O$ | a) - $25.2^o$ (c 3.1, $CHCl_3$) (t = $22^o$)<br>b) + $27.5^o$ (c 0.5, $CHCl_3$) (t = $23^o$) | K 4<br>K 4 |
| $C_{10}H_{20}O$ | $CH_3CH(CH_3)(CH_2)_3CH(CH_3)CH_2CH{=}O$ | (+)-Tartramidic acid hydrazide in $MeOH\text{-}H_2O$ | + $7.48^o$ (neat) (t = $20^o$) | K 19 |
| $C_{11}H_{12}O$ | 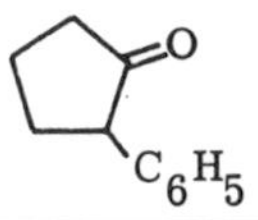 | (+)-Tartramidic acid hydrazide in MeOH | + $4440^o$ (c 1.042, toluene) (t = $28^o$) | K 20 |
| $C_{11}H_{16}O$ | 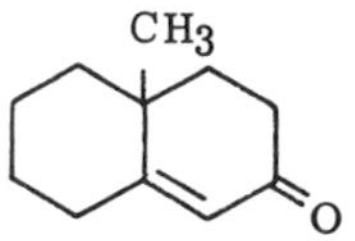 | (+)-Camphor-10-sulfonic acid on the pyrrolidine enamine in EtOAc | a) + $26.4^o$ (c 1, 95% EtOH) (t = $27^o$)<br>b) - $3.4^o$ (c 1, 95% EtOH) (t = $27^o$) | K 21 |
| $C_{11}H_{20}O$ | 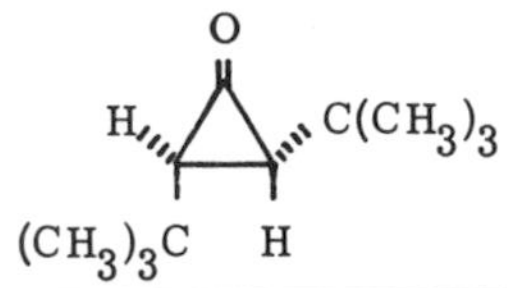 | (+)-Amphetamine *via* asymmetric destruction<br>Partial resolution | + $76^o$ (c 0.5, $CCl_4$) (t = $25^o$) (436 nm) | K 22 |
| $C_{12}H_{10}O$ | 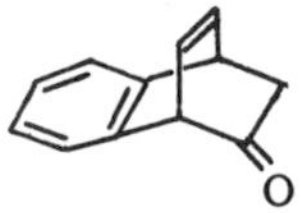 | (-)-Menthydrazide in EtOH-HOAc-NaOAc | + 87.5 ± $1.2^o$ (c 1.038, $CHCl_3$) (t = $22^o$) | A 283 (p.136) |
| $C_{12}H_{12}O$ | 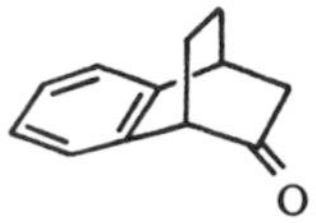 | (-)-Menthydrazide in EtOH-HOAc-NaOAc | - 356.2 ± $4^o$ (c 0.957, $CHCl_3$) (t = $22^o$) | A 283 |

TABLE 11 (continued)

| Formula | Substance resolved | Resolving agent and solvent | Rotation $[\alpha]_D$ of resolved compound | Reference |
|---|---|---|---|---|
| $C_{13}H_{20}O$ | 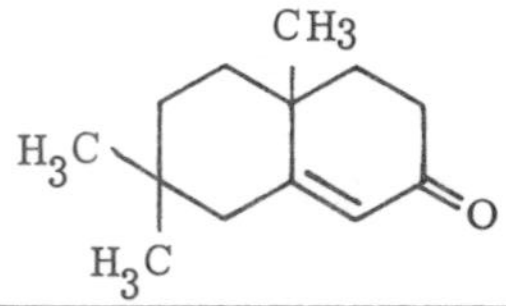 | See $C_{13}H_{22}O$ in Table 10b | | |
| $C_{13}H_{20}O$ | 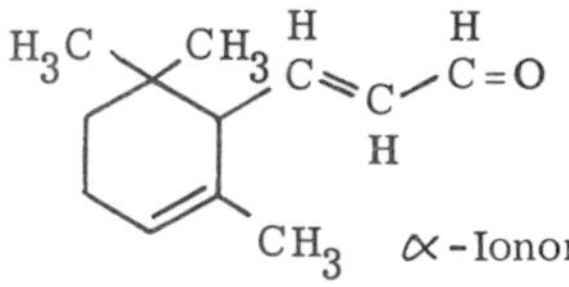 $\alpha$-Ionone | (-)-Menthydrazide in EtOH-HOAc-NaOAc | a) - 406° (t = 27°)<br>b) + 347° (t = 23°) | K 23 |
| $C_{14}H_{12}O_2$ | | (+)-4-($\alpha$-Methylbenzyl)-semicarbazide · HCl in py-$H_2O$ | a) + 118.3° (c 1.26, acetone) (t = 11°)<br>b) - 118.5° (c 1.15, acetone) (t = 11°) | K 24 |
| | | Also, resolution with (+)-4-(1-phenylpropyl)-semicarbazide hydrochloride (K 25) and with quinidine on the H Phthal derivative (Table 10c). See also Table 10c for the resolution of substituted benzoins. | | |
| $C_{14}H_{14}ORu$ | See Table 12 | | | |
| $C_{14}H_{16}O_2$ | 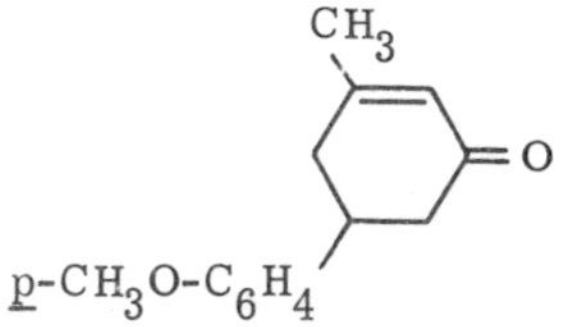 p-$CH_3O$-$C_6H_4$ | Potassium (+)-camphor-10-sulfonate on pyrrolidine iminium perchlorate salt. Recr.: $CH_3CN$-$Et_2O$ | a) + 55.6° (c 1, 95% EtOH) (t = 27°)<br>b) - 26.8° (c 1, 95% EtOH) (t = 27°) | K 21 |
| $C_{14}H_{20}O_2$ | 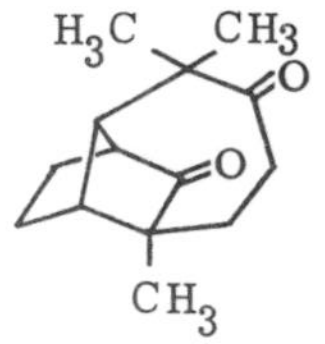 | (+)-2,3-Butanedithiol *via* chromatographic separation of the mono-thioketal derivatives on alumina | + 500° (c 0.0647, cyclohexane) (302 nm) on $C_{14}H_{22}O$ (Longicamphenylone) | K 26 |

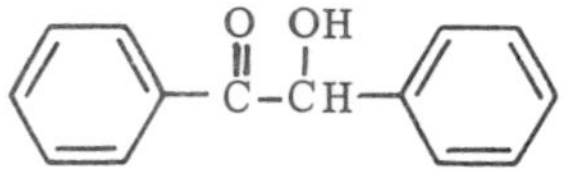

TABLE 11 (continued)

| | Substance resolved | Resolving agent and solvent | Rotation $[\alpha]_D$ of resolved compound | Reference |
|---|---|---|---|---|
| $C_{14}H_{22}O$ | 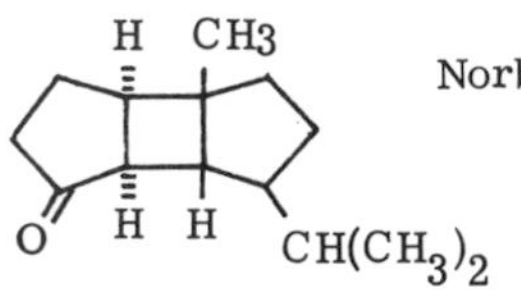 Norbourbonone (structure: H, CH3, H, H, O, $CH(CH_3)_2$) | (-)-2,3-Butanediol *via* gas chromatography on the ketal derivatives | a) $+205^{\circ}$ (c 0.83, $CHCl_3$) (t = $25^{\circ}$)<br>b) $-143^{\circ}$ (c 0.28, $CHCl_3$) (t = $25^{\circ}$) | K 27 |
| $C_{14}H_{20}O$ | (structure: $CH_3$, $H_3C$, $CH_3$, O) | (-)-2,3-Butanediol *via* gas chromatography on the ketal derivatives | a) $+157^{\circ}$ (c 1.67, $CHCl_3$) (t = $24^{\circ}$)<br>b) $-159^{\circ}$ (c 1.34, $CHCl_3$) (t = $23^{\circ}$) | K 28 |
| | | Also resolved with (+)-2,3-butanediol | | K 28 |
| $C_{14}H_{22}O$ | (structure: $CH_3$, $H_3C$, $CH_3$, O) | 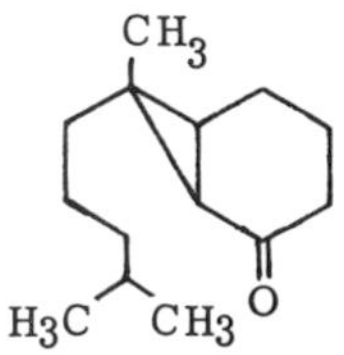 (-)-2,3-Butanediol *via* gas chromatography on the ketal derivatives | a) $+147^{\circ}$ (c 0.61, $CHCl_3$) (t = $23^{\circ}$)<br>b) $-147^{\circ}$ (c 0.57, $CHCl_3$) (t = $23^{\circ}$) | K 28 |
| $C_{15}H_{12}O$ | 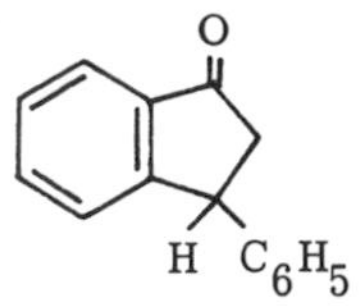 (structure: O, H, $C_6H_5$) | Mandelic acid hydrazide | $+210^{\circ}$ (dioxane) (t = $20^{\circ}$) on the oxime | K 29 |
| $C_{15}H_{12}O_2$ | 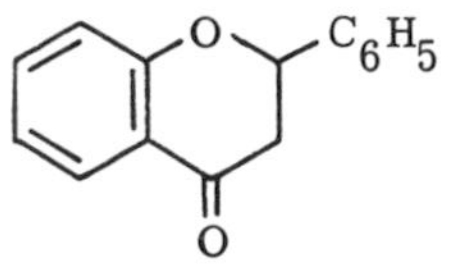 Flavanone (structure: O, $C_6H_5$, O) | (+)-2,3-Butanedithiol *via* the thioketal derivatives. Recr.: $C_6H_6$<br>Also, resolution with 5-methyl-5-($\alpha$-methyl-$\beta$-phenylethyl)-semioxamazide, "Hosemiazide" (K 31). | a) $-53.5^{\circ}$ (c 2.27, $CHCl_3$) (t = $25^{\circ}$)<br>b) $+52^{\circ}$ (c 0.75, $CHCl_3$) (t = $25^{\circ}$) | K 30 |

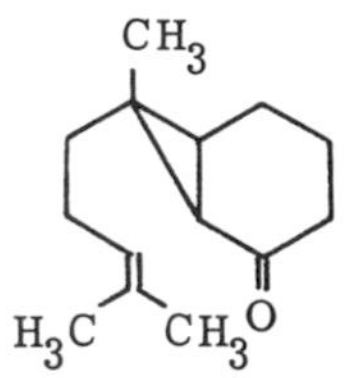

TABLE 11 (continued)

| Formula | Substance resolved | Resolving agent and solvent | Rotation $[\alpha]_D$ of resolved compound | Reference |
|---|---|---|---|---|
| $C_{15}H_{16}O$ | 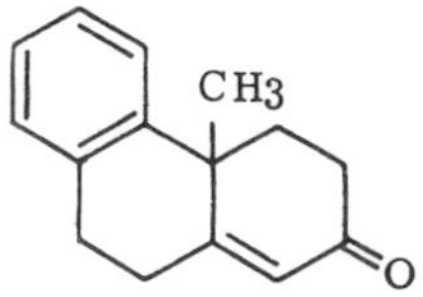 CH3 | (+)-Camphor-10-sulfonic acid on the pyrrolidine enamine in acetone | a) - 332° (c 1, 95% EtOH) (t = 27°)<br>b) + 332° (c 1, 95% EtOH) (t = 27°) | K 21 |
| $C_{15}H_{16}O$ | 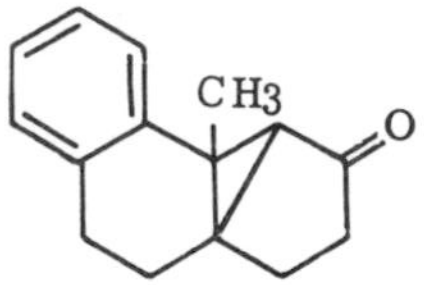 CH3 | Potassium (+)-camphor-10-sulfonate on pyrrolidine iminium perchlorate salt in MeOH. Recr.: EtOAc-MeOH | a) + 42.5° (c 1, 95% EtOH) (t = 27°)<br>b) - 42.3° (c 1, 95% EtOH) (t = 27°) | K 21 |
| $C_{15}H_{22}O_4$ | trans-Humulinic acid | See Table 10d | | |
| $C_{16}H_{19}NO_2$ | 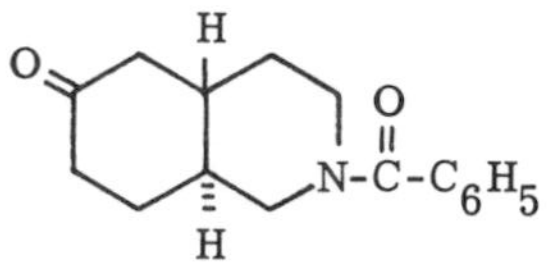 H, H, N-C(=O)-$C_6H_5$ | (-)-2,3-Butanediol via the ketal derivative | a) - 62.6° (c 1.005, $CHCl_3$) (t = 25°)<br>b) + 61.8° (c 1.01, $CHCl_3$) (t = 26°) | B 229 (p.195) |
| $C_{18}H_{16}O$ | 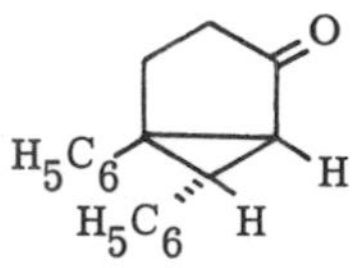 $H_5C_6$, $H_5C_6$, H, H | See $C_{20}H_{23}N$ Table 9a | | |
| $C_{18}H_{18}O$ | 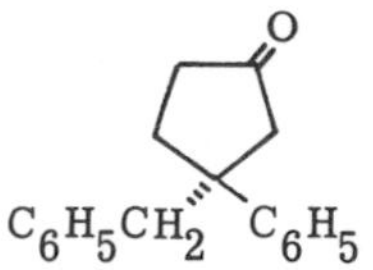 $C_6H_5CH_2$, $C_6H_5$ | See $C_{20}H_{25}N$ Table 9a | | |
| $C_{18}H_{22}O_2$ | Estrone | See $C_{19}H_{26}O_2$ Table 10c | | |
| $C_{19}H_{22}O_5$ | Zearalenone | See Table 10d | | |

TABLE 11 (continued)

| | Substance resolved | Resolving agent and solvent | Rotation $[\alpha]_D$ of resolved compound | Reference |
|---|---|---|---|---|
| $C_{20}H_{26}O_2$ | 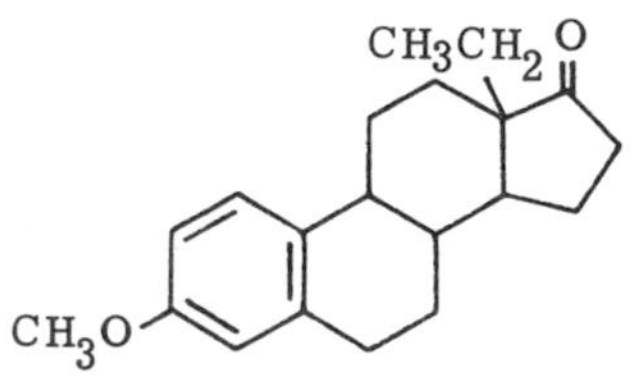 | (-)-2-Amino-1-butanol *via* formation of imine hydrochloride salts Recr.: EtOH-HCl | a) - 94° ($CHCl_3$) (t = 23°)<br>b) + 101° ($CHCl_3$) (t = 23°) | Al 121 (p. 224) |
| $C_{21}H_{26}O_3$ | Methyl-3-keto-$\Delta^{4,9(11),16}$-etiocholatrienate | See $C_{21}H_{28}O_3$ | Table 10c | |
| $C_{21}H_{28}O_3$ | Ketoprogesterone | See $C_{23}H_{32}O_4$ below | | |
| $C_{23}H_{32}O_4$ | 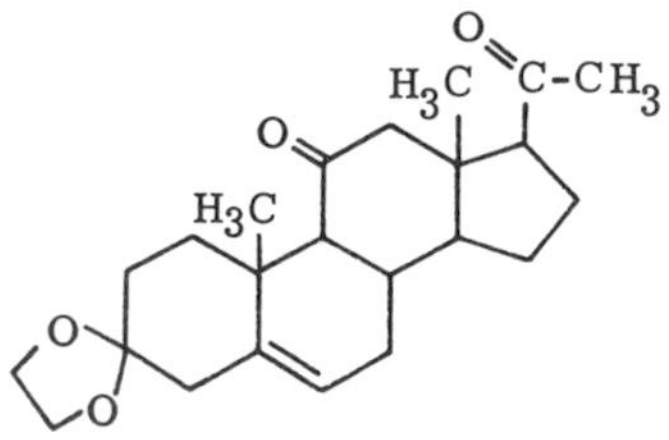 | Strychnine on the 21-oxalyl derivative | + 52 ± 2° ($CHCl_3$) (t = 25°) | K 32 |
| $C_{28}H_{14}O_2$ | 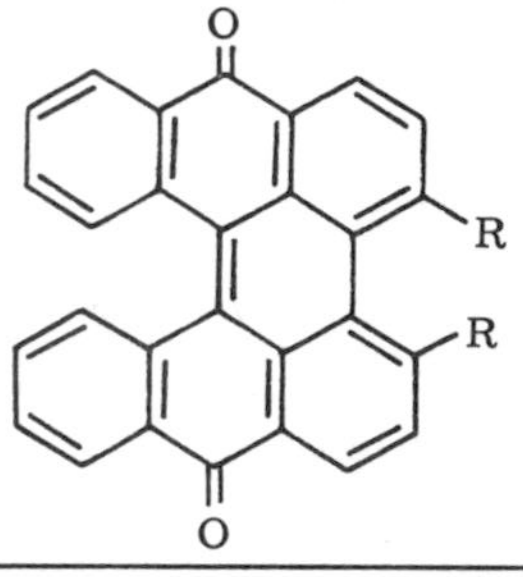 R = H Helianthrone | (-)-Menthoxyacetyl chloride on the hydroquinone derivative. Esters separated by column chromatography | + 7.0° (c 0.27, $C_6H_5NO_2$ or $C_6H_6$) (t = 20°) | K 33 |
| $C_{30}H_{18}O_2$ | R = $CH_3$ | Resolved similarly | + 40.2° (c 0.24, $C_6H_6$) (t = 20°) | K 33 |

## References to Table 11
## Resolutions of Aldehydes and Ketones


(K 1) M.K. Hargreaves and B. Modarai, J. Chem. Soc. C, 1971, 1013.

(K 2) R. Adams, C.M. Smith and S. Loewe, J. Amer. Chem. Soc., 64, 2087 (1942).

(K 3) R. Adams and J.D. Garber, J. Amer. Chem. Soc., 71, 522 (1949).

(K 4) G. Adolphen, E.J. Eisenbraun, G.W. Keen and P.W.K. Flanagan, Org. Prep. Proc., 2, 93 (1970).

(K 5) N.J. Leonard and J.H. Boyer, J. Org. Chem., 15, 42 (1950).

(K 6) J.K. Shillington, G.S. Denning, Jr., W.B. Greenough, III, T. Hill, Jr., and O.B. Ramsey, J. Amer. Chem. Soc., 80, 6551 (1958).

(K 7) E.J. Eisenbraun, G.H. Adolphen, K.S. Schorno and R.N. Morris, J. Org. Chem., 36, 414 (1971).

(K 8) M. Sanz-Burata, J. Irrure-Peréz and S. Juliá-Arechaga, Afinidad, 27, 705 (1970).

(K 9) M. Caserio, H.E. Simmons, Jr., A.E. Johnson and J.D. Roberts, J. Amer. Chem. Soc., 82, 3102 (1960).

(K 10) E.F. Jenny and J.D. Roberts, J. Amer. Chem. Soc., 78, 2005 (1956).

(K 11) M.S. Newman and R.M. Layton, J. Org. Chem., 33, 2338 (1968).

(K 12) U.A. Huber and A.S. Dreiding, Helv. Chim. Acta, 53, 495 (1970).

(K 13) L. Friedman and J.G. Miller, Science, 172, 1045 (1971).

(K 14) W. Acklin, V. Prelog, F. Schenker, B. Serdarević and P. Walter, Helv. Chim. Acta, 48, 1725 (1965).

(K 15) R.B. Woodward, T.P. Kohman and G.C. Harris, J. Amer. Chem. Soc., 63, 120 (1941).

(K 16) A.M.T. Finch, Jr., and W.R. Vaughan, J. Amer. Chem. Soc., 91, 1416 (1969).

(K 17) J. Casanova and E.J. Corey, Chem. Ind. (London), 1961, 1664.

(K 18) D.S. Tarbell and F.C. Loveless, J. Amer. Chem. Soc., 80, 1963 (1958).

(K 19) F. Nerdel and E. Henkel, Chem. Ber., 85, 1138 (1952).

(K 20) K. Mislow and C.L. Hamermesh, J. Amer. Chem. Soc., 77, 1590 (1955).

(K 21) W.R. Adams, O.L. Chapman, J.B. Sieja and W.J. Welstead, Jr., J. Amer. Chem. Soc., 88, 162 (1966).

(K 22) D.B. Sclove, J.F. Pazos, R.L. Camp and F.D. Greene, J. Amer. Chem. Soc., 92, 7488 (1970).

(K 23) H. Sobotka, E. Bloch, H. Cahnmann, E. Feldblau and E. Rosen, J. Amer. Chem. Soc., 65, 2061 (1943).

(K 24) I.V. Hopper and F.J. Wilson, J. Chem. Soc., 1928, 2483.

(K 25) A.J. Little, J. M'Lean and F.J. Wilson, J. Chem. Soc., 1940, 336.

(K 26) E.J. Corey, M. Ohno, R.B. Mitra and P.A. Vatakencherry, J. Amer. Chem. Soc., 86, 478 (1964).

(K 27) J.D. White and D.N. Gupta, J. Amer. Chem. Soc., 90, 6171 (1968).

(K 28) J.J. Plattner and H. Rapoport, J. Amer. Chem. Soc., 93, 1758 (1971).

(K 29) K. Bott, Tetrahedron Lett., 1965, 4569.

(K 30) E.J. Corey and R.B. Mitra, J. Amer. Chem. Soc., 84, 2938 (1962).

(K 31) M. Kotake and G. Nakaminami, Proc. Japan Acad., 29, 56 (1953).

(K 32) L.H. Sarett, G.E. Arth, R.M. Lukes, R.E. Beyler, G.I. Poos, W.F. Johns and J.M. Constantin, J. Amer. Chem. Soc., 74, 4974 (1952).

(K 33) W. Theilacker and W. Thomas, Justus Liebigs Ann. Chem., 632, 115 (1960).

TABLE 12 RESOLUTIONS OF ORGANOMETALLIC AND ORGANOMETALLOIDAL COMPOUNDS, INCLUDING PHOSPHORUS AND SULFUR COMPOUNDS *)

| Formula | Substance resolved | Resolving agent and solvent | Rotation $[\alpha]_D$ of resolved compound | Reference |
|---|---|---|---|---|
| | Antimony Compounds | | | |
| $C_{19}H_{16}NSb$ | 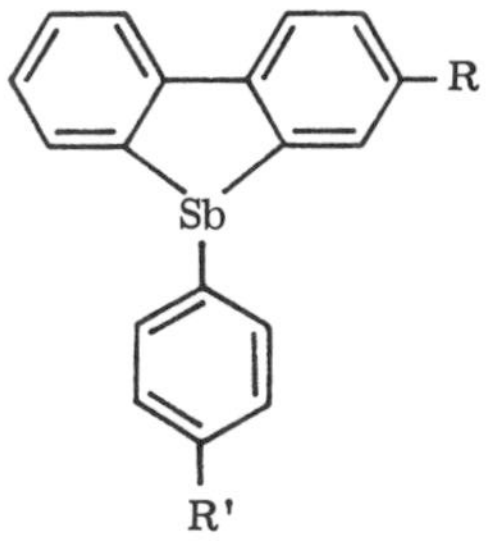 R = $NH_2$, R' = $CH_3$ | (+)-Tartaric acid<br>(-)-Tartaric acid<br>both in EtOH | a) + 250.5 ± 1° (c 0.530, $C_6H_6$) (t = 22°)<br>b) - 248.0 ± 2° (c 0.248, $C_6H_6$) (t = 23°) | M 1 |
| $C_{20}H_{15}O_2Sb$ | R = COOH, R' = $CH_3$ | (-)-α-Methylbenzylamine<br>(+)-α-Methylbenzylamine<br>both in EtOH | a) + 245.2° (c 0.25, py) (t = 20°)<br>b) - 245.7° (c 0.25, py) (t = 20°) | M 2 |
| $C_{20}H_{15}O_2Sb$ | R = $CH_3$, R' = COOH | (-)-α-Methylbenzylamine in EtOAc<br>(-)-Ephedrine in EtOH-$CHCl_3$<br>(Partial resolutions) | a) - 69.2° (py)<br>b) + 78.0° (py) | M 3<br>M 3 |
| $C_{20}H_{15}O_3Sb$ | R = $OCH_3$, R' = COOH | (+)-α-Methylbenzylamine<br>(-)-Ephedrine<br>both in EtOH | a) - 145.5 ± 2° (py)<br>b) + 153.0 ± 1° (py) | M 3<br>M 3 |
| $C_{20}H_{15}O_3Sb$ | 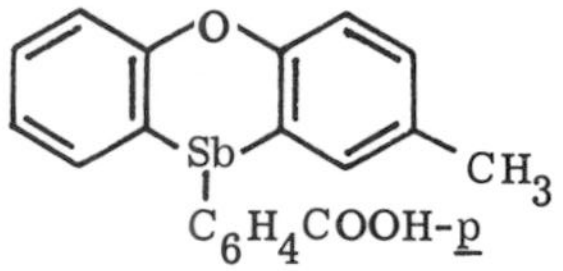 (phenoxantimonin structure: Sb–$C_6H_4COOH$-p, $CH_3$) | Strychnine in EtOH-$CHCl_3$ | a) + 77.50° (c 0.555, $CHCl_3$)<br>b) - 77.2° (c 0.518, $CHCl_3$) | M 4 |
| $C_{21}H_{17}O_3Sb$ | See $C_{19}H_{16}NSb$ above<br>R = $OCH_2COOH$<br>R' = $CH_3$ | (-)-Ephedrine in EtOH | a) - 112.0° (c 0.3, py)<br>b) + 112.0° (c 0.3, py) | M 3 |

Structure for the first four rows: 5-(4-R'-phenyl)dibenzostibole with R on the ring (labels: Sb, R, R').

*) References to Table 12 begin on p. 261. See also comments on pp. 47-48.

TABLE 12 (continued)

| Substance resolved | Resolving agent and solvent | Rotation $[\alpha]_D$ of resolved compound | Reference |
|---|---|---|---|
| $C_{23}H_{17}O_2Sb$ (Sb, $C_6H_5$, COOH) | 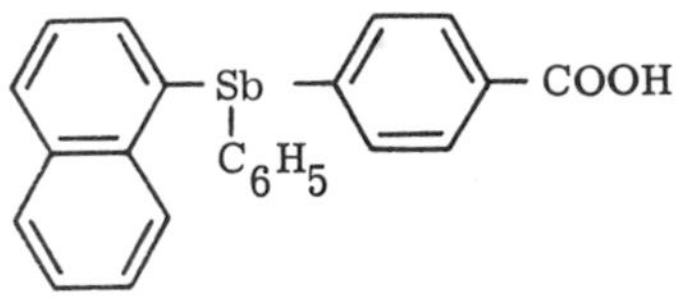 (+)-α-Methylbenzylamine<br>(-)-α-Methylbenzylamine<br>both in EtOH | a) $-35.5 \pm 1.5^\circ$ (c 0.324, $CHCl_3$) (t = $18^\circ$)<br>b) $+35.1 \pm 1.6^\circ$ (c 0.314, $CHCl_3$) (t = $18^\circ$) | M 5 |
| $C_{25}H_{19}O_2Sb$ (HOOC, Sb, $C_6H_5$) | 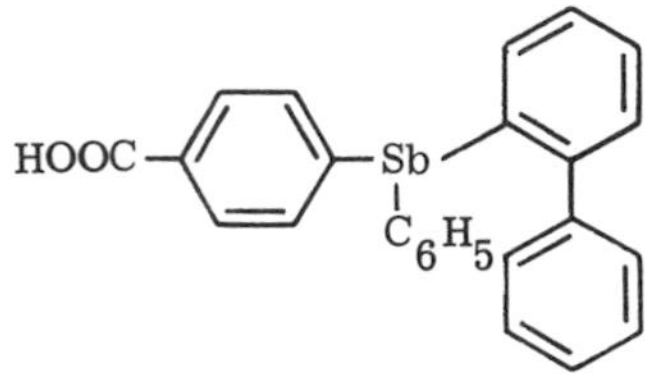 (+)-α-Methylbenzylamine<br>in acetone | a) $-46.9^\circ$ (c 0.245, py) (t = $20^\circ$)<br>b) $+47.2^\circ$ (c 0.223, py) (t = $20^\circ$) | M 6 |
| Arsenic Compounds | | | |
| $C_9H_{13}As$ ($CH_3$, As-$CH_2CH_3$) | 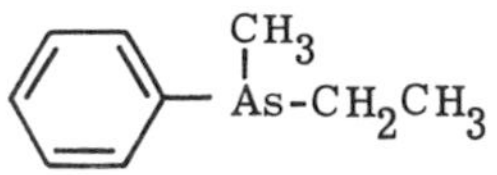 via formation and crystallization ($CHCl_3$) of the complex chloromethylethylphenylarsine-(+)-stilbenediamine platinum(II) chloride<br>See also $C_{16}H_{20}O_4AsCl$ below | a) $+3.26 \pm 0.1^\circ$ (c 4.193, $Et_2O$) (t = $20^\circ$)<br>b) $-3.05 \pm 0.1^\circ$ (c 3.69, $Et_2O$) (t = $20^\circ$) | M 7 |
| $C_{10}H_{13}O_2As$ (As-$CH_2COOH$, $CH_2CH_3$) | 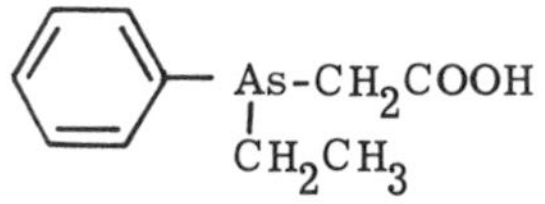 (-)-α-Methylbenzylamine<br>in EtOH | $+2.65^\circ$ (c 1.4300, EtOH) (t = $20^\circ$) | M 8 |
| $C_{11}H_{17}As$ See $C_{18}H_{24}O_4AsCl$ | | | |

TABLE 12 (continued)

| | Substance resolved | Resolving agent and solvent | Rotation $[\alpha]_D$ of resolved compound | Reference |
|---|---|---|---|---|
| $C_{12}H_{17}O_2As$ | R-As(R')-C₆H₄-COOH<br>R = $CH_3CH_2CH_2$  R' = $CH_3CH_2$ | Quinine<br>(+)-α-Methylbenzylamine<br>both in EtOH | a) $+3.01^o$ (c 2.230, $C_6H_6$) (t = $20^o$)<br>b) $-3.04^o$ (c 1.2436, $C_6H_6$) (t = $20^o$) | M 9<br>M 9 |
| $C_{13}H_{19}O_2As$ | R = $CH_3CH_2CH_2CH_2$  R' = $CH_3CH_2$ | Quinine<br>in EtOH | $-3.96^o$ (c 4.4559, $C_6H_6$) (t = $20^o$) | M 9 |
| $C_{14}H_{12}O_2BrAs$ | See $C_{12}H_{17}O_2As$<br>R = $CH_3$  R' = p-Br$C_6H_4$ | Quinine | a) $+8.57^o$ (c 1.6528) (t = $20^o$)<br>b) $-7.12^o$ (c 0.9634) (t = $20^o$) | M 10 |
| $C_{14}H_{13}O_2As$ | See $C_{12}H_{17}O_2As$<br>R = $CH_3$  R' = $C_6H_5$ | Quinine | a) $+8.71^o$ (c 1.5140) (t = $20^o$)<br>b) $-6.2^o$ (c 2.9800) (t = $20^o$) | M 10 |
| $C_{15}H_{14}O_2BrAs$ | See $C_{12}H_{17}O_2As$<br>R = $CH_3CH_2$  R' = p-Br$C_6H_4$ | Quinine | a) $+13.54^o$ (c 0.8184) (t = $20^o$)<br>b) $-13.0^o$ (c 2.8820) (t = $20^o$) | M 10 |
| $C_{15}H_{15}O_2As$ | $C_6H_5$-As($CH_2CH_3$)-$C_6H_4$-COOH<br>meta | Quinine<br>in EtOH | a) $+16.63^o$ (c 1.5488, $C_6H_6$) (t = $20^o$)<br>b) $-16.28^o$ (c 0.5834, $C_6H_6$) (t = $20^o$) | M 11 |
| | ortho | Quinine<br>in EtOH | a) $+18.36^o$ (c 1.9042, $C_6H_6$) (t = $20^o$)<br>b) $-18.41^o$ (c 2.4616, $C_6H_6$) (t = $20^o$) | M 11, M 10 |
| | para | Quinine<br>in EtOH | a) $+12.75^o$ (c 1.8886, $C_6H_6$) (t = $20^o$)<br>b) $-12.56^o$ (c 1.2130, $C_6H_6$) (t = $20^o$) | M 11, M 12 |

TABLE 12 (continued)

| | Substance resolved | Resolving agent and solvent | Rotation $[\alpha]_D$ of resolved compound | Reference |
|---|---|---|---|---|
| $C_{16}H_{20}O_4AsCl$ | $C_6H_5-\overset{\oplus}{As}(R)(R')CH_2-C_6H_5\ X^{\ominus}$<br>R = $CH_3$ R' = $CH_3CH_2$<br>$X^{\ominus}$ = $ClO_4^{\ominus}$ | Silver (-)-O,O'-dibenzoyl-tartrate | a) + 21.4$^o$ (MeOH) (t = 20$^o$)<br>b) - 19.3$^o$ (MeOH) (t = 20$^o$)<br>Reduction of salt at Hg cathode yields ethylmethylphenylarsine $[\alpha]_D^{20}$ +1.9$^o$ and - 1.15$^o$ (both MeOH). | M 13 |
| | | Resolution of chloride salt by chromatography on (+)-lactose | | M 14 |
| $C_{17}H_{22}As_2Br_2$ | $(CH_3)_2\overset{\oplus}{As}$ … $\overset{\oplus}{As}(CH_3)_2$ $2\,Br^{\ominus}$ (biphenyl, bridged by $(CH_2)_n$)<br>n = 1 | Silver (+)-O,O'-dibenzoyl-tartrate in $H_2O$<br>Silver (-)-O,O'-dibenzoyl-tartrate in $H_2O$ | a) + 29.5$^o$ (c 0.814 %, $H_2O$) (t = 20-23$^o$)<br>b) - 30.3$^o$ (c 0.388 %, $H_2O$) (t = 20-23$^o$) | M 15 |
| $C_{18}H_{14}NAs$ | (dibenzarsole, As–$C_6H_4R'$-p, ring substituent R)<br>R = $NH_2$<br>R' = H | (+)-Tartaric acid<br>(-)-Tartaric acid<br>both in EtOH | a) + 255 ± 1$^o$ (c 0.45, EtOH) (t = 25$^o$)<br>b) - 251 ± 1$^o$ (c 0.45, EtOH) (t = 25$^o$) | M 16 |
| $C_{18}H_{24}As_2Br_2$ | See $C_{17}H_{22}As_2Br_2$<br>n = 2 | Silver (+)-3-bromocamphor-9-sulfonate in $H_2O$<br>Silver (-)-3-bromocamphor-9-sulfonate in $H_2O$ | a) + 50.7$^o$ (c 0.799 %, $H_2O$) (t = 20-23$^o$)<br>b) - 50.8$^o$ (c 0.7040 %, $H_2O$) (t = 20-23$^o$) | M 15 |
| $C_{18}H_{24}O_4AsCl$ | See $C_{16}H_{20}O_4AsCl$<br>R = $CH_3$ R' = $CH_3CH_2CH_2CH_2$ | Silver (-)-O,O'-dibenzoyl-tartrate | + 29$^o$ (MeOH) (t = 20$^o$)<br>Reduction of salt at Hg cathode yields butylmethylphenylarsine $[\alpha]_D^{20}$ + 12.4$^o$ (MeOH) | M 13 |

TABLE 12 (continued)

| Formula | Substance resolved | Resolving agent and solvent | Rotation $[\alpha]_D$ of resolved compound | Reference |
|---|---|---|---|---|
| $C_{19}H_{12}O_3AsCl$ | Z, As, R; $C_6H_4COOH$-p; Z = O, R = Cl | Strychnine in EtOH-$CHCl_3$<br>Cinchonine in EtOH | a) $+162.1^o$ (c 0.944, $CHCl_3$) (t = $20^o$) (579 nm)<br>b) $-161.3^o$ (c 0.930, $CHCl_3$) (t = $20^o$) (579 nm) | M 17<br>M 17 |
| $C_{19}H_{26}AsCl$ | See $C_{16}H_{20}O_4AsCl$<br>R = $CH_3CH_2$ R' = $CH_3CH_2CH_2CH_2$ X = $Cl^-$ | Chromatography on<br>(-)-quartz<br>(+)-lactose | a) $-12.8^o$ ($CHCl_3$) (t = $20^o$)<br>b) $+14.4^o$ (EtOAc) (t = $20^o$) | M 14<br>M 14 |
| $C_{19}H_{26}As_2Br_2$ | See $C_{17}H_{22}As_2Br_2$<br>n = 3 | Silver (+)-3-bromocamphor-9-sulfonate in $H_2O$<br>Silver (-)-3-bromocamphor-9-sulfonate in $H_2O$ | a) $\alpha_D^{20-23}$ $+1.130^o$ (l 1 dm, c 1.354 %, $H_2O$)<br>b) $\alpha_D^{20-23}$ $-1.095^o$ (l 1 dm, c 1.272 %, $H_2O$) | M 15 |
| $C_{20}H_{15}O_3As$ | See $C_{18}H_{14}NAs$<br>R = $OCH_3$ R' = COOH | (-)-α-Methylbenzylamine<br>(+)-α-Methylbenzylamine<br>both in EtOH | a) $+160.7 \pm 1^o$ (py) (t = $21^o$)<br>b) $-160.2 \pm 1^o$ (py) (t = $21^o$) | M 16 |
| $C_{20}H_{22}AsBr$ | $CH_3$, $\oplus$, $AsCH_2$, $CH_2CH_3$, $Br^\ominus$ | Silver (+)-3-bromocamphor-9-sulfonate in acetone | $+22.8^o$ (t = $25^o$) | M 18 |
| $C_{20}H_{28}AsBr$ | See $C_{16}H_{20}O_4AsCl$<br>R = $CH_3CH_2$ R' = $CH_3(CH_2)_4$ X = Br | Chromatography on<br>(+)-quartz<br>(-)-quartz | a) $+22.1^o$ ($CHCl_3$) (t = $20^o$)<br>b) $-19.56^o$ ($CHCl_3$) (t = $20^o$) | M 19, M 14 |

TABLE 12 (continued)

| | Substance resolved | Resolving agent and solvent | Rotation $[\alpha]_D$ of resolved compound | Reference |
|---|---|---|---|---|
| $C_{21}H_{17}O_2As$ | See $C_{19}H_{12}O_3AsCl$<br>$Z = CH_2$ $R = CH_3$ | (+)-Amphetamine<br>(-)-$\alpha$-Methylbenzylamine<br>both in EtOH | a) $-83.9^o$ (c 0.8, MeOH) ($t = 25.5^o$); $-65^o$ (dioxane)<br>b) $+84.1^o$ (c 0.7, MeOH) ($t = 25^o$); $+62^o$ (dioxane) | M 20<br>M 20 |
| $C_{21}H_{19}O_2As$ | See $C_{12}H_{17}O_2As$<br>$R = CH_3CH_2$ $R' = \underline{p}\text{-}C_6H_5C_6H_4$ | Quinine<br>in $EtOH\text{-}CCl_4$ | a) $+12.84^o$ (c 1.5213, $CHCl_3$) ($t = 20^o$)<br>b) $-11.30^o$ (c 1.2145, $CHCl_3$) ($t = 20^o$) | M 10 |
| $C_{23}H_{17}O_2As$ | 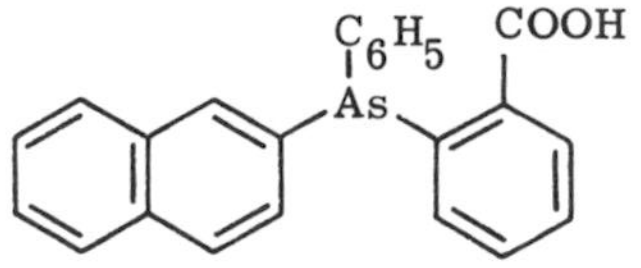 [structure: $C_6H_5$, As, COOH] | Quinine<br>in MeOH | a) $+6.04^o$ (c 0.7119, $C_6H_6$) ($t = 20^o$)<br>b) $-4.93^o$ (c 1.3871, $C_6H_6$) ($t = 20^o$) | M 9 |
| $C_{24}H_{28}As_2Br_2$ | 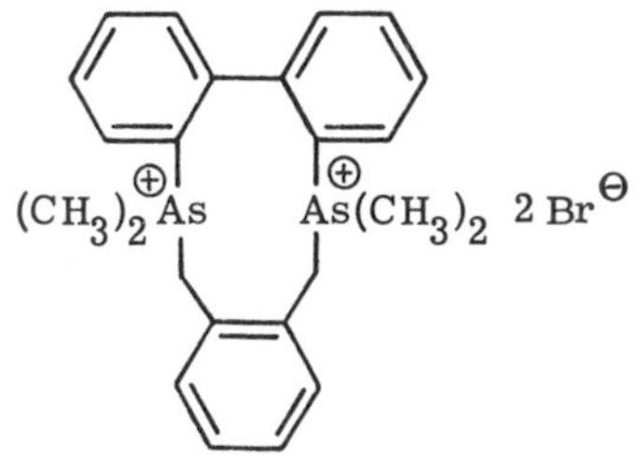 [structure: $(CH_3)_2\overset{\oplus}{A}s$, $\overset{\oplus}{A}s(CH_3)_2$ $2\,Br^{\ominus}$] | Silver (+)-3-bromocamphor-9-sulfonate in $H_2O$<br>Silver (-)-3-bromocamphor-9-sulfonate in $H_2O$ | a) $\alpha_D^{20-23}$ $+1.495^o$ (1 2 dm, c 0.889 %, $H_2O$)<br>b) $\alpha_D^{20-23}$ $-0.970^o$ (1 2 dm, c 0.589 %, $H_2O$) | M 15 |
| $C_{25}H_{19}O_2As$ | 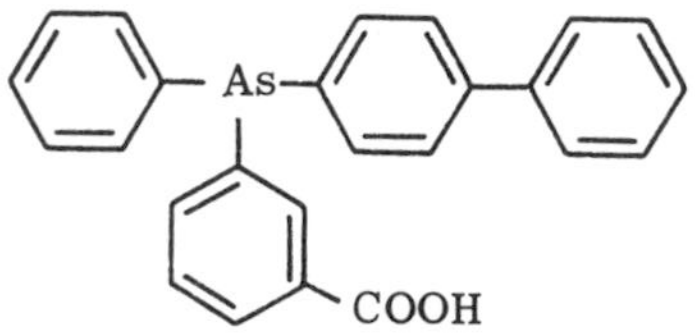 [structure: As, COOH] | (-)-$\alpha$-Methylbenzylamine<br>(+)-Amphetamine | a) $+4^o$ (dioxane); $[\phi]_{600}^{31}$ $+16^o$ (c 2.0, dioxane)<br>b) $-6^o$ (dioxane); $[\phi]_{600}^{26}$ $-24^o$ (c 2.1, dioxane) | M 20<br>M 20 |
| $C_{30}H_{34}As_2I_2$ | 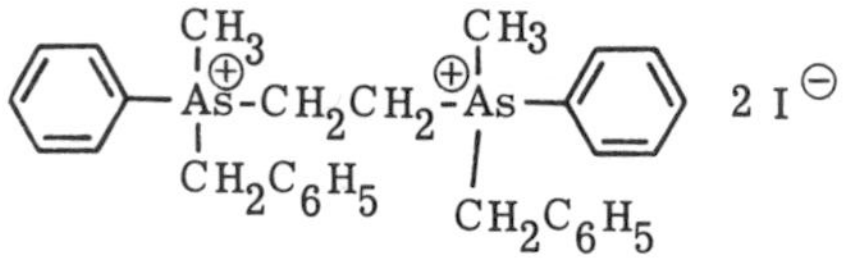 [structure: $CH_3$, $\overset{\oplus}{A}s\text{-}CH_2CH_2\text{-}\overset{\oplus}{A}s$, $CH_3$, $CH_2C_6H_5$, $CH_2C_6H_5$, $2\,I^{\ominus}$] | Silver (-)-O,O'-dibenzoyl-tartrate in $H_2O\text{-}MeOH$ | a) $+5.22^o$ (c 4.003, acetone) ($t = 20^o$)<br>b) $-7.74^o$ (c 4.006, acetone) ($t = 20^o$)<br>both on the 2 $PF_6^-$ salts | M 7 |

TABLE 12 (continued)

| Formula | Substance resolved | Resolving agent and solvent | Rotation $[\alpha]_D$ of resolved compound | Reference |
|---|---|---|---|---|
| | **Boron Compounds** | | | |
| $C_8H_{11}O_2B$ | $C_6H_5$-CH($CH_3$)-B(OH)-OH | Bisdemethylbrucine sulfate + $NaHCO_3$ on the dibutyl ester of the acid | a) - 23.9° ($C_6H_6$) (t = 18-23°) and<br>- 10.4° (neat) (t = 23°) on the dibutyl ester<br>b) + 4 to + 5° (neat) (t = 23°) on the dibutyl ester | M 21<br>M 22<br><br>M 22 |
| $C_9H_{17}N_2BBrF_6P$ | [$CH_3$-$C_5H_4$N-B(H)(X)-N($CH_3$)$_3$]$^{\oplus}$ $PF_6^{\ominus}$ X = Br | (-)-KAs($C_6H_4O_2$)$_3$, an anionic catechol complex of As(V), in $H_2O$-acetone on the bromide salt | a) - 60.9° (c 0.025 M, acetone) (t = 25°)<br>b) + 35.1° (c 0.025 M, acetone) (t = 25°) | M 23 |
| $C_9H_{17}N_2BClF_6P$ | X = Cl | | - 41.2° (c 0.025 M, acetone) (t = 25°) | M 23 |
| $C_{16}H_{27}O_2B$ | See $C_8H_{11}O_2B$ | | | |
| | **Chromium Compounds** | | | |
| $C_{10}H_5O_5CrF$ | (R'-$C_6H_4$-R)Cr(CO)$_3$ R = COOH, R' = 2-F | Quinidine<br>Quinine<br>both in $CH_3CN$ | a) + 71° (c 0.83, $CHCl_3$) (t = 25°)<br>b) - 71° (c 0.81, $CHCl_3$) (t = 22°) | M 24<br>M 24 |
| | R = COOH, R' = 3-F | Quinine in $CH_3CN$ | - 94° (c 0.73, $CHCl_3$) (t = 25°) | M 24 |
| $C_{10}H_9NO_3Cr$ | See $C_{10}H_5O_5CrF$<br>R = $NH_2$, R' = 2-$CH_3$ | (+)-Camphor-10-sulfonic acid in EtOAc | a) + 15.2 ± 0.6° (c 1.0, EtOAc) (t = 22°)<br>b) - 13.6 ± 0.8° (c 0.7, EtOAc) (t = 22°) | M 25 |
| | R = $NH_2$, R' = 3-$CH_3$ | (+)-Camphor-10-sulfonic acid in EtOAc-acetone (1:1) | a) + 5.5 ± 0.4° (c 1.2, EtOAc) (t = 23°)<br>b) - 5.2 ± 0.7° (c 0.7, EtOAc) (t = 23°) | M 25 |

TABLE 12 (continued)

| Substance resolved | | Resolving agent and solvent | Rotation $[\alpha]_D$ of resolved compound | Reference |
|---|---|---|---|---|
| $C_{12}H_{10}O_5Cr$ | 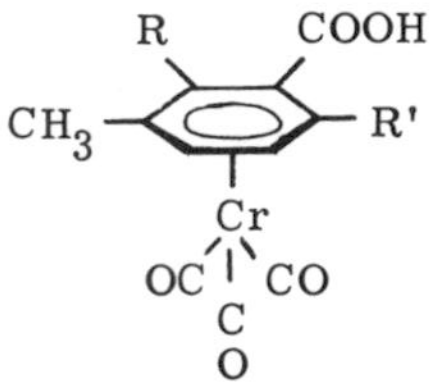 R = $CH_3$, R' = H | (-)-α-Methylbenzylamine on the carboxylic anhydride in py via iterative kinetic (partial) resolution | + 1.90° (EtOH) (t = 20°) | M 26 |
| | R = H, R' = $CH_3$ | (-)-α-Methylbenzylamine on the carboxylic anhydride in py via iterative kinetic resolution | + 58.0° (EtOH) (t = 20°) | M 26 |
| $C_{18}H_{19}NO_3Cr$ | 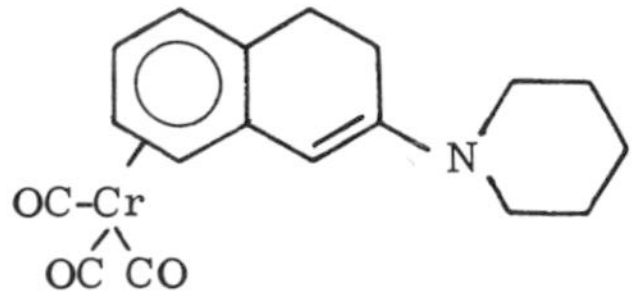 | (+)-Camphor-10-sulfonic acid in EtOAc | a) + 6.5 ± 0.3° (c 1.7, EtOAc) (t = 23°)<br>b) - 5.9 ± 0.3° (c 1.7, EtOAc) (t = 23°) | M 25 |
| Germanium Compounds | | | | |
| $C_{17}H_{16}GeO$ | 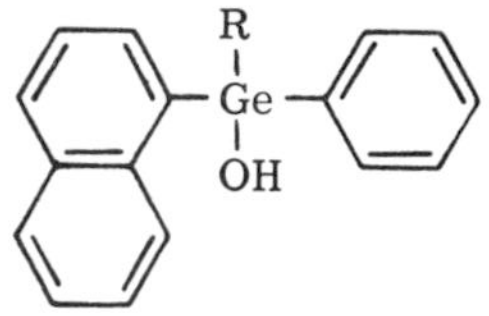 R = $CH_3$ | (-)-Menthol by exchange reaction with the $OCH_3$ derivative | a) + 26.7° (c 10.6, cyclohexane) (t = 25°)<br>b) - 25.5° (c 11.3, cyclohexane) (t = 25°)<br>both on the -GeH compound obtained upon $LiAlH_4$ reduction of the menthoxygermane | M 27 |
| $C_{18}H_{18}GeO$ | See $C_{17}H_{16}GeO$, R = $CH_3CH_2$ | Sodium (-)-methoxide on on the -GeBr derivative in $Et_2O$-$C_6H_6$ | a) + 23.6° (c 5.5, $C_6H_6$) (t = 25°)<br>b) - 9.53° (t = 20°)<br>both on the -GeH compound obtained upon $LiAlH_4$ reduction of the menthoxygermane | M 28, M 29<br>M 29 |

TABLE 12 (continued)

| Formula | Substance resolved | Resolving agent and solvent | Rotation $[\alpha]_D$ of resolved compound | Reference |
|---|---|---|---|---|
| | Iron Compounds* | | | |
| $C_8H_4O_8Fe$ | 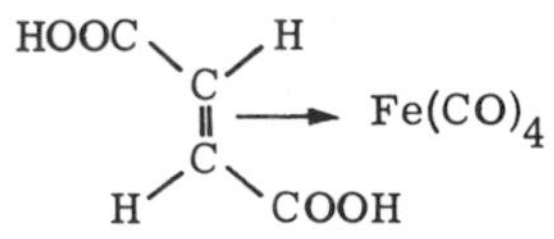 HOOC(H)C=C(H)COOH → $Fe(CO)_4$ | Brucine in acetone-$H_2O$ | a) - 593° (c 0.848, acetone) (t = 25°)<br>b) + 587° (c 0.921, acetone) (t = 25°) | M 30 |
| $C_{10}H_{11}NFe$ |  [structure: Fe, N, $CH_3$] | (-)-6,6'-Dinitro-2,2'-diphenic acid in EtOH | a) - 62° (c 0.04, cyclohexane) (t = 20°)<br>b) + 33° (cyclohexane) (t = 20°) | M 31 |
| $C_{12}H_{14}OFe$ | Fc-CH(OH)$CH_3$ | Partition thin layer chromatography with (+)-diethyl tartrate | + 2.5 ± 0.5° (c 0.35, EtOH) (t = 20°) | M 32 |
| $C_{13}H_{16}OFe$ | Fc-$CH_2$CH(OH)$CH_3$ | Strychnine on the HPhthal ester in $CHCl_3$-acetone | $\alpha_D^{26}$ + 3.99° (c 15.2, $CCl_4$) | M 33 |
| $C_{14}H_{14}OFe$ | [structure: Fe, =O] | Countercurrent distribution in cyclohexane-(+)-diethyl tartrate (Partial resolution) | + 234° after recrystallization from cyclohexane | M 34 |
| | | For resolution with (-)-menthydrazide see ref. M35. | | |

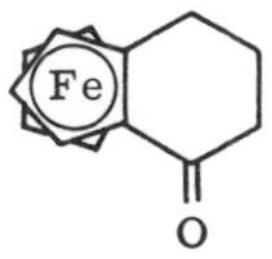

* In this section, the symbol Fc is used to represent the ferrocenyl group [structure: ferrocenyl, Fe]

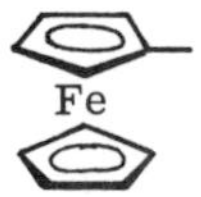

TABLE 12 (continued)

| Formula | Substance resolved | Resolving agent and solvent | Rotation $[\alpha]_D$ of resolved compound | Reference |
|---|---|---|---|---|
| $C_{14}H_{14}OFe$ | R = $CH_3$, R' = H | (-)-Menthydrazide in NaOAc-HOAc-EtOH (Both are partial resolutions) | - $44^o$ (MeOH) (t = $20^o$) | M 36 |
| | R = H, R' = $CH_3$ | | - $6.2^o$ (MeOH) (t = $20^o$) | M 36 |
| $C_{14}H_{14}O_2Fe$ | R = COOH, R' = H | (-)-$\alpha$-Methylbenzylamine in $Et_2O$ | + $20.3^o$ (EtOH) (t = $20^o$) | M 36 |
| | R = H, R' = COOH | (-)-$\alpha$-Methylbenzylamine in $Et_2O$ | + $155^o$ (EtOH) (t = $20^o$) | M 36 |
| $C_{14}H_{19}NFe$ | Fc-CHCH$_3$ with N(CH$_3$)$_2$ on the CH | (+)-Tartaric acid in MeOH | a) - $14.1^o$ (c 1.6, EtOH) (t = $25^o$)<br>b) + $14.0^o$ (EtOH) (t = $25^o$) | M 37 |
| $C_{14}H_{21}NFeSi$ | 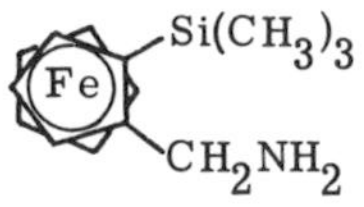 Fc with $Si(CH_3)_3$ and $CH_2NH_2$ | (R)-O,O'-Di-$\alpha$-naphthoyl-tartaric acid in MeOH | - $58.5^o$ (c 2.2, EtOH) (t = $20^o$) | M 38 |
| $C_{15}H_{16}O_3Fe$ | Fc-C(=O)CH$_2$CH(CH$_3$)COOH | (-)-$\alpha$-Methylbenzylamine | a) - 104 ± $3^o$ (c 2.1, $CHCl_3$)<br>b) + 105 ± $3^o$ (c 2.2, $CHCl_3$) | M 39 |
| $C_{15}H_{16}O_3Fe$ | Fc-C(=O)CH(CH$_3$)CH$_2$COOH | Brucine | a) + 173 ± $4^o$ (c 2.4, $CHCl_3$)<br>b) - 175 ± $4^o$ (c 2.2, $CHCl_3$) | M 39 |
| $C_{16}H_{20}O_2Fe$ | FcCH$_2$CH(CH(CH$_3$)$_2$)COOH | $\alpha$-Methylbenzylamine | a) + 8 ± $1^o$ (c 5, $CHCl_3$) (t = $24^o$)<br>b) - 8 ± $1^o$ (c 5, $CHCl_3$) (t = $24^o$) | M 40 |
| $C_{17}H_{18}O_2Fe$ |  COOH | (-)-$\alpha$-Methylbenzylamine on the carboxylic anhydride in py; asymmetric destruction (Partial resolution) | - $15^o$ (EtOH) (t = $20^o$) | M 36 |

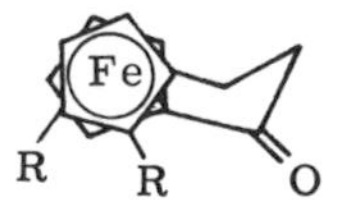

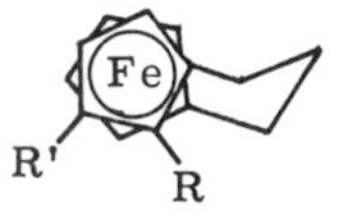

TABLE 12 (continued)

| Formula | Substance resolved | Resolving agent and solvent | Rotation $[\alpha]_D$ of resolved compound | Reference |
|---|---|---|---|---|
| $C_{19}H_{18}O_2Fe$ | $FcCH_2CH(C_6H_5)COOH$ | (-)-α-Methylbenzylamine in $Et_2O$ | $-1.48 \pm 0.10^o$ (c 1.4-1.6, EtOH) (t = $20^o$) | M 32 |
| $C_{22}H_{20}OFe_2$ | (biferrocenyl structure) R = CHO, R' = $CH_3$, R" = H | (-)-Menthydrazide in NaOAc-HOAc-EtOH (Partial resolution) | $+145^o$ (c ~ 0.1, $C_6H_6$) (t = $20^o$) | M 41 |
| $C_{24}H_{27}NFe_2$ | R = $CH_2N(CH_3)_2$, R' = $CH_3$, R = H | (-)-O,O'-Dibenzoyltartaric acid in $Et_2O$ | $+760^o$ (c ~ 0.1, $C_6H_6$) (t = $20^o$) | M 41 |
| $C_{26}H_{32}N_2Fe_2$ | R = R" = $CH_2N(CH_3)_2$, R' = H | (-)-O,O'-Dibenzoyltartaric acid in $Et_2O$ | $+696^o$ (c ~ 0.1, $C_6H_6$) (t = $20^o$) | M 41 |
| $C_{25}H_{20}O_2F_6FeP_2$ | $[(C_5H_5)Fe(CO)(CO)P(C_6H_5)_3]^{\oplus} PF_6^{\ominus}$ | Sodium (-)-mentholate in $Et_2O$ *via* ester formation | a) $-75^o$ ($C_5H_{12}$)<br>b) $+30^o$ ($C_5H_{12}$)<br>both on the menthyl esters | M 42 |
| | **Manganese Compounds** | | | |
| $C_{24}H_{20}NO_2F_6MnP_2$ | $[(C_5H_5)Mn(NO)(CO)P(C_6H_5)_3]^{\oplus} PF_6^{\ominus}$ | Sodium (-)-mentholate in THF *via* ester formation | a) $+460^o$ (c 0.1, $C_6H_6$) (t = $25^o$) (578 nm)<br>b) $-450^o$ (c 0.1, $C_6H_6$) (t = $25^o$) (578 nm) | M 43 |
| | **Mercury Compounds** | | | |
| $C_4H_9BrHg$ | $CH_3CH_2CH(HgBr)CH_3$ | (-)-Mandelic acid in MeOH on the -HgOH derivative | a) $-25.9^o$ (c 3, acetone) (t = $22^o$)<br>$-25.8^o$ (c 5, EtOH) (t = $22^o$)<br>b) $+18.8^o$ (c ~ 5, acetone) (t = $20^o$) | M 44, M 45<br>M 45<br>M 44, M 45 |

TABLE 12 (continued)

| | Substance resolved | Resolving agent and solvent | Rotation $[\alpha]_D$ of resolved compound | Reference |
|---|---|---|---|---|
| $C_6H_{13}ClHg$ | $CH_3(CH_2)_3CH(HgCl)CH_3$ | (-)-Mandelic acid<br>(+)-Mandelic acid<br>both in THF on the -HgBr derivatives | a) $-36.0 \pm 0.2^o$ (c 3, acetone) ($t = 20.5^o$)<br>b) $+29.5 \pm 0.2^o$ ($t = 21^o$) | M 46 |
| $C_7H_{15}BrHg$ | $(CH_3)_2CHCH_2CH_2CH(HgBr)CH_3$ | Mercury monoethyl (+)-tartrate<br>on the $R_2Hg$ derivative | a) $-41.6^o$ ($t = 18^o$)<br>b) $+25^o$ ($t = 18^o$) | M 47 |
| $C_8H_9BrHg$ | $C_6H_5CH(HgBr)CH_3$ | (-)-Mandelic acid<br>in THF-heptane<br>on the -HgOH derivative<br>(Partial resolution) | $+5.8 \pm 0.25^o$ (c 4, THF) | M 48 |
| | Phosphorus Compounds | | | |
| $CH_5O_2P$ | $CH_3-P(=O)(OH)-OH$ | (-)-Menthol<br>via ester formation<br>Ester Recr.: hexane | $-96.6^o$ (c 1.0, $C_6H_6$) ($t = 25^o$)<br>on the menthyl ester | M 49 |
| | O-Alkyl Alkylphosphonothioic Acids<br>$R-P(=S)(OR')-OH$ | | | |
| $C_2H_7O_2PS$ | $R = CH_3$ $R' = CH_3$ | (+)-$\alpha$-Methylbenzylamine<br>(-)-$\alpha$-Methylbenzylamine<br>both in $Et_2O$ | a) $+6.37^o$ (c 5, MeOH) ($t = 25^o$)<br>b) $-6.35^o$ (c 5, MeOH) ($t = 25^o$)<br>both on the dicyclohexylamine salts | M 50 |

TABLE 12 (continued)

| Formula | Substance resolved | Resolving agent and solvent | Rotation $[\alpha]_D$ of resolved compound | Reference |
|---|---|---|---|---|
| | O-Alkyl Alkylphosphonothioic Acids (continued) | | | |
| $C_3H_9O_2PS$ | R = $CH_3$ R' = $CH_3CH_2$ | (+)-α-Methylbenzylamine<br>(-)-α-Methylbenzylamine<br>both in $Et_2O$ | a) + 8.47° (c 5, MeOH) (t = 25°)<br>b) - 8.47° (c 5, MeOH) (t = 25°)<br>both on the dicyclohexylamine salts | M 50 |
| $C_3H_9O_2PS$ | R = $CH_3CH_2$ R' = $CH_3$ | Quinine<br>in acetone | a) - 7.0° (MeOH) (t = 26-28°)<br>b) + 4.1° (MeOH) (t = 26-28°)<br>both on the dicyclohexylamine salts | M 51 |
| $C_4H_{11}O_2PS$ | R = $CH_3$ R' = $(CH_3)_2CH$ | (+)-α-Methylbenzylamine<br>(-)-α-Methylbenzylamine<br>both in $Et_2O$<br>Also, resolution with quinine | a) + 7.76° (c 5, MeOH) (t = 25°)<br>b) - 7.72° (c 5, MeOH) (t = 25°)<br>both on the dicyclohexylamine salts | M 50<br>M 50, M 51 |
| $C_4H_{11}O_2PS$ | R = $CH_3CH_2$ R' = $CH_3CH_2$ | Quinine<br>in acetone | a) - 7.11 ± 0.23° (c 2.150, MeOH)<br>b) + 6.85 ± 0.25° (c 3.230, MeOH) (t = 25°)<br>both on the dicyclohexylamine salts | M 52 |
| $C_5H_{13}O_2PS$ | R = $CH_3$ R' = $CH_3(CH_2)_3$ | (+)-α-Methylbenzylamine<br>(-)-α-Methylbenzylamine<br>both in $Et_2O$ | a) + 7.25° (c 5, MeOH) (t = 25°)<br>b) - 7.21° (c 5, MeOH) (t = 25°)<br>both on the dicyclohexylamine salts | M 50 |
| $C_5H_{13}O_2PS$ | R = $(CH_3)_2CH$ R' = $CH_3CH_2$ | Quinine<br>in acetone | a) - 7.0° (MeOH) (t = 26-28°)<br>b) + 6.5° (MeOH) (t = 26-28°)<br>both on the dicyclohexylamine salts | M 51 |
| $C_7H_8NO_4PS$ | R = $CH_3$ R' = $O_2N-C_6H_4-$ | Strychnine<br>Recr.: MeOH | a) + 8.2 ± 0.2° ($C_6H_6$)<br>b) + 8.0 ± 0.2° ($C_6H_6$)<br>both on the S-methyl esters | M 53 |

TABLE 12 (continued)

| Formula | Substance resolved | Resolving agent and solvent | Rotation $[\alpha]_D$ of resolved compound | Reference |
|---|---|---|---|---|
| $C_3H_7O_4P$ | epoxide: H, $CH_3$ / O / H, $P(O)(OH)_2$ | (-)-Menthol via pyridinium menthyl ester | - 14° (c 5%, $H_2O$) (t = 28°) (405 nm) on the disodium salt | M 54 |
| | | Resolutions with (+)-α-methylbenzylamine (M 55), with calcium (+)-gluconate and with calcium (-)-lactate (M 56). | | |
| $C_3H_8O_4ClP$ | $CH_3CH(OH)$-$CH(Cl)PO_3H_2$ threo-isomer | (-)-α-Methylbenzylamine | + 19.03° (c 3.415, $H_2O$) (405 nm) | M 57 |
| $C_3H_9O_2P$ | $CH_3CH_2P(=O)(CH_3)$-OH | (-)-Menthol via ester formation in py-$C_6H_6$ on -POCl derivative Partial separation | - 6.7° (MeOH) on ethylmethylphenylphosphine oxide by reaction of menthylester with $C_6H_5MgBr$ | M 58 |
| $C_3H_9O_2PS$ | See above, under O-alkyl alkylphosphonothioic acids | | | |
| $C_4H_{11}OPSSe$ | $CH_3CH_2$-P(=Se)(-SH)$OCH_2CH_3$ | Quinine in $C_6H_6$ (Partial resolution) | a) - 6.7° (neat) (t = 20°)<br>b) + 4.9° (neat) (t = 20°) | M 59 |
| $C_4H_{11}O_2PS$<br>$C_5H_{13}O_2PS$ | See above, under O-alkyl alkylphosphonothioic acids | | | |
| $C_5H_{13}O_3PS$ | $CH_3CH_2O$-P(=S)(-OH)OR R = $(CH_3)_2CH$ | Quinine in acetone | a) + 2.1° (neat)<br>b) - 2.0° (neat)<br>both on the S-methyl esters | M 60 |
| $C_6H_{15}O_3PS$ | R = $CH_3(CH_2)_3$ | Cinchonidine in acetone | a) - 1.1° (neat)<br>b) + 0.8° (neat)<br>both on the S-methyl esters | M 60 |

TABLE 12 (continued)

| Formula | Substance resolved | Resolving agent and solvent | Rotation $[\alpha]_D$ of resolved compound | Reference |
|---|---|---|---|---|
| $C_6H_7O_2P$ | $C_6H_5$-P(=O)(OH)H | (-)-Menthol<br>Ester Recr.: hexane<br>(Partial separation) | - 21.0° ($C_6H_6$) (t = 22°)<br>and - 89.6° ($C_6H_6$) (t = 22°) on two menthyl ester mixtures. Esters converted to phosphine oxides | M 61 |
| $C_7H_8NO_4PS$ | See above, under O-alkyl alkylphosphonothioic acids | | | |
| $C_7H_8NO_5PS$ | $O_2N$-$C_6H_4$-O-P(=S)(OH)$OCH_3$ | Strychnine methiodide on the silver salt of the acid in MeOH | a) - 35.6 ± 0.1° (c 1.25, MeOH) (t = 21°)<br>b) + 35.6 ± 0.1° (c 1.25, MeOH) (t = 21°)<br>both on the S-methyl esters | M 62 |
| $C_7H_9OPS$ | $C_6H_5$-P(=O)(SH)$CH_3$ | Quinine<br>Brucine<br>both in acetone | a) - 9.22° (c 1.75, MeOH) (t = 25°) (578 nm)<br>b) + 9.25° (t = 25°) (578 nm)<br>both on the dicyclohexylamine salts | M 63 |
| $C_7H_9O_2P$ | $C_6H_5$-P(=O)(OH)$CH_3$ | (-)-Menthol<br>via ester formation<br>on -POCl derivative | - 93.8° (c 1.45, $CHCl_3$) (t = 25°) on one of the menthyl esters; reaction with $CH_3CH_2MgBr$ yields ethylmethylphenylphosphine oxide $[\alpha]_D^{22}$ + 21.1° ($H_2O$) | M 64<br>M 65 |
| | | Also with (-)-cholesterol | | M 64 |
| $C_9H_{13}OP$ | $C_6H_5$-P(=O)($CH_3$)$CH_2CH_3$<br>See also $C_3H_9O_2P$ and $C_7H_9O_2P$ above | (+)-3-Bromocamphor-9-sulfonic acid. Recr.: EtOAc | + 22.6° (c 1, $H_2O$) (t = 20°) | M 66 |

TABLE 12 (continued)

| | Substance resolved | Resolving agent and solvent | Rotation $[\alpha]_D$ of resolved compound | Reference |
|---|---|---|---|---|
| $C_9H_{13}O_2P$<br><br>$C_{10}H_{15}OP$ | O, P-OH, $CH_2CH_2CH_3$ | (-)-Menthol<br>via ester formation<br>on -POCl derivative | $-14^o$ ($C_6H_6$) (t = $23^o$)<br>and $-81.3^o$ ($C_6H_6$) (t = $23^o$) on the two menthyl esters; reaction with $CH_3MgCl$ yields methylphenyl-n-propylphosphine oxide $[\alpha]_D^{23}$ $+17.3^o$ (MeOH) | M 65<br><br>M 65 |
| $C_{11}H_{10}O_3BrPS$ | 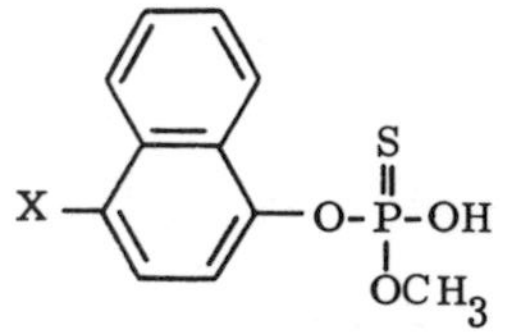 X = Br | (-)-Ephedrine · HCl | $+43.4^o$ (c 2.44, $CHCl_3$)<br>on the S-methyl esters | M 67 |
| $C_{11}H_{11}O_3PS$ | X, S, O-P-OH, $OCH_3$; X = H | (-)-Ephedrine · HCl<br>on the $(CH_3)_4N^+$ salt of the acid in $H_2O$ | a) $+53.4^o$ (c 6.34, $CHCl_3$) (t = $24^o$)<br>b) $-51^o$ (c 1.75, $CHCl_3$) (t = $24^o$)<br>both on the S-methyl esters | M 67 |
| $C_{11}H_{17}P$ | 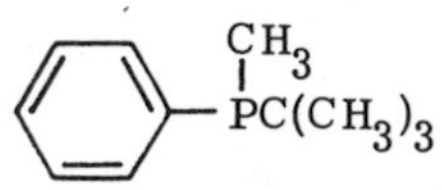 $CH_3$, $PC(CH_3)_3$ | via formation and crystallization of the complex trans-dichloro [(+)-deoxyephedrine] (methylphenyl-t-butylphosphine)platinum (II) | $+10^o$ (578 nm)<br>on the phosphine oxide | M 68 |
| $C_{12}H_{13}O_3P$ | 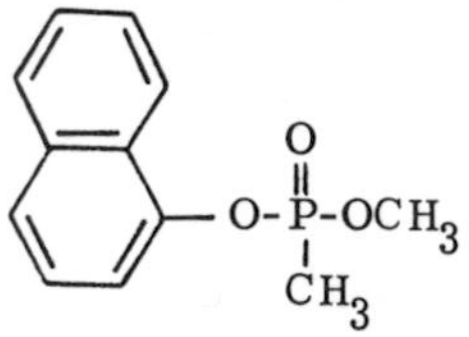 O, O-P-$OCH_3$, $CH_3$ | (-)-$\alpha$-(2,4,5,7-Tetranitrofluorenylideneaminoöxy)-propionic acid in propionic acid-Pet E (1:1)<br>(Partial resolution) | $+44^o$ (c 1.5, dioxane) (t = $20^o$) | M 69 |
| $C_{12}H_{19}P$ | See $C_{15}H_{23}NIP$ | | | |
| $C_{13}H_{13}O_2P$<br><br>$C_{13}H_{13}OP$ | O, P-OH, $CH_2C_6H_5$ | (-)-Menthol<br>Ester Recr.: hexane | $-65.4^o$ (c 0.0314, $CHCl_3$) (t = $23^o$) on one of the menthyl esters; reduction with $LiAlH_4$ yields benzylphenylphosphine oxide $[\alpha]_D^{23}$ $-0.52^o$ (c 0.0735, MeOH) | M 70<br><br>M 70 |

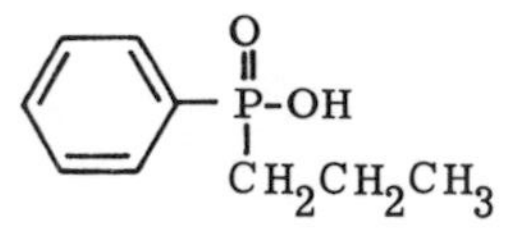

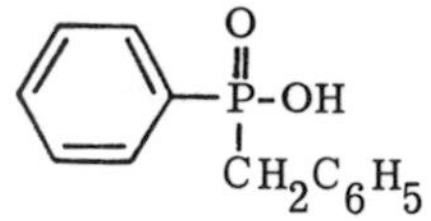

TABLE 12 (continued)

| | Substance resolved | Resolving agent and solvent | Rotation $[\alpha]_D$ of resolved compound | Reference |
|---|---|---|---|---|
| $C_{14}H_{20}N_3PS$ | 2-methylimidazol-1-yl–$P(=S)(C_6H_5)N(CH_2CH_3)_2$ | Methyl (+)-camphor-3-sulfonate in $Et_2O$ via formation and crystallization ($C_6H_6$) of the $CH_3\!-\!\overset{\oplus}{N}=$ salt | $+ 7.9^o$ (c 1.3, $CHCl_3$) ($t = 25^o$) | M 71 |
| $C_{15}H_{22}NOP$ | 3-methyl-3-phospholene 1-oxide, P–$C_6H_4N(CH_2CH_3)_2$-p | (+)-Camphor-9-sulfonic acid in $CH_2Cl_2$ | $- 69.4^o$ (c 0.6 M, $C_6H_6$) ($t = 20^o$) | M 72 |
| $C_{15}H_{23}NIP$ | $CH_3\text{-}\overset{\oplus}{P}(CH_2CH_2CN)(CH_2CH_2CH_2CH_3)(CH_2C_6H_5)$ $I^{\ominus}$ | Silver (-)-O,O'-dibenzoyl hydrogen tartrate<br>Silver (+)-O,O'-dibenzoyl hydrogen tartrate<br>Recr.: PrOH | a) $+ 24 \pm 2^o$ (c 2.9, $C_6H_6$) ($t = 27^o$)<br>b) $- 28 \pm 2^o$ (c 2.9, $C_6H_6$) ($t = 27^o$) both on benzylbutylmethylphosphine (from salts + $NaOCH_3$) | M 73<br>M 73 |
| $C_{16}H_{13}O_2P$<br>$C_{16}H_{13}OP$ | 2-naphthyl–P(=O)(OH)–phenyl | (-)-Menthol via ester formation on -POCl derivative | $- 0.59^o$ (c 0.476, $CHCl_3$) ($t = 25^o$) on $\beta$-naphthylphenylphosphine oxide (ester + $LiAlH_4$) | M 74 |
| $C_{16}H_{20}XP$ | $C_6H_5\text{-}\overset{\oplus}{P}(CH_3)(CH_2C_6H_5)(CH_2CH_3)$ $X^{\ominus}$<br>X = I<br><br>X = Br | Silver (-)-O,O'-dibenzoyl hydrogen tartrate in MeOH<br>Silver (+)-O,O'-dibenzoyl hydrogen tartrate in MeOH<br>Analogous resolutions | a) $+ 25.0 \pm 1.0^o$ (c 2.815, MeOH) ($t = 25^o$)<br>b) $- 24.5 \pm 1.0^o$ (c 2.625, MeOH) ($t = 25^o$)<br>$+ 23.07$ (c 1.517, $CHCl_3$) ($t = 20^o$) | M 75<br>M 75<br>M 76, M 77 |
| $C_{16}H_{20}IP$ | $C_6H_5\text{-}\overset{\oplus}{P}(CH_3)(CH_2CH_3)\text{-}C_6H_4\text{-}CH_3$ $I^{\ominus}$ | Silver (-)- and (+)-O,O'-dibenzoyl hydrogen tartrate in MeOH | a) $0.00 \pm 1.00^o$ (MeOH) ($t = 25^o$)<br>b) $0.00 \pm 1.00^o$ (MeOH) ($t = 25^o$) | M 75 |

TABLE 12 (continued)

| | Substance resolved | Resolving agent and solvent | Rotation $[\alpha]_D$ of resolved compound | Reference |
|---|---|---|---|---|
| $C_{17}H_{20}BrP$ | 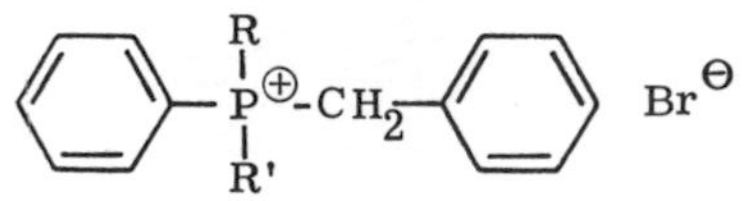 R = $CH_3$ R' = $CH_2 = CHCH_2$ | Silver (-)-O, O'-dibenzoyl hydrogen tartrate | + 15.7 ± 1° (c 1.908, MeOH) | M 77 |
| $C_{17}H_{22}BrP$ | R = $CH_3$ R' = $CH_3CH_2CH_2$ | | + 36.8 ± 1° (c 1.507, MeOH) | M 77 |
| $C_{17}H_{22}XP$ | 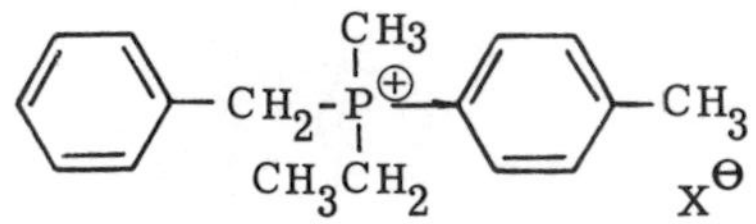 | Silver (-)-O, O'-dibenzoyl hydrogen tartrate | X = Cl + 20.87° (c 0.995, MeOH) (t = 20°)<br>X = Br + 24.42° (c 2.093, $CHCl_3$) (t = 20°)<br>X = I + 22.20° (c 1.1178, $CHCl_3$) (t = 20°) | M 76 |
| $C_{20}H_{19}N_2OP$<br><br>$C_{20}H_{19}N_2P$ | 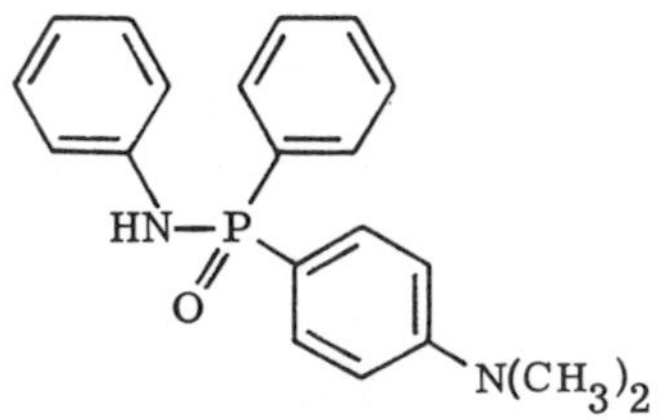 | (+)-Camphor-10-sulfonic acid<br>(-)-Camphor-10-sulfonic acid<br>both in EtOH-EtOAc (1:2) | a) + 152.6 ± 1.2° (EtOH) (t = 22°)<br>b) - 152.8 ± 1.2° (EtOH) (t = 21°)<br>Reduction with $LiAlH_4$ yields the azaphosphines $[\alpha]_D^{21}$ - 128.0° (c 0.2265, EtOAc) and + 141.6° ± 2.3 (c 0.219, EtOAc), resp. | M 78<br><br>M 78 |
| $C_{23}H_{19}P$ | 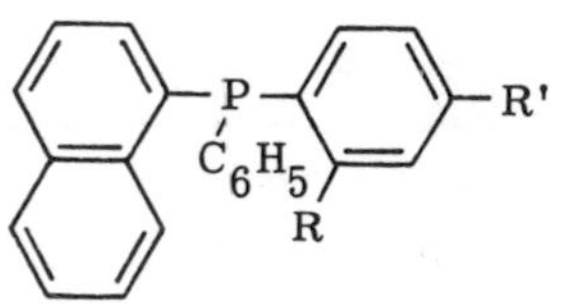 R = $CH_3$, R' = H | Di-μ-chloro-bis[(-)-N, N, α-trimethylbenzylamine-2C, N]dipalladium(II) in $C_6H_6$ | a) - 1.04° (c 7.72, $CH_2Cl_2$) (t = 26°)<br>b) + 2.39° (c 1.78, $CH_2Cl_2$) (t = 26°) | M 79 |
| $C_{24}H_{21}OP$ | R = H, R' = $OCH_2CH_3$ | | a) + 0.91° (c 6.06, $CH_2Cl_2$) (t = 26°)<br>b) - 2.57° (c 6.80, $CH_2Cl_2$) (t = 26°) | M 79 |
| $C_{28}H_{21}P$ | 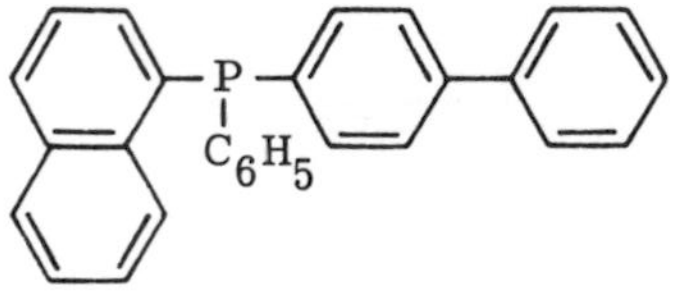 | (+)-Camphor-10-sulfonic acid on the $-\overset{\oplus}{P}-CH_2OH$ derivative in $C_6H_6$ | a) - 5.2° (c 3.38, $CH_2Cl_2$) (t = 24°) (578 nm)<br>b) + 8.7° (c 3.11, $CH_2Cl_2$) (t = 25.4°) (578 nm) | M 80 |

TABLE 12 (continued)

| | Substance resolved | Resolving agent and solvent | Rotation $[\alpha]_D$ of resolved compound | Reference |
|---|---|---|---|---|
| $C_{30}H_{26}OBrP$ | See $C_{17}H_{20}BrP$<br>R = α-naphthyl R' = p-anisyl | Silver O, O'-dibenzoyl hydrogen tartrate<br>Recr.: tert-BuOH | + 11.0° (c 0.137, $C_6H_5NO_2$-DMF, 1:2) (578 nm) | M 81 |
| $C_{30}H_{26}OClP$ | Corresponding phosphonium chloride | | + 15.7° (c 1.4992, MeOH) | M 81 |
| $C_{30}H_{34}Br_2P_2$ | $2\ Br^{\ominus}$ 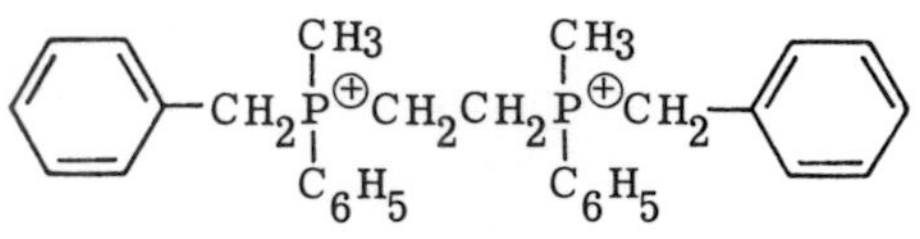<br>$C_6H_5CH_2P^{\oplus}(CH_3)(C_6H_5)CH_2CH_2P^{\oplus}(CH_3)(C_6H_5)CH_2C_6H_5$ | Silver (-)-O, O'-dibenzoyl hydrogen tartrate | a) + 64.05° (MeOH)<br>b) - 65.4° (MeOH)<br>both on the $OSO_3CH_3^-$ salts | M 82 |
| $C_{36}H_{24}KP$ | [P(biphenylyl)₃ anion, R] $^{\ominus}$ $K^{\oplus}$<br>R = H | Brucine methiodide in MeOH on the $Li^+$ salt | a) - 1930 ± 20° (acetone) (t = 24.5°) (578 nm)<br>b) + 1930 ± 20° (acetone) (t = 24.5°) (578 nm) | M 83 |
| $C_{37}H_{26}KP$ | R = $CH_3$ | Brucine methiodide in MeOH-acetone-$H_2O$ | a) - 1870 ± 30° (acetone) (t = 24°) (578 nm)<br>b) + 1870 ± 30° (acetone) (t = 24°) (578 nm) | M 84 |
| $C_{37}H_{27}P$ 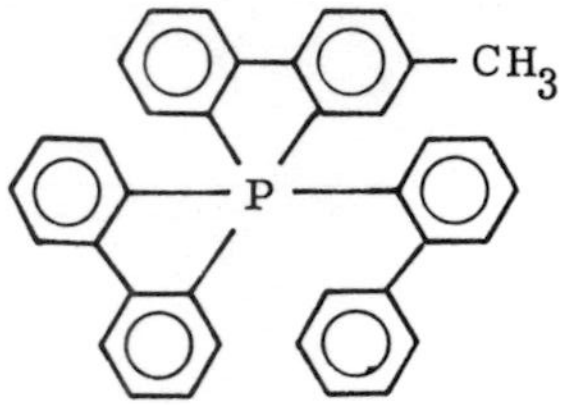 | [structure: P with biphenylyl groups, $CH_3$] | Spontaneous resolution by crystallization from acetone in presence of (-)-phosphorane seeds<br>Partial resolution | - 52.9° (c 1.22, THF) (t = 24°) (578 nm) | M 84 |

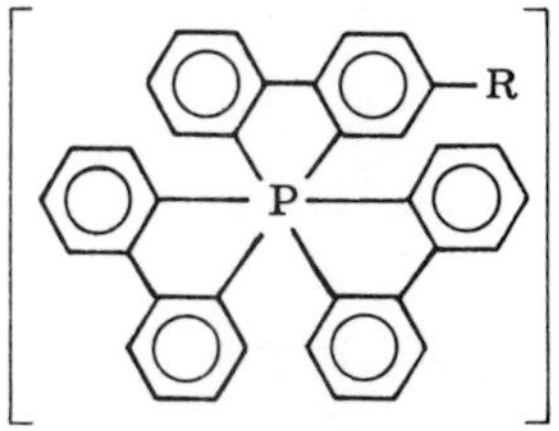

TABLE 12 (continued)

| | Substance resolved | Resolving agent and solvent | Rotation $[\alpha]_D$ of resolved compound | Reference |
|---|---|---|---|---|
| $C_{42}H_{36}KP$ | $[(CH_3-C_6H_4-C_6H_4-CH_3)_3P]^{\ominus} K^{\oplus}$ | Brucine methiodide in MeOH-acetone | - 1450 ± 30° (acetone) (t = 24°) (578 nm) | M 85 |
| | Ruthenium Compounds | | | |
| $C_{12}H_{12}O_2Ru$ | (Ru) COOH, $CH_3$ | (-)-α-Methylbenzylamine in EtOH | a) + 47.3° (EtOH) (t = 20°)<br>b) - 38.5° (EtOH) (t = 20°) | M 86 |
| $C_{14}H_{14}ORu$ | (Ru), O | (-)-Menthydrazide in NaOAc-HOAc-EtOH | + 375° (EtOH) (t = 20°) | M 86 |
| | Selenium Compounds | | | |
| $C_4H_{11}OPSSe$ | See Phosphorus Compounds | | | |
| $C_7H_{10}O_4Se$ | HOOC, Se, COOH | Quinine in EtOH-aq. $NaHCO_3$<br>Brucine in $H_2O$ | a) + 112.3° (c 1.730, acetone) (t = 25°)<br>b) - 112.0° (c 1.689, acetone) (t = 25°) | M 87<br>M 87 |
| $C_{13}H_{12}O_2Se$ | Se-CH($CH_3$)COOH | (-)-α-Methylbenzylamine<br>(-)-α-Methylbenzylamine<br>both in EtOAc | a) + 98.0° (c 1.010, EtOH) (t = 25°)<br>b) - 98.3° (c 1.017, EtOH) (t = 25°) | M 88 |
| | Silicon Compounds | | | |
| $C_{14}H_{21}NFeSi$ | See Iron Compounds | | | |

TABLE 12 (continued)

| | Substance resolved | Resolving agent and solvent | Rotation $[\alpha]_D$ of resolved compound | Reference |
|---|---|---|---|---|
| $C_{16}H_{14}OSi$ | (α-naphthyl)(phenyl)Si(R)(OH) | (-)-Menthol via menthoxysilane formed from -SiCl derivative | Reaction of one of the menthoxysilanes with $CH_3MgI$ yields methyl-α-naphthylphenylsilane $[\alpha]_D$ - 32.9° ($C_5H_{12}$) | M 89 |
| $C_{16}H_{18}O_2Si$ | R-$C_6H_4$-Si($CH_3$)($CH_2CH_3$)-$C_6H_4$-COOH; R = H | Quinine in EtOH | a) + 2.75 ± 0.2° (c 2.8, $C_6H_6$) (t = 20°)<br>b) - 1.65 ± 0.2° (c 3.2, $C_6H_6$) (t = 20°) | M 90 |
| $C_{17}H_{16}OSi$<br><br>$C_{17}H_{16}Si$ | See $C_{16}H_{14}OSi$<br>R = $CH_3$ | (-)-Menthol + KOH in toluene; via menthoxysilane formed from -$SiOCH_3$ derivative<br>Recr.: $C_5H_{12}$ | - 53.9° (c 11.32, cyclohexane) and -47° (c 1.40, cyclohexane) on the two menthoxysilanes. Reduction with $LiAlH_4$ yields methyl-α-naphthylphenylsilanes $[\alpha]_D$ + 33.7° (c 4.00, cyclohexane) and -32.8° (c 6.07, cyclohexane), resp. | M 91<br><br>M 91 |
| $C_{17}H_{20}O_3Si$ | See $C_{16}H_{18}O_2Si$<br>R = $OCH_3$ | Quinine in MeOH | + 1.2° (c 3.7, $C_6H_6$) (t = 25°) | M 92 |
| $C_{18}H_{16}OSi$<br><br>$C_{18}H_{16}Si$ | See $C_{16}H_{14}OSi$<br>R = $CH_2$=CH | (-)-Menthol + $NaOCH_3$ in toluene; via menthoxysilane formed from -$SiOCH_2CH_3$ derivative | - 43.5° and -34.0°, both (heptane) (t = 20°) on the two menthoxysilanes. Reduction with $LiAlH_4$ yields α-naphthylphenylvinylsilanes $[\alpha]_D^{20}$ + 13° and - 10.3°, both (heptane), resp. | M93, M94<br><br>M 93 |
| $C_{18}H_{18}OSi$ | (α-naphthyl)(phenyl)Si($CH_3$)($OCH_3$) | (+)-Quinidine-$LiAlH_4$ complex in $Et_2O$ via asymmetric destruction (Partial resolution) | - 0.93° (c 14.34, EtOH) (t = 20°) | M 95 |

TABLE 12 (continued)

| | Substance resolved | Resolving agent and solvent | Rotation $[\alpha]_D$ of resolved compound | Reference |
|---|---|---|---|---|
| $C_{25}H_{22}OSi$ | $C_6H_5$-CH(OH)-Si$(C_6H_5)_3$ | Brucine · 4 $H_2O$ on the H Phthal derivative in acetone-$H_2O$ | a) - 49.5 ± 1.1° (c 1.775, $CHCl_3$) (t = 22°)<br>b) + 48.6 ± 1.1° (c 1.915, $CHCl_3$) (t = 22°) | M 96 |
| $C_{25}H_{24}OSi_2$ | $(C_6H_5)_3$Si-Si($C_6H_5$)($CH_3$)-OH | (-)-Menthol<br>Menthoxysilane Recr.: $C_5H_{12}$ and hexane | + 20.0° (c 2.0, $C_5H_{12}$) and - 55.5° (c 1.0, $C_5H_{12}$) on the two menthoxysilanes. | M 97 |
| $C_{25}H_{24}Si_2$ | | | Reduction with $LiAlH_4$ yields the corresponding disilanes $[\alpha]_D$ - 14.3 (c 4.2, $C_5H_{12}$) and + 7.7° (c 4.0, $C_5H_{12}$), resp. | M 97 |
| | Sulfur Compounds | | | |
| $CH_3O_2Cl$ | $CH_3O$-S(=O)-Cl | (+)-Diphenacyl malate<br>Ester Recr.: acetone-$H_2O$ | - 50.2° (acetone) and - 4.4° (acetone) on the two diastereomeric methyl(diphenacyl malate) sulfites | M 98 |
| $C_4H_{11}NOS$ | $CH_3$S(=O)$CH_2CH_2CH_2$-$NH_2$ | (+)-3-Bromocamphor-9-sulfonic acid in EtOH | a) - 115° (c 0.735 %, EtOH) (t = 18.5°) | M 99 |
| | | (+)-O,O'-Dibenzyltartaric acid in MeOH | b) + 111° (t = 18.5°) | M 99 |
| $C_5H_{13}NOS$ | $CH_3$S(=O)$CH_2CH_2CH_2CH_2$-$NH_2$ | (+)-Tartaric acid<br>(-)-Tartaric acid<br>both in MeOH-EtOH (1:1) | a) + 125.5 ± 1° (c 0.797, MeOH) (t = 19°)<br>b) - 124.7 ± 1° (c 0.369, MeOH) (t = 19°) | M 100 |
| $C_7H_8O_2S$ | $CH_3$S(=O)-$C_6H_4$-OH ortho | 2,3,4,6-Tetra-O-acetylglucopyranosyl bromide via glucoside formation<br>Recr.: MeOH | a) + 273° ($H_2O$) (t = 20°)<br>b) - 271° ($H_2O$) (t = 20°) | M 101 |
| | para | | a) - 110.4° ($H_2O$) (t = 20°)<br>b) + 99.3° ($H_2O$) (t = 20°) | M 101 |

TABLE 12 (continued)

| Formula | Substance resolved | Resolving agent and solvent | Rotation $[\alpha]_D$ of resolved compound | Reference |
|---|---|---|---|---|
| $C_7H_9NOS$ | $C_6H_5-S(\rightarrow O)(\rightarrow NH)-CH_3$ | (+)-Camphor-10-sulfonic acid in acetone | a) + 29° (t = 20°)<br>b) - 29° (t = 20°)<br>both on the HCl salt | M 102 |
| $C_7H_{17}O_4ClS$ | $CH_3CH_2\overset{\oplus}{S}(CH_3)-C(CH_3)_3\ ClO_4^{\ominus}$ | *via* formation of the (-)-O,O'-dibenzoyl hydrogen tartrate | - 34.6° (c 1.5, MeOH) (t = 25°) | M 103 |
| $C_8H_{10}O_2S$ | $CH_3CH_2S(=O)-C_6H_4-OH$, *ortho* | 2,3,4,6-Tetra-O-acetylgluco-pyranosyl bromide *via* glucoside formation | a) - 282.3° ($H_2O$) (t = 20°)<br>b) + 270° ($H_2O$) (t = 20°) | M 101 |
| | *para* | Recr.: MeOH | + 148° ($H_2O$) (t = 20°) | M 101 |
| $C_9H_{12}OS$ | $CH_3CH_2S(=O)-C_6H_4-CH_3$ (*para*) | (-)-*trans*-Dichloro(ethylene)-(α-methylbenzylamine)platinum(II)<br>(+)-Enantiomer<br>both in $CH_2Cl_2$ | a) + 203.2° (c 0.60, acetone) (t = 25°)<br>b) - 203.6° (c 0.93, acetone) (t = 25°) | M 104<br>M 104 |
| $C_{10}H_{14}NO_4ClS$ | $R,R'-C_6H_3-CH_2\overset{\oplus}{S}(CH_3)CH_2CH_3\ ClO_4^{\ominus}$ (R at the para position, R' at the meta position)<br>R = $NO_2$, R' = H | *via* formation of the (-)-O,O'-dibenzoyl hydrogen tartrate | + 7.6° (c 0.9, MeOH) (t = 25°) | M 105 |
| $C_{10}H_{15}O_4ClS$ | R = R' = H | | - 8.20° (c 3.0, MeOH) (t = 25°) | M 105 |
| $C_{11}H_{15}O_4ClS$ | $C_6H_5-C(=O)CH_2\overset{\oplus}{S}(CH_3)CH_2CH_3\ ClO_4^{\ominus}$ | *via* formation of the (-)-O,O'-dibenzoyl hydrogen tartrate | - 10.7° (c 0.88, MeOH) (t = 25°) | M 105 |

TABLE 12 (continued)

| | Substance resolved | Resolving agent and solvent | Rotation $[\alpha]_D$ of resolved compound | Reference |
|---|---|---|---|---|
| $C_{11}H_{16}NO_7ClS$ | See $C_{10}H_{14}NO_4ClS$<br>$R = OCH_3$ $R' = NO_2$ | via formation of the (-)-O,O'-dibenzoyl hydrogen tartrate | $+10.6^o$ (c 2.93, acetone) ($t = 27^o$) | M 105 |
| $C_{11}H_{17}O_5ClS$ | $R = OCH_3$ $R' = H$ | | $-11.8^o$ (c 0.62, MeOH) ($t = 25^o$) | M 105 |
| $C_{13}H_{11}O_3ClS_2$ | 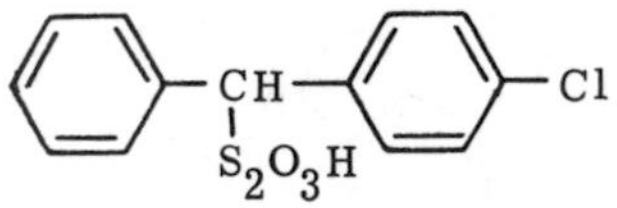 $CH$, $S_2O_3H$, $Cl$ | Quinine·HCl in $H_2O$ on the sodium salt | $-21.3^o$ (c 4.21, $CCl_4$) ($t = 25^o$) on the thiocyanate derivative $Ar_2CHSCN$ | M 106 |
| $C_{13}H_{23}O_4ClS$ | 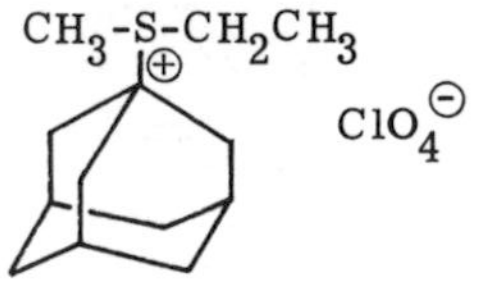 $CH_3$-S-$CH_2CH_3$ ⊕ $ClO_4^{\ominus}$ | (-)-Malic acid in $H_2O$-MeOH on the sulfonium hydroxide | $+5.93^o$ (c 2.5, MeOH) | M 107 |
| $C_{14}H_{10}NClS$ | See $C_{13}H_{11}O_3ClS_2$ above | | | |
| $C_{14}H_{12}S_2$ | 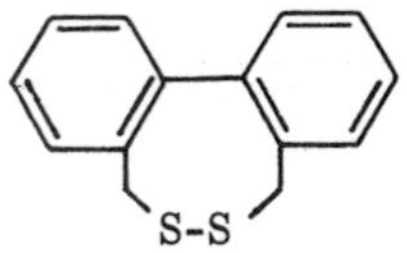 S-S | via chromatography on cellulose 2.5-acetate | a) $+498^o$ (c 0.62 %, $CHCl_3$) ($t = 20^o$)<br>b) $-498^o$ (c 0.62 %, $CHCl_3$) ($t = 20^o$)<br>both at 578 nm | M 108 |
| $C_{14}H_{13}NO_4S_2$ | $S = NSO_2C_6H_5$, $CH_3$, HOOC | Brucine in acetone-MeOH | a) $+254.0^o$ (c 1.00, MeOH) ($t = 20^o$) | M 109 |
| | | (-)-α-Methylbenzylamine in MeOH | b) $-253.5^o$ (c 1.00, MeOH) ($t = 20^o$) | M 109 |
| $C_{14}H_{13}NO_5S_2$ | 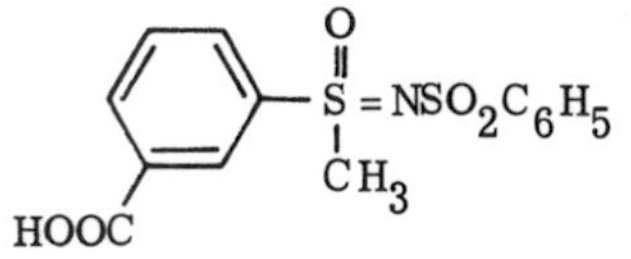 O, $S = NSO_2C_6H_5$, $CH_3$, HOOC | (-)-α-Methylbenzylamine<br>(+)-α-Methylbenzylamine<br>both in MeOH | a) $+143.5^o$ (c 1.00, acetone-$H_2O$) ($t = 20^o$)<br>b) $-143^o$ (c 1.00, acetone-$H_2O$) | M 109 |

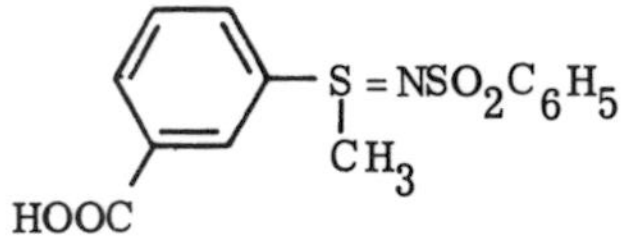

TABLE 12 (continued)

| | Substance resolved | Resolving agent and solvent | Rotation $[\alpha]_D$ of resolved compound | Reference |
|---|---|---|---|---|
| $C_{14}H_{14}O_2S_2$ | 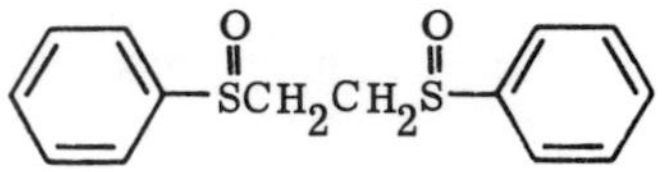 $C_6H_5S(O)CH_2CH_2S(O)C_6H_5$ | via chromatography on (+)-lactose (Partial resolution) | a) $+14^{\circ}$ ($CHCl_3$) (t = $20^{\circ}$) (350 nm)<br>b) $-36^{\circ}$ ($CHCl_3$) (t = $20^{\circ}$) (350 nm) | M 110 |
| $C_{24}H_{12}S_3$ | 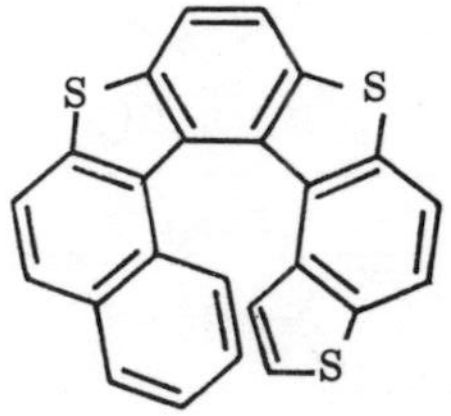 | via crystallization in (-)-$\alpha$-pinene (Partial resolution) | $+2400^{\circ}$ ($CHCl_3$) (t = $25^{\circ}$) (436 nm) | M 111 |

## References to Table 12

## Resolutions of Organometallic and Organometalloidal Compounds, including Phosphorus and Sulfur Compounds

(M 1) I.G.M. Campbell, J. Chem. Soc., 1952, 4448.

(M 2) I.G.M. Campbell, J. Chem. Soc., 1950, 3109.

(M 3) I.G.M. Campbell and D.J. Morrill, J. Chem. Soc., 1955, 1662.

(M 4) I.G.M. Campbell, J. Chem. Soc., 1947, 4.

(M 5) I.G.M. Campbell and A.W. White, J. Chem. Soc., 1958, 1184.

(M 6) I.G.M. Campbell, J. Chem. Soc., 1955, 3116.

(M 7) B. Bosnich and S.B. Wild, J. Amer. Chem. Soc., 92, 459 (1970).

(M 8) Y.F. Gatilov, L.B. Ionov and F.D. Yambushev, Zh. Obshch. Khim., 40, 2250 (1970) [Engl. transl., p. 2237].

(M 9) Y.F. Gatilov and L.B. Ionov, Zh. Obshch. Khim., 38, 2561 (1968) [Engl. transl., p. 2476].

(M 10) Y.F. Gatilov and L.B. Ionov, Zh. Obshch. Khim., 39, 1064 (1969) [Engl. transl., p. 1036].

(M 11) Y.F. Gatilov, L.B. Ionov and F.D. Yambushev, Zh. Obshch. Khim., 41, 570 (1971) [Engl. transl., p. 565].

(M 12) Y.F. Gatilov, G.K. Kamai and L.B. Ionov, Zh. Obshch. Khim., 38, 370 (1968) [Engl. transl., p. 368].

(M 13) L. Horner and H. Fuchs, Tetrahedron Lett., 1962, 203.

(M 14) G. Kamai and Y.F. Gatilov, Zh. Obshch. Khim., 34, 782 (1964) [Engl. transl., p. 781].

(M 15) M.H. Forbes, F.G. Mann, I.T. Millar and E.A. Moelwyn-Hughes, J. Chem. Soc., 1963, 2833.

(M 16) I.G.M. Campbell and R.C. Poller, J. Chem. Soc., 1956, 1195.

(M 17) M. Lesslie, J. Chem. Soc., 1949, 1183.

(M 18) G. Kamai and Y.F. Gatilov, Zh. Obshch. Khim., 32, 3170 (1962).

(M 19) G. Kamai, E.I. Klabunovskii, Y.F. Gatilov and G.S. Khodakov, Dokl. Akad. Nauk SSSR, 139, 1112 (1961) [Chem. Abstr., 56, 41f (1962)].

(M 20) K. Mislow, A. Zimmerman and J.T. Melillo, J. Amer. Chem. Soc., 85, 594 (1963).

(M 21) A.G. Davies and B.P. Roberts, J. Chem. Soc. B, 1967, 17.

(M 22) D.S. Matteson and R.A. Bowie, J. Amer. Chem. Soc., 87, 2587 (1965).

(M 23) G.E. Ryschkewitsch and J.M. Garrett, J. Amer. Chem. Soc., 90, 7234 (1968).

(M 24) G. Jaouen and R. Dabard, J. Organometal. Chem., 21, P43 (1970).

(M 25) S. Rosca and C.D. Nenitzescu, Rev. Roum. Chim., 15, 259 (1970).

(M 26) H. Falk and K. Schlögl, Monatsh. Chem., 99, 578 (1968).

(M 27) A.G. Brook and G.J.D. Peddle, J. Amer. Chem. Soc., 85, 1869 (1963).

(M 28) C. Eaborn, P. Simpson and I.D. Varma, J. Chem. Soc. A, 1966, 1133.

(M 29) R.W. Bott, C. Eaborn and I.D. Varma, Chem. Ind. (London) 1963, 614.

(M 30) G. Paiaro, R. Palumbo, A. Musco and A. Panunzi, Tetrahedron Lett., 1965, 1067.

(M 31) K. Bauer, H. Falk and K. Schlögl, Angew. Chem., 81, 150 (1969); Angew. Chem. Int. Ed. Engl., 8, 135 (1969).

(M 32) H. Falk, C. Krasa and K. Schlögl, Monatsh. Chem., 100, 254 (1969).

(M 33) M.J. Nugent, R.E. Carter and J.H. Richards, J. Amer. Chem. Soc., 91, 6145 (1969).

(M 34) K. Bauer, H. Falk and K. Schlögl, Monatsh. Chem., 99, 2186 (1968).

(M 35) K. Schlögl, M. Fried and H. Falk, Monatsh. Chem., 95, 576 (1964).

(M 36) H. Falk, O. Hofer and K. Schlögl, Monatsh. Chem., 100, 624 (1969).

(M 37) D. Marquarding, H. Klusacek, G. Gokel, P. Hoffmann and I. Ugi, J. Amer. Chem. Soc., 92, 5389 (1970).

(M 38) G. Gokel, P. Hoffmann, H. Kleimann, H. Klusacek, D. Marquarding and I. Ugi, Tetrahedron Lett., 1970, 1771.

(M 39) B. Gautheron, C.R. Acad. Sci., Paris, Ser. C, 267, 1359 (1968).

(M 40) B. Gautheron and J.-C. Leblanc, C.R. Acad. Sci., Paris, Ser. C, 269, 431 (1969).

(M 41) K. Schlögl and M. Walser, Monatsh. Chem., 100, 1515 (1969).

(M 42) H. Brunner and E. Schmidt, J. Organometal. Chem., 21, P53 (1970).

(M 43) H. Brunner, Angew. Chem., 81, 395 (1969); Angew. Chem. Int. Ed. Engl., 8, 382 (1969).

(M 44) H.B. Charman, E.D. Hughes and C. Ingold, J. Chem. Soc., 1959, 2523.

(M 45) F.R. Jensen, L.D. Whipple, D.K. Wedegaertner and J.A. Landgrebe, J. Amer. Chem. Soc., 82, 2466 (1960); see also J.A. Landgrebe and R.D. Mathis, ibid., 88, 3545 (1966).

(M 46) E.V. Uglova, T.B. Svetlanova, V.G. Smirnova and O.A. Reutov, Zh. Org. Khim., 5, 19 (1969) [Engl. transl., p. 16].

(M 47) O.A. Reutov and E.V. Uglova, Izv. Akad. Nauk SSSR, Otdel Khim. Nauk, 1959, 757 [Chem. Abstr., 53, 21626a (1959)].

(M 48) E.V. Uglova, E.Y. Fishilevich and O.A. Reutov, Zh. Org. Khim., 3, 834 (1967) [Engl. transl., p. 802].

(M 49) H.P. Benschop, D.H.J.M. Platenburg, F.H. Meppelder and H.L. Boter, Chem. Commun., 1970, 33.

(M 50) H.L. Boter and D.H.J.M. Platenburg, Rec. Trav. Chim. Pays-Bas, 86, 399 (1967).

(M 51) H.S. Aaron, J. Braun, T.M. Shryne, H.F. Frack, G.E. Smith, R.T. Uyeda and J.I. Miller, J. Amer. Chem. Soc., 82, 596 (1960).

(M 52) H.S. Aaron, T.M. Shyrne and J.I. Miller, J. Amer. Chem. Soc., 80, 107 (1958).

(M 53) A. J. J. Ooms and H. L. Boter, Biochem. Pharmacol., 14, 1839 (1965).

(M 54) R. F. Shuman, Ger. Offen. 1,924,085, Jan. 15, 1970 [Chem. Abstr., 72, 90634y (1970)].

(M 55) E. J. Glamkowski, G. Gal, R. Purick, A. J. Davidson and M. Sletzinger, J. Org. Chem., 35, 3510 (1970).

(M 56) R. F. Shuman, Ger. Offen., 1,924,084, Jan. 1970 [Chem. Abstr., 73, 4027u (1970)].

(M 57) N. N. Girotra and N. L. Wendler, Tetrahedron Lett., 1969, 4647.

(M 58) R. A. Lewis and K. Mislow, J. Amer. Chem. Soc., 91, 7009 (1969).

(M 59) C. Krawiecki, J. Michalski and Z. Tulimowski, Chem. Ind. (London), 1965, 34.

(M 60) J. Michalski, A. Ratajczak and Z. Tulimowski, Bull. Acad. Polon. Sci., Ser. Sci. Chim., 11, 237 (1963) [Chem. Abstr., 59, 8579f (1963)].

(M 61) W. B. Farnham, R. K. Murray, Jr., and K. Mislow, J. Amer. Chem. Soc., 92, 5809 (1970).

(M 62) G. Hilgetag and G. Lehmann, J. Prakt. Chem., 8, 224 (1959).

(M 63) H. P. Benschop and G. R. Van den Berg, Rec. Trav. Chim. Pays-Bas, 87, 362 (1968).

(M 64) A. Nudelman and D. J. Cram, J. Amer. Chem. Soc., 90, 3869 (1968).

(M 65) O. Korpium, R. A. Lewis, J. Chickos and K. Mislow, J. Amer. Chem. Soc., 90, 4842 (1968).

(M 66) O. Červinka and O. Kříž, Collect. Czech. Chem. Commun., 31, 1910 (1966).

(M 67) C. Donninger and D. H. Hutson, Tetrahedron Lett., 1968, 4871.

(M 68) T. H. Chan, Chem. Commun., 1968, 895.

(M 69) M. Green and R. F. Hudson, J. Chem. Soc., 1958, 3129.

(M 70) T. L. Emmick and R. L. Letsinger, J. Amer. Chem. Soc., 90, 3459 (1968).

(M 71) J. N. Seiber and H. Tolkmith, Tetrahedron Lett., 1967, 3333.

(M 72) G. Ostrogovich and F. Kerek, Angew. Chem. 83, 496 (1971); Angew. Chem. Int. Ed. Engl., 10, 498 (1971).

(M 73) D.P. Young, W.E. McEwen, D.C. Velez, J.W. Johnson and C.A. Van der Werf, Tetrahedron Lett., 1964, 359.

(M 74) O. Cervinka, O. Belovsky and M. Hepnerova, Chem. Commun., 1970, 562.

(M 75) W.E. McEwen, K.F. Kumli, A. Blade-Font, M. Zanger and C.A. Van der Werf, J. Amer. Chem. Soc., 86, 2378 (1964).

(M 76) G. Kamai and G.M. Usacheva, Zh. Obshch. Khim., 34, 785 (1964) [Engl. transl., p. 784].

(M 77) L. Horner, H. Winkler, A. Rapp, A. Mentrup, H. Hoffmann and P. Beck, Tetrahedron Lett., 1961, 161.

(M 78) I.G.M. Campbell and J.K. Way, J. Chem. Soc., 1960, 5034.

(M 79) S. Otsuka, A. Nakamura, T. Kano and K. Tani, J. Amer. Chem. Soc., 93, 4301 (1971).

(M 80) G. Wittig, H.J. Cristau and H. Braun, Angew. Chem., 79, 721 (1967); Angew. Chem. Int. Ed. Engl., 6, 700 (1967).

(M 81) L. Horner, F. Schedlbauer and P. Beck, Tetrahedron Lett., 1964, 1421.

(M 82) L. Horner, J.P. Bercz and C.V. Bercz, Tetrahedron Lett., 1966, 5783.

(M 83) D. Hellwinkel, Chem. Ber., 99, 3628 (1966).

(M 84) D. Hellwinkel, Chem. Ber., 99, 3642 (1966).

(M 85) D. Hellwinkel, Chem. Ber., 99, 3660 (1966).

(M 86) O. Hofer and K. Schlögl, J. Organometal. Chem., 13, 457 (1968).

(M 87) A. Fredga and K. Styrman, Ark. Kemi, 14, 461 (1959).

(M 88) B. Sjöberg, Ark. Kemi, 15, 397 (1960).

(M 89) R.J.P. Corriu and G.F. Lanneau, Tetrahedron Lett., 1971, 2771.

(M 90) C. Eaborn and C.G. Pitt, J. Chem. Soc. C, 1966, 1524.

(M 91) L.H. Sommer, C.L. Frye, G.A. Parker and K.W. Michael, J. Amer. Chem. Soc., 86, 3271 (1964).

(M 92) R.W. Bott, C. Eaborn and P.W. Jones, J. Organometal. Chem., 6, 484 (1966).

(M 93) S.L. Davydova, Y.A. Purinson, B.D. Lavrukhin and N.A. Platé, Izv. Akad. Nauk SSSR, Ser. Khim., 1965, 387 [Engl. transl., p. 373].

(M 94) R. Corriu and G. Royo, J. Organometal. Chem., 14, 291 (1968).

(M 95) A. Holt, A.W.P. Jarvie and G.J. Jervis, Tetrahedron Lett., 1968, 4087.

(M 96) M.S. Biernbaum and H.S. Mosher, J. Org. Chem., 36, 3168 (1971).

(M 97) L.H. Sommer and K.T. Rosborough, J. Amer. Chem. Soc., 91, 7067 (1969).

(M 98) M.K. Hargreaves, P.G. Modi and J.G. Pritchard, Chem. Commun., 1968, 1306.

(M 99) P. Karrer, E. Scheitlin and H. Siegrist, Helv. Chim. Acta, 33, 1237 (1950).

(M 100) H. Schmid and P. Karrer, Helv. Chim. Acta, 31, 1497 (1948).

(M 101) G. Wagner and S. Böhme, Arch. Pharm. (Weinheim), 297, 257 (1964).

(M 102) R. Fusco and F. Tenconi, Chim. Ind. (Milan), 47, 61 (1965).

(M 103) D. Darwish and G. Tourigny, J. Amer. Chem. Soc., 88, 4303 (1966).

(M 104) A.C. Cope and E.A. Caress, J. Amer. Chem. Soc., 88, 1711 (1966).

(M 105) D. Darwish, S.H. Hui and R. Tomlinson, J. Amer. Chem. Soc., 90, 5631 (1968).

(M 106) U. Tonellato, O. Rossetto and A. Fava, J. Org. Chem., 34, 4032 (1969).

(M 107) R. Scartazzini and K. Mislow, Tetrahedron Lett., 1967, 2719.

(M 108) A. Lüttringhaus, U. Hess and H.J. Rosenbaum, Z. Naturforsch., B22, 1296 (1967).

(M 109) G. Kresze and B. Wustrow, Chem. Ber., 95, 2652 (1962).

(M 110) M. Cinquini, S. Colonna and F. Taddei, Boll. Sci. Fac. Chim. Ind. Bologna, 27, 231 (1969).

(M 111) H. Wynberg and M.B. Groen, J. Amer. Chem. Soc., 92, 6664 (1970).

TABLE 13 RESOLUTIONS OF AMINO ACIDS, AMINO ACID DERIVATIVES AND RELATED COMPOUNDS *)

| | Substance resolved | Derivative used +) | Resolving agent and solvent | Rotation $[\alpha]_D$ of resolved compound | Reference |
|---|---|---|---|---|---|
| $C_3H_6N_2O_2$ | Cycloserine<br>see Table 9d | | (-)-Tartaric acid<br>(+)-Tartaric acid | a) - 115° (c 1, $H_2O$) (t = 22°)<br>b) + 115° (c 1, $H_2O$) (t = 22°) | AA 1 |
| $C_3H_7N_5$ | 5-(1-aminoethyl)tetrazole: tetrazole ring (N-H) bearing $-CH(R)-NH_2$, R = $CH_3$<br>Tetrazole analog of alanine | Z | (+)-Tyrosine hydrazide in MeOH | + 34.5° (c 1, MeOH) (t = 20°)<br>on the Z-derivative | AA 2 |
| $C_3H_7NO_2$ | $CH_3CH(NH_2)COOH$ Alanine | Z | (+)-Tyrosine hydrazide in EtOH | a) + 13.8° (c 2, HOAc) (t = 23°)<br>b) - 13.8° (c 2, HOAc) (t = 23°)<br>both on the Z-derivatives | AA 3 |
| | | Z | Also, resolutions with (-)-ephedrine (B 86) and (-)-cytisine (AA 4). | | |
| | | Tos | (-)-Menthol<br>*via* ester formation | - 46.5° (c 1%, MeOH)<br>on the ester Tos-derivative | AA 5, AA 6 |
| $C_3H_7N_5S$ | See $C_3H_7N_5$ above<br>R = Btm<br>Tetrazole analog of S-Benzylcysteine | Z | (+)-Tyrosine hydrazide in MeOH | + 35.5° (c 1, MeOH) (t = 20°)<br>on the S-Benzyl Z-derivative | AA 2 |
| $C_3H_7NO_2S$ | $CH_2(SH)CH(NH_2)COOH$ Cysteine | Z<br>S-Benzyl | (+)-Tyrosine hydrazide in MeOH | + 44.5° (c 2, acetone) (t = 20°)<br>on the S-Benzyl Z-derivative | AA 7 |

*) References to Table 13 begin on p. 291. See also comment on p. 48.

+) For abbreviations used in this table see pp. 42-43.

TABLE 13 (continued)

| | Substance resolved | Derivative used | Resolving agent and solvent | Rotation $[\alpha]_D$ of resolved compound | Reference |
|---|---|---|---|---|---|
| $C_3H_7NO_3$ | $CH_2CHCOOH$ (OH, $NH_2$) Serine | Z, O-Cbzo | (-)-Ephedrine<br>(+)-Ephedrine<br>both in EtOAc | a) - 4.3° (EtOH) (t = 22-26°)<br>b) + 4.4° (EtOH) (t = 22-26°)<br>both on the O-Cbzo Z-derivatives | AA 8 |
| | | Tos | (+)-threo-2-Amino-1-(p-nitrophenyl)-1,3-propanediol in $H_2O$ | a) - 54° (1 N, 5% aq. NaOH) (t = 20°)<br>b) + 54° (1 N, 5% aq. NaOH) (t = 20°)<br>both on the Tos-derivatives | AA 9 |
| | | Tos | Also, resolution with brucine and quinine (AA 10) | | |
| | | N-Formyl, O-Benzyl | Brucine in EtOH | a) - 23.3° (c 2, 80% aq. AcOH + 1 equiv HCl)<br>b) + 22.8° (c 2, 80% aq. AcOH + 1 equiv HCl)<br>both on the O-Benzyl-derivatives at 25° | AA 11, AA 12 |
| | | N-Formyl, O-Benzyl | (-)-α-Methylbenzylamine<br>(+)-α-Methylbenzylamine<br>both in isoPrOH | a) - 47.2° (c 3.5, 80% EtOH) (t = 20°)<br>b) + 48.1° (c 3.5, 80% EtOH) (t = 20°)<br>both on the O-Benzyl N-formyl-derivatives | AA 13 |
| | | Z, O-Benzyl | (-)-α-Methylbenzylamine<br>(+)-α-Methylbenzylamine<br>both in isoPrOH | a) - 16.8° (c 1, EtOH) (t = 30°)<br>b) + 17.3° (c 1, EtOH) (t = 32°)<br>both on the O-Benzyl Z-derivatives | AA 14 |
| | | Also, resolutions of serine methyl ester with O,O'-dibenzoyl-D-tartaric acid (AA 15), and of O-methylserineamide with L-tartaric acid (AA 16). | | | |
| $C_3H_8N_2O_2$ | $CH_2$-$CHCOOH$ ($NH_2$, $NH_2$) | N,N'-bis-Bz | (+)-α-Methylbenzylamine<br>(-)-α-Methylbenzylamine<br>both in EtOAc | a) - 33° (c 2, 0.1 N aq. NaOH) (t = 20°)<br>b) + 32.8° (c 2, 0.1 N aq. NaOH) (t = 20°)<br>both on the bis-Bz-derivatives | AA 17 |
| $C_4H_6NO_2F_3$ | $CF_3CH_2CHCOOH$ ($NH_2$) | Tfa | Dimethyl L-(+)-glutamate on the oxazolone cyclization product in THF; via amide formation | - 6.3° (c 16.4, 1 N aq. HCl) (t = 25°) (546 nm) | AA 18 |

TABLE 13 (continued)

| | Substance resolved | Derivative used | Resolving agent and solvent | Rotation $[\alpha]_D$ of resolved compound | Reference |
|---|---|---|---|---|---|
| $C_4H_7NO_2$ | N, H, H, COOH | Z | (+)-Tyrosine hydrazide in MeOH | a) + 98.5° (c 3.9, $CHCl_3$) (t = 20°)<br>b) - 99.0° (c 3.9, $CHCl_3$) (t = 20°)<br>both on the Z-derivative | AA 19 |
| $C_4H_7NO_2$ | $HOOCCH_2CH(NH_2)COOH$<br>Aspartic acid | - | (-)-α-Methylbenzylamine<br>(+)-α-Methylbenzylamine<br>both in $H_2O$-MeOH-acetone | a) + 25.0° (c 1.97, 6 N aq. HCl) (t = 27°)<br>b) - 23.0° (c 2.30, 6 N aq. HCl) (t = 27°) | AA 20 |
| | | - | Also, partial resolution with (-)-2-amino-1-butanol (B 58). | | |
| $C_4H_7NO_5$ | $HOOCCH(OH)-CH(NH_2)COOH$<br>β-Hydroxyaspartic acid<br>erythro | - | (+)-Ornithine<br>(-)-Ornithine<br>both in $H_2O$-MeOH | a) + 52.0° (c 1.0, 5 N aq. HCl) (t = 20°)<br>b) - 51.8° (c 1.0, 5 N aq. HCl) (t = 20°) | AA 21 |
| | erythro | Bzl | L-Histidine in $H_2O$ - EtOH | a) - 17.1° (c 2.9, 5 N aq. HCl) (t = 20°)<br>b) + 15.2° (c 1.6, 5 N aq. HCl) (t = 20°)<br>both on the Bzl-derivatives | AA 22 |
| | threo | - | (+)-Lysine in $H_2O$ | a) + 6.4° (c 1.0, 5 N aq. HCl) (t = 20°)<br>b) - 6.5° (c 1.0, 5 N aq. HCl) (t = 20°) | AA 21 |
| | threo | Bzl | (-)-Ephedrine in EtOH - $H_2O$ | a) + 5.3° (c 15, 5 N aq. HCl) (t = 16°)<br>b) - 5.9° (c 15, 5 N aq. HCl) (t = 20°)<br>both on the Bzl-derivatives | AA 23 |
| $C_4H_8N_2O_4$ | $HOOCCH(NH_2)-CH(NH_2)COOH$<br>threo-isomer | N-COOEt<br>N, N'-bis-Bzl | (+)-threo-2-Amino-1-(p-nitrophenyl)-1,3-propanediol | a) + 47.1° (c 0.9, MeOH) (t = 20°) | AA 24 |
| | | | (-)-Enantiomer<br>both in $H_2O$ | b) - 47.9° (c 0.4, MeOH) (t = 20°)<br>both on the N, N'-bis Bzl N-COOEt derivatives | AA 24 |

TABLE 13 (continued)

| | Substance resolved | Derivative used | Resolving agent and solvent | Rotation $[\alpha]_D$ of resolved compound | Reference |
|---|---|---|---|---|---|
| $C_4H_9NO_2$ | $CH_3CH_2CH(NH_2)COOH$ Butyrine | OBzl | (-)-O,O'-Dibenzoyltartaric acid in PrOH | a) - 14.0° (c 2.01, $H_2O$) (t = 20°)<br>b) + 14.1° (c 2.21, $H_2O$) (t = 20°)<br>both on the hydrochlorides | AA 25 |
| | | Tos | (-)-Menthol *via* ester formation | - 47.2° (c 1%, MeOH)<br>on the ester Tos-derivative | AA 5 |
| | | Ac | Resolution with (-)- and (+)-α-methylbenzylamine (AA 26). | | |
| $C_4H_9NO_2S$ | $CH_2(SH)CH_2CH(NH_2)COOH$ Homocysteine | S-Carboxy-methyl | Brucine in $H_2O$ - EtOH (Partial resolution) | a) - 5.1° (c 1, 1 N aq. HCl) (t = 25°)<br>b) + 3.5° (c 1, 1 N aq. HCl) (t = 25°)<br>both on the S-carboxymethyl derivatives | AA 27 |
| $C_4H_9NO_3$ | $CH_3CH(OH)CH(NH_2)COOH$ Threonine | Pht | Brucine in methyl cellosolve | a) - 28.2 ± 1° (c 2, $H_2O$) (t = 20°)<br>b) + 27.6 ± 1° (c 2, $H_2O$) (t = 20°) | AA 28 |
| | | N-Formyl O-Ethyl | (+)-α-Methylbenzylamine in MeOH-acetone | + 14.3° (c 1, $H_2O$) (t = 25°)<br>on the O-ethyl N-formyl derivative | AA 29 |
| | | Oxazolidone | Brucine in MeOH | a) - 40.5° (c 3.07, $H_2O$) (t = 24.5°)<br>b) + 41.7° (c 3.07, $H_2O$) (t = 25°)<br>both on the oxazolidone derivatives | AA 30 |
| $C_4H_9NO_3$ | $CH_3CH(OH)-CH(NH_2)COOH$ Allothreonine | Pht | Brucine in methyl cellosolve | a) - 32.4° (c 1, 1 N aq. HCl) (t = 20°)<br>b) + 32.8° (c 1, 1 N aq. HCl) (t = 20°) | AA 31 |
| | | Oxazolidone | Brucine | a) - 19.2° (c 3.29, $H_2O$) (t = 16°) | AA 30 |
| | | Oxazolidone | Quinine | b) + 19.0° (c 3.21, $H_2O$) (t = 19.5°)<br>both on the oxazolidone derivatives | AA 30 |

TABLE 13 (continued)

| | Substance resolved | Derivative used | Resolving agent and solvent | Rotation $[\alpha]_D$ of resolved compound | Reference |
|---|---|---|---|---|---|
| $C_4H_9NO_3$ | $CH_2$-CHCH$_2$COOH (NH$_2$, OH) | Bz | Strychnine in aq. NaOH | a) + 20.2° (c 2, 1 N aq. NaOH) (t = 23°) | AA 32 |
| | | Bz | Brucine in $H_2O$ | b) - 21.45° (c 2, 1 N aq. NaOH) (t = 23°) both on the Bz-derivatives | AA 32 |
| $C_4H_9NO_4$ | $CH_2$-CH-CHCOOH (OH, OH, NH$_2$) threo-isomer | Z, O-Benzyl | (+)-Tyrosine hydrazide in MeOH | a) + 10.9° (c 2.0, EtOH) (t = 23°)<br>b) - 10.3° (c 2.0, EtOH) (t = 23°)<br>both on the O-benzyl Z-derivatives (cyclohexylamine salts) | AA 33 |
| $C_5H_6N_2O_4$ | 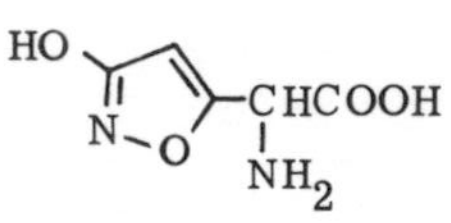 HO-isoxazolyl-CHCOOH (NH$_2$) Ibotenic acid | Tfa | Quinine in EtOH-$H_2O$ | a) + 8.6° (c 0.2, $H_2O$)<br>b) - 10.2° (c 0.2, $H_2O$) | AA 34 |
| $C_5H_7NO_2$ | 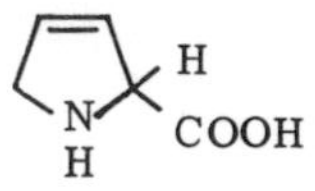 Dehydroproline | Amide | Ammonium (+)-3-bromo-camphor-9-sulfonate in $H_2O$ | + 300° (c 2, $H_2O$) (t = 20°) on the amide | AA 35 |
| $C_5H_9NO_2$ | 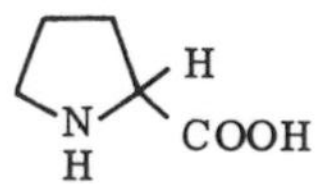 Proline | Z | (+)-Tyrosine hydrazide in MeOH | a) + 61.2° (c 5.3, HOAc) (t = 23°)<br>b) - 61.7° (c 5.3, HOAc) (t = 22°)<br>both on the Z-derivatives | AA 3 |
| $C_5H_9NO_4$ | HOOCCH—CHCOOH (CH$_3$, NH$_2$) β-Methylaspartic acid | Z | (-)-Ephedrine<br>(+)-Ephedrine<br>both in EtOAc | a) + 17.8° (py) (t = 22-26°)<br>b) - 16.5° (py) (t = 22-26°)<br>both on the Z-derivatives | AA 8 |
| $C_5H_9NO_4$ | Glutamic acid | See Table 3, ref. 19, p. 30. | | | |

TABLE 13 (continued)

| | Substance resolved | Derivative used | Resolving agent and solvent | Rotation $[\alpha]_D$ of resolved compound | Reference |
|---|---|---|---|---|---|
| $C_5H_9NO_4$ | $HOOCCH_2CH(NH_2)CH_2COOH$ | Ac<br>mono-OEt | Cinchonidine<br>Strychnine | a) + 5.8° (c 2.5%, acetone) (t = 25°)<br>b) - 5.9° (c 2.5%, acetone) (t = 25°)<br>both on the Ac OEt-derivatives | AA 36<br>AA 36 |
| $C_5H_{11}N_5$ | See $C_3H_7N_5$ above<br>R = $(CH_3)_2CH$<br>Tetrazole analog of valine | Z | (+)-Tyrosine hydrazide<br>in MeOH | + 33° (c 1, MeOH) (t = 20°)<br>on the Z-derivative | AA 2 |
| $C_5H_{11}NO_2$ | $CH_3CH_2C(CH_3)(NH_2)COOH$<br>Isovaline | Bz | Sodium (-)-menthoxide on the oxazolin-5-one cyclization product in $C_6H_6$; *via* ester formation | a) - 10.4° (c 1.82, MeOH)<br>b) + 10.3° (c 1.44, MeOH)<br>both on the Bz-derivatives | AA 37 |
| | | N-Formyl | Brucine | + 7.17° (3/4 N aq. KOH) (t = 23°)<br>on the N-formyl derivative | AA 38 |
| $C_5H_{11}NO_2$ | $CH_2(NH_2)CH_2CH(CH_3)COOH$ | Pht | Quinine<br>in EtOAc | a) - 21.5 ± 0.4° (c 1.75, $C_6H_6$) (t = 24°)<br>b) + 20.3 ± 0.4° (c 2.288, $C_6H_6$) (t = 23°)<br>both on the Pht-derivatives | A 240 |
| $C_5H_{11}NO_2$ | $CH_3CH(CH_3)$—$CH(NH_2)COOH$<br>Valine | Ac | (+)-*threo*-2-Amino-1-(*p*-nitrophenyl)-1,3-propanediol<br>in EtOH | - 19.9° (c 4, $H_2O$) (t = 20°) | AA 39 |
| | | Z | (-)-Ephedrine<br>(+)-Ephedrine<br>both in EtOAc | a) + 0.7° (EtOH) (t = 22-26°)<br>b) - 0.53° (EtOH) (t = 22-26°)<br>both on the Z-derivatives | AA 8 |
| | | Tos | (-)-Menthol<br>*via* ester formation | - 54.4° (c 1%, MeOH)<br>on the ester Tos-derivative | AA 5, AA 6 |
| | | — | Also, resolution with (-)-pantolactone (AA 55). | | |

TABLE 13 (continued)

| | Substance resolved | Derivative used | Resolving agent and solvent | Rotation $[\alpha]_D$ of resolved compound | Reference |
|---|---|---|---|---|---|
| $C_5H_{11}NO_2$ | $CH_3CH_2CH_2CH(NH_2)COOH$<br>Norvaline | Tos | (-)-Menthol<br>*via* ester formation | - 49.9° (c 1%, MeOH)<br>on the ester Tos-derivative | AA 5 |
| $C_5H_{11}NO_2S$ | $CH_2(SCH_3)CH_2CH(NH_2)COOH$<br>Methionine | Z | (-)-Ephedrine<br>(+)-Ephedrine<br>both in EtOAc | a) + 17.2° (EtOH) (t = 22-26°)<br>b) - 19.4° (EtOH) (t = 22-26°)<br>both on the Z-derivatives | AA 8 |
| | | Tos | (-)-Menthol<br>*via* ester formation | - 30.7° (c 1%, MeOH)<br>on the ester Tos-derivative | AA 5 |
| | | Amide | Resolution with (-)-2-pyrrolidone-5-carboxylic acid (AA 40). | | |
| $C_5H_{11}NO_2S$ | $CH_3C(CH_3)(SH)-CH(NH_2)COOH$<br>Penicillamine | N-Formyl<br>S-Benzyl | Brucine | a) + 63° (c 1%, 1 N aq. NaOH) (t = 22°)<br>b) - 56° (c 1%, 1 N aq. NaOH) (t = 22°)<br>+ 64° (c 1%, py) (t = 22°) on the<br>N-formyl derivative | AA 41 |
| $C_5H_{11}NO_3$ | $CH_2(OH)CH_2C(CH_3)(NH_2)COOH$<br>$\alpha$-Methylhomoserine | Ac<br>O-Benzyl | Sodium (-)-menthoxide on the oxazolin-5-one cyclization product in $C_6H_6$; *via* ester formation | a) - 7.3° (c 1.98, MeOH)<br>b) + 8.0° (c 1.88, MeOH)<br>both on the Ac-derivatives | AA 37 |
| $C_5H_{11}NO_3$ | $CH_3C(CH_3)(OH)-CH(NH_2)COOH$<br>$\beta$-Hydroxyvaline | N-Formyl<br>O-Methyl | Brucine<br>in EtOH | a) - 5° (c 1.0, $H_2O$) (t = 20°)<br>b) + 5° (c 1.0, $H_2O$) (t = 20°)<br>both on the N-formyl O-methyl derivative | AA 10 |
| | | Bz | (-)-$\alpha$-Methylbenzylamine<br>in MeOH-acetone | - 30.5° (c 2.01, EtOH) (t = 25°)<br>on the Bz-derivative | AA 42 |

TABLE 13 (continued)

| | Substance resolved | Derivative used | Resolving agent and solvent | Rotation $[\alpha]_D$ of resolved compound | Reference |
|---|---|---|---|---|---|
| $C_5H_{12}N_2O_3$ | $CH_2CH_2OCH_2CHCOOH$ ($NH_2$, $NH_2$)<br>4-Oxalysine | N, N'-bis-Pht | Brucine<br>in methyl cellosolve | a) + 65.5° (c 1, DMF) (t = 25°)<br>b) - 65.1° (c 1, DMF) (t = 22°)<br>both on the N, N'-bis-Pht-derivatives | AA 43 |
| $C_6H_5N_3O_2S$ | 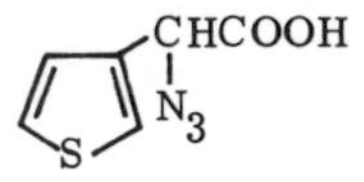 S, CHCOOH, $N_3$ | — | Cinchonine and<br>(-)-$\alpha$-methylbenzylamine<br>(-)-Ephedrine<br>all in EtOAc | a) - 108° (acetone) (t = 25°)<br>b) + 107° (c 0.877, acetone) (t = 25°) | AA 44<br>AA 44 |
| $C_6H_7NO_2S$ | 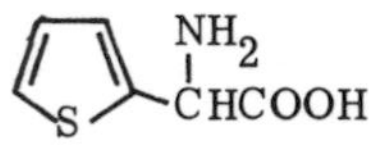 S, $NH_2$, CHCOOH | N-Formyl | Dehydroabietylamine<br>in $MeOH-H_2O$ | - 78.0° (c 0.423, $H_2O$) (t = 25°) | A 187 |
| $C_6H_7NO_2S$ | 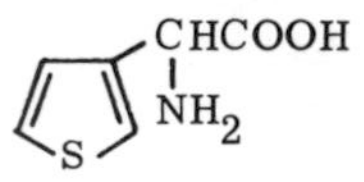 S, CHCOOH, $NH_2$ | —<br>— | (+)-3-Bromocamphor-9-sulfonic acid in EtOAc<br>(+)-Camphor-10-sulfonic acid in isobutyl alcohol | a) + 100° (c 0.250, $H_2O$) (t = 25°)<br>b) - 100° (c 0.270, $H_2O$) (t = 25°) | AA 44<br>AA 44 |
| | | N-Formyl | Dehydroabietylamine<br>in $MeOH-H_2O$ | - 132° (c 1, 1 N aq. HCl) (t = 25°) | A 187 |
| $C_6H_9N_3O_2$ | 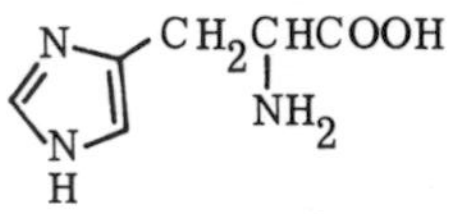 N, N, H, $CH_2CHCOOH$, $NH_2$<br>Histidine | — | (-)-Tartaric acid<br>(+)-Tartaric acid<br>both in $H_2O$ | a) - 39.0° (c 2, $H_2O$) (t = 25°)<br>b) + 40.0° (c 2.8, $H_2O$) | AA 45<br>AA 46 |
| $C_6H_9N_3O_4$ | $HOOCCH_2CH_2CH_2CHCOOH$ ($N_3$) | — | Dehydroabietylamine<br>in EtOH<br>or (-)-ephedrine in acetone<br>(+)-ephedrine in acetone | a) + 54° (c 3, acetone) (t = 25°)<br>+ 53.3° (c 2, acetone) (t = 25°)<br>b) - 53° (c 1, acetone) (t = 25°) | AA 47<br>AA 47<br>AA 47 |

TABLE 13 (continued)

| Formula | Substance resolved | Derivative used | Resolving agent and solvent | Rotation $[\alpha]_D$ of resolved compound | Reference |
|---|---|---|---|---|---|
| $C_6H_{11}NO_2$ | Pipecolic acid (piperidine-2-COOH) | Z | (+)-Tyrosine hydrazide in MeOH | a) + 57.2 ± $1^o$ (c 5.3, HOAc) (t = $25^o$)<br>b) - 57 ± $1^o$ (c 5.3, HOAc) (t = $25^o$)<br>both on the Z-derivatives | AA 48 |
| | | — | (+)-Tartaric acid<br>(-)-Tartaric acid<br>both in 96 % EtOH | a) + 26.2 ± $1^o$ (c 3.106, $H_2O$) (t = $19^o$)<br>b) - 25.4 ± $1^o$ (c 2.85, $H_2O$) (t = $20^o$) | AA 49, AA 50<br>AA 49, AA 50 |
| $C_6H_{11}NO_3$ | (Isopropylideneaminooxy)-propionic acid | | See Table 8b | | |
| $C_6H_{11}NO_4$ | $HOOCCH_2CH_2CH_2CH(NH_2)COOH$ | Z | (-)-threo-2-Amino-1-(p-nitro-phenyl)-1,3-propanediol<br>(+)-Enantiomer<br>both in EtOH | a) + 17 ± $1^o$ (c 2, EtOH-2 N aq. NaOH 9:1)<br>b) - $15^o$ (c 2, EtOH-2 N aq. NaOH 9:1)<br>both on the Z-derivatives at t = $25^o$ | AA 51<br>AA 51 |
| | | Bz | Resolution with the same resolving agents or with brucine (AA 52). | | |
| $C_6H_{12}N_2O$ | α-Aminocaprolactam | See Table 9d | See also refs. AA 53 and AA 54. | | |
| $C_6H_{13}N_5$ | See $C_3H_7N_5$ above<br>R = $(CH_3)_2CHCH_2$<br>Tetrazole analog of leucine | Z | (+)-Tyrosine hydrazide in MeOH | + 39.5$^o$ (c 1, MeOH) (t = $20^o$)<br>on the Z-derivative | AA 2 |
| $C_6H_{13}NO_2$ | $(CH_3)_3C-CH(NH_2)COOH$ tert-Leucine | Ac | Cinchonidine in EtOH | - $40^o$ (c 1, $H_2O$) (t = $20^o$)<br>on the Ac-derivative | A 122 (p.127) |
| | | Amide | (+)-Tartaric acid in $H_2O$ | - 9.0$^o$ (c 4, $H_2O$) (t = $20^o$) | A 122 |
| | | N-Formyl | Resolution with brucine (A 15) (p. 121). | | |

TABLE 13 (continued)

| Formula | Substance resolved | Derivative used | Resolving agent and solvent | Rotation $[\alpha]_D$ of resolved compound | Reference |
|---|---|---|---|---|---|
| $C_6H_{13}NO_2$ | $CH_3CH_2CH(CH_3)-CH(NH_2)COOH$ Isoleucine | Z | (+)-Tyrosine hydrazide in MeOH | a) $-4.3^o$ (c 2, EtOH) ($t = 23^o$)<br>b) $+4.1^o$ (c 2, EtOH) ($t = 23^o$)<br>both on the dicyclohexylamine salts of Z-derivatives | AA 3 |
| | | Z | (-)-Ephedrine<br>(+)-Ephedrine<br>both in $Et_2O$ | a) $+40.8^o$ (6 N aq. HCl) ($t = 22\text{-}25^o$)<br>b) $-40.3^o$ (6 N aq. HCl) ($t = 22\text{-}25^o$) | AA 8 |
| $C_6H_{13}NO_2$ | $CH_3CH(CH_3)CH_2CH(NH_2)COOH$ Leucine | Z | (-)-Ephedrine<br>(+)-Ephedrine<br>both in EtOAc-Pet E (2:3) | a) $-15.9^o$ (6 N aq. HCl) ($t = 22\text{-}25^o$)<br>b) $+14.6^o$ (6 N aq. HCl) ($t = 22\text{-}25^o$) | AA 8 |
| | | Tos | (-)-Menthol<br>*via* ester formation | $-44.2^o$ (c 1%, MeOH)<br>on the ester Tos-derivative | AA 5, AA 6 |
| | | — | Also, resolution with (-)-pantolactone (AA 55). | | |
| $C_6H_{13}NO_2$ | $CH_3CH_2CH_2CH_2CH(NH_2)COOH$ Norleucine | Tos | (-)-Menthol<br>*via* ester formation | $-43.1^o$ (c 1%, MeOH)<br>on the ester Tos-derivative | AA 5 |
| $C_6H_{14}N_2O_2$ | $CH_2(NH_2)CH_2CH_2CH_2CH(NH_2)COOH$ Lysine<br>See also $C_6H_{12}N_2O$ $\alpha$-aminocaprolactam above and Table 9 d | — | (+)-2-Phenoxypropionic acid | a) $+26.0^o$ (c 10, 6 N aq. HCl) ($t = 20^o$) | AA 56, AA 57 |
| | | — | (-)-Menthyl dihydrogen phosphate in $H_2O$-EtOH<br>Partial resolution | $+10.2^o$ (0.6 N aq. HCl) ($t = 20^o$)<br>b) $-13.8^o$ (0.6 N aq. HCl) ($t = 20^o$) | AA 58 |
| | | | Also, resolutions with 2-pyrrolidone-5-carboxylic acid (AA 59), with (+)- and (-)-glutamic acids (AA 60), and with D-homocysteinic acid (AA 61). | | |

TABLE 13 (continued)

| Formula | Substance resolved | Derivative used | Resolving agent and solvent | Rotation $[\alpha]_D$ of resolved compound | Reference |
|---|---|---|---|---|---|
| $C_7H_{15}NO_2$ | $CH_3CH(CH_3)-CH(CH_3)-CH(NH_2)COOH$ erythro-β-Methylleucine | Z | (-)-Ephedrine<br>(+)-Ephedrine<br>both in $Et_2O$ | a) + 38.9° (c 0.485, 5 N aq. HCl) (t = 24°)<br>b) - 39.4° (c 0.38, 5 N aq. HCl) (t = 25°) | AA 8, AA 62 |
| $C_8H_6N_3O_2F$ | 3-R-$C_6H_4$-$CH(N_3)COOH$, R = F | – | Pseudoephedrine in MeOH-$Et_2O$ | - 101.1° (c 0.9, EtOH) (t = 20°) | AA 63 |
| $C_8H_7N_3O_2$ | R = H | – | (+)-threo-2-Amino-1-(p-nitrophenyl)-1,3-propanediol | a) - 145.8° (c 1.000%, EtOH) | AA 64 |
| | | – | (-)-Enantiomer<br>both in acetone-$CH_2Cl_2$ | b) + 145.6° (c 1.013%, EtOH) | AA 64 |
| | | | Also, resolutions with (-)-ephedrine and (-)-α-(2-naphthyl)-ethylamine (AA 65). | | |
| $C_8H_9NO_2$ | 4-R-$C_6H_4$-$CH(NH_2)COOH$, R = H<br>α-Phenylglycine | Z | Dehydroabietylamine in MeOH-$H_2O$ | - 116.5° (c 1, EtOH) (t = 25°) on the Z-derivative | A 187 (p.131) |
| $C_8H_9NO_3$ | R = OH | Z, O-Cbzo | Quinine · 3 $H_2O$ in EtOH | a) - 120.2° (c 1, MeOH) (t = 21°) | AA 66 |
| | | Z | (-)-Ephedrine in EtOH | b) + 120.2° (c 1, MeOH) (t = 21°)<br>both on the Z-derivatives | AA 66 |
| $C_8H_{12}N_2O_3S$ | 6-Aminopenicillanic acid | $C_6H_5OCH_2C(=O)-$, $K^{\oplus}$ salt | (-)-erythro-1,2-Diphenyl-2-methylaminoethanol · HCl in $H_2O$<br>(+)-Enantiomer in butyl acetate | a) $\alpha_D^{25}$ + 92° (c 0.2, N-methylpyrrolidone)<br>b) $\alpha_D^{25}$ - 86° (c 0.4, N-Methylpyrrolidone)<br>both on the diastereomeric salts | AA 67 |

TABLE 13 (continued)

| | Substance resolved | Derivative used | Resolving agent and solvent | Rotation $[\alpha]_D$ of resolved compound | Reference |
|---|---|---|---|---|---|
| $C_8H_{15}NO_2$ | (cyclohexyl)(H)C–CHCOOH, with $NH_2$ on the CH | Bz | Quinine<br>in $H_2O$-EtOH (3:2) | a) - 26.1° (c 0.803, 0.5 N aq. NaOH) (t = 23°)<br>b) + 25.9° (c 1.082, aq. NaOH) (t = 23°)<br>both on the Bz-derivatives | AA 68 |
| $C_8H_{15}NO_4$ | $HOOC(CH_2)_5CH(NH_2)COOH$ | Z | (-)-Tyrosine hydrazide<br>(+)-Tyrosine hydrazide<br>both in MeOH-EtOAc | a) - 9.1° (c 4, DMF) (t = 25°)<br>b) + 9.1° (c 4, DMF) (t = 28°)<br>both on the Z-derivatives | AA 69 |
| $C_8H_{17}NO_2$ | $CH_3CH(CH_3)$–$CH(CH_3)$–$CH(NHCH_3)COOH$<br>threo-N, β-Dimethylleucine | Z | (-)-Ephedrine<br>(+)-Ephedrine<br>both in EtOAc | a) - 75.9° (EtOH) (t = 22-26°)<br>b) + 75.3° (EtOH) (t = 22-26°)<br>both on the Z-derivatives | AA 8, AA 62 |
| $C_9H_8NO_2Cl_3$ | See $C_9H_{11}NO_2$, N-Phenylalanine | | | | |
| $C_9H_9N_3O_2$ | $CH_3CHCOOH$ on N-1 of benzotriazole | — | Cinchonidine<br>in EtOAc | a) - 51.7° (t = 25°)<br>b) + 51.3° (t = 25°) | AA 70 |
| $C_9H_9N_3O_2$ | $CH_3CHCOOH$ on N-2 of benzotriazole | — | Cinchonidine<br>in MeOH-$H_2O$<br>(+)-α-Methylbenzylamine<br>in EtOAc | a) + 32.9° (acetone) (t = 25°)<br>b) - 33.1° (acetone) (t = 25°) | AA 71<br>AA 71 |
| $C_9H_9N_3O_7$ | See $C_9H_{11}NO_2$ Substituted β-phenylalanines | | | | |
| $C_9H_{10}N_2O_5$ | See $C_9H_{11}NO_3$ below | | | | |
| $C_9H_{11}N_5$ | See $C_3H_7N_5$<br>R = $C_6H_5CH_2$<br>Tetrazole analog of β-phenylalanine | Z | (+)-Tyrosine hydrazide<br>in MeOH | - 47° (c 1, DMF) (t = 20°)<br>on the Z-derivative | AA 2 |

TABLE 13 (continued)

| Formula | Substance resolved | Derivative used | Resolving agent and solvent | Rotation $[\alpha]_D$ of resolved compound | Reference |
|---|---|---|---|---|---|
| $C_9H_{11}NO_2$ | $C_6H_5$-CH(COOH)-NHCH$_3$<br>$\alpha$-Phenylsarcosine | Boc | (-)-Ephedrine<br>(+)-Ephedrine<br>both in $C_6H_6$-hexane | a) + 134.8° (c 0.804, EtOH) (t = 24°)<br>b) - 134.1° (c 0.682, EtOH) (t = 29°)<br>both on the Boc-derivatives | AA 72 |
| $C_9H_{11}NO_2$ | R = H: 2,4,6-R$_3$C$_6$H$_2$-NH-CH(CH$_3$)COOH<br>N-Phenylalanine | — | Quinine<br>in MeOH-acetone | + 71° (c 2%, EtOH) (t = 23°) | AA 73 |
| | R = Cl<br>$C_9H_8NO_2Cl_3$ | — | (-)-$\alpha$-Methylbenzylamine<br>(+)-$\alpha$-Methylbenzylamine<br>both in acetone-$H_2O$ | a) + 34.3° (c 2.286, acetone) (t = 25°)<br>b) - 34.2° (c 2.384, acetone) (t = 25°) | AA 74 |
| $C_9H_{11}NO_2$ | $C_6H_5$-C(CH$_3$)(NH$_2$)COOH | N-Formyl<br>N-Formyl | Quinine<br>Cinchonidine<br>both in $H_2O$ | a) + 90.4° (c 3.3, EtOH) (t = 25°)<br>b) - 89.3° (c 3.4, EtOH) (t = 26°)<br>both on the N-formyl derivatives | AA 75, AA 76<br>AA 75 |
| | | OEt | Tartaric acid<br>in EtOH | + 25.02° (c 9.43, EtOH) (t = 18°)<br>on the ethyl ester | AA 77 |
| | | Amide | (+)-Tartaric acid<br>in acetone-$H_2O$ | + 6.7 ± 1.1° (c 1.08, EtOH) (t = 25°) | AA 78 |
| $C_9H_{11}NO_2$ | $C_6H_5$-CH$_2$CH(NH$_2$)COOH<br>$\beta$-Phenylalanine | Z | (-)-Ephedrine<br>(+)-Ephedrine<br>both in $Et_2O$ | a) + 8.4° (EtOH) (t = 22-26°)<br>b) - 8.3° (EtOH) (t = 22-26°)<br>both on the Z-derivatives | AA 8 |
| | | Z | Also resolved with (-)- and (+)-$\alpha$-methylbenzylamines (B 86). | | |

TABLE 13 (continued)

| | Substance resolved | Derivative used | Resolving agent and solvent | Rotation $[\alpha]_D$ of resolved compound | Reference |
|---|---|---|---|---|---|
| $C_9H_{11}NO_2$ | β-Phenylalanine (continued) | N-Formyl | (-)-α-Methylbenzylisothiouronium acetate | a) - 72° (c 2.5, EtOH) (t = 20°) | A 103 (p.126) |
| | | N-Formyl | (+)-Enantiomer<br>both in $MeOH-H_2O$ | b) + 74° (c 5, EtOH) (t = 20°)<br>both on the N-formyl derivatives | A 103 |
| | | N-Formyl | Also resolved with (-)-2-amino-1-butanol (B 58) (p. 185). | | |
| | | O-tertBu | (+)-Carbobenzoxyphenylalanine in $Et_2O$ | a) + 43.8° (c 2, EtOH) (t = 25°)<br>b) - 44.4° (c 2, EtOH) (t = 25°)<br>both on the tert-Bu esters HCl salts | AA 79 |
| | | — | Also, resolutions with (-)-pantolactone (AA 55) and with cinchonine methohydroxide (AA 80). | | |
| | Substituted β-Phenylalanines | | | | |
| $C_9H_9N_3O_7$ | 3,5-Dinitro-4-hydroxy (3,5-Dinitrotyrosine) | Ac | (-)-α-Methylbenzylamine<br>(+)-α-Methylbenzylamine<br>both in EtOH | a) - 12.2° (t = 21°)<br>b) + 12.2° (t = 21°)<br>both on the Ac-derivatives | AA 81 |
| $C_9H_{11}NO_3$ | 3-Hydroxy (m-Tyrosine) | N-Formyl | Brucine in 95% EtOH | a) - 44.7° (c 1%, $H_2O$) (t = 27°)<br>b) + 45.7° ($H_2O$)<br>both on the N-formyl derivatives | AA 82 |
| $C_{10}H_{11}NO_4$ | 3,4-Methylenedioxy (Methylene DOPA) | Ac | Cinchonine in EtOH | a) - 53.4° (c 1.841, EtOH) (t = 18°)<br>b) + 53.4° (c 2.262, EtOH) (t = 13°)<br>both on the Ac-derivatives | AA 83 |
| | | Ac | Also, resolutions with (+)-threo-2-amino-1-(p-nitrophenyl)-1,3-propanediol (AA 84) and with (+)-ephedrine (AA 85). | | |
| $C_{10}H_{13}NO_4$ | 3-Methoxy-4-hydroxy | Bz | Resolution with dehydroabietylamine | | AA 86 |

TABLE 13 (continued)

| | Substance resolved | Derivative used | Resolving agent and solvent | Rotation $[\alpha]_D$ of resolved compound | Reference |
|---|---|---|---|---|---|
| | Substituted β-Phenylalanines (continued) | | | | |
| $C_{11}H_{15}NO_4$ | 3,4-Dimethoxy | Ac | Resolution with (+)- and (-)-α-methylbenzylamines (AA 87, AA 88), with (+)- and (-)-threo-2-amino-1-(p-nitrophenyl)-1,3-propanediol (AA 89a), and with (+)-ephedrine (AA 85). | | |
| $C_{12}H_{15}NO_5$ | 3-Methoxy-4-acetoxy | Ac | Resolution with (+)- and (-)-α-methylbenzylamines (AA 88), and with (+)- and (-)-threo-2-amino-1-(p-nitrophenyl)-1,3-propanediol (AA 89a, AA 89b). | | |
| $C_{12}H_{17}NO_5$ | 2,4,5-Trimethoxy | Bz | (-)-α-Methylbenzylamine<br>(+)-α-Methylbenzylamine<br>both in isoPrOH | a) $-53.1^{o}$ (c 1, MeOH) ($t = 25^{o}$)<br>b) $+52.7^{o}$ (c 1, MeOH) ($t = 24^{o}$)<br>both on the Bz-derivatives | AA 90 |
| $C_{24}H_{25}NO_5$ | 3,5-bis(Benzyloxy)-4-methoxy | Ac | Resolution with (+)-ephedrine | | AA 91 |
| $C_9H_{11}NO_2$ | $C_6H_5$-CH($NH_2$)$CH_2$COOH | N-Formyl | Quinidine<br>in MeOH | $+103^{o}$ (c 0.64%, EtOH) ($t = 22^{o}$)<br>on the N-formyl derivative | A 151 (p.129) |
| $C_9H_{11}NO_3$ | R-$C_6H_4$-CH(OH)-CH($NH_2$)COOH  R = H<br>threo-β-phenylserine | Ac | (+)-threo-2-Amino-1-(p-nitrophenyl)-1,3-propanediol | a) $+18.0 \pm 3^{o}$ (c 2.02, MeOH) | AA 92 |
| | | Ac | (-)-Enantiomer<br>both in $H_2O$ | b) $-18.5 \pm 1^{o}$ (c 2.0, MeOH)<br>both on the Ac-derivatives | AA 92 |
| | | Ac | Resolution with quinine | | AA 93 |
| | | — | Resolution on the cyclohexyl ester with (-)-2-pyrrolidone-5-carboxylic acid (AA 92). | | |
| $C_9H_{10}N_2O_5$ | R = $NO_2$<br>erythro-β-(p-nitrophenyl)serine | OEt | (-)-O,O'-Dibenzoyltartaric acid in EtOH | a) $+51.5 \pm 2^{o}$ (c 1.03%, EtOH) ($t = 20^{o}$)<br>b) $-51.2 \pm 2^{o}$ (c 1.102%, EtOH) ($t = 20^{o}$)<br>both on the ethyl esters | AA 94 |

TABLE 13 (continued)

| | Substance resolved | Derivative used | Resolving agent and solvent | Rotation $[\alpha]_D$ of resolved compound | Reference |
|---|---|---|---|---|---|
| $C_9H_{15}NO_3S$ | S; O=; N; H; H; $(CH_2)_5COOH$ | — | Brucine<br>in EtOH | a) $- 51.4^o$ (c 1, MeOH) ($t = 25^o$)<br>b) $+ 57^o$ ($t = 25^o$) | AA 95 |
| $C_{10}H_9N_2O_3I_3$ | $NH_2$; I; I; I; COOH; $CH_3$; $CH_3$; NC; O | — | Strychnine | a) $+ 13.0^o$ (DMF) ($t = 25^o$)<br>b) $- 12.3^o$ (DMF) ($t = 25^o$) | AA 96 |
| $C_{10}H_{10}N_2O_2$ | $CH_3$; N-CHCOOH; N | — | (+)-$\alpha$-Methylbenzylamine<br>in EtOAc<br>Cinchonidine<br>in EtOAc-EtOH (3 : 1) | a) $+ 21.3^o$ (c 0.869, DMSO) ($t = 25^o$)<br>b) $- 22.2^o$ (c 0.586, DMSO) ($t = 25^o$) | AA 97<br>AA 97 |
| $C_{10}H_{10}N_2O_2S$ | See Table 8e | | | | |
| $C_{10}H_{10}N_2O_3$ | $CH_3$; COOH; N; NO | — | (-)-Ephedrine<br>Brucine<br>both in EtOAc | a) $+ 12^o$ (c 2, EtOH) ($t = 24^o$)<br>b) $- 12^o$ (c 2, EtOH) ($t = 24^o$) | B 162 (p.191)<br>B 162 |
| $C_{10}H_{11}NO_4$ | See $C_9H_{11}NO_2$ | Substituted $\beta$-phenylalanines | | | |
| $C_{10}H_{13}NO_2$ | $CH_3$; $CH_2CCOOH$; $NH_2$<br>$\alpha$-Methylphenylalanine | Ac<br>Ac | Sodium (-)-menthoxide on the oxazolin-5-one cyclization product in $C_6H_6$; *via* ester formation<br>Resolutions with (-)- and (+)-$\alpha$-methylbenzylamines (AA 99) and with cinchonidine and cinchonine (AA 100). | a) $+ 79.3^o$ (c 1.082, MeOH) ($t = 20^o$)<br>b) $- 78.4^o$ (c 1.036, MeOH) ($t = 19^o$)<br>both on the Ac-derivatives | AA 98<br>AA 98 |

TABLE 13 (continued)

| | Substance resolved | Derivative used | Resolving agent and solvent | Rotation $[\alpha]_D$ of resolved compound | Reference |
|---|---|---|---|---|---|
| | Substituted $\alpha$-Methylphenylalanines | | | | |
| $C_{10}H_{13}NO_3$ | 4-Hydroxy ($\alpha$-Methyltyrosine) | Ac | (-)-$\alpha$-Methylbenzylamine<br>(+)-$\alpha$-Methylbenzylamine<br>both in acetone | a) $-61.1^{\circ}$ (c 1-2 %, MeOH) (t = $25^{\circ}$)<br>b) $+61.0^{\circ}$ (c 1-2 %, MeOH) (t = $25^{\circ}$)<br>both on the Ac-derivatives | AA 99<br>AA 99 |
| $C_{10}H_{13}NO_4$ | 3,4-Dihydroxy ($\alpha$-Methyl DOPA) | N-CH(OH)$C_6H_5$ OMe | D-Tartaric acid in acetone | a) $-32^{\circ}$ (c 1 %, MeOH)<br>b) $+32^{\circ}$ (c 1 %, MeOH)<br>both on the N-CH(OH)$C_6H_5$ OMe derivatives | AA 101 |
| $C_{11}H_{13}NO_4$ | 3,4-Methylenedioxy | | Resolution with tartaric acid | | AA 102 |
| $C_{12}H_{17}NO_4$ | 3,4-Dimethoxy | Ac | (-)-$\alpha$-Methylbenzylamine | a) $-56^{\circ}$ (c 1, MeOH) (t = $25^{\circ}$)<br>on the Ac-derivative<br>$-4^{\circ}$ (c 2, 0.1 N aq. HCl) (t = $25^{\circ}$) | AA 103 |
| | | Ac | (+)-$\alpha$-Methylbenzylamine<br>both in MeOH | b) $+4^{\circ}$ (c 2, 0.1 N aq. HCl) (t = $25^{\circ}$) | AA 103 |
| | | Ac | Sodium (-)-menthoxide on the oxazolin-5-one cyclization product in $C_6H_6$; *via* ester formation | a) $+60.6^{\circ}$ (c 1.024, MeOH)<br>b) $-58.2^{\circ}$ (c 1.11, MeOH)<br>both on the Ac-derivatives | AA 37 |
| $C_{14}H_{17}NO_6$ | 3,4-Diacetoxy | Ac | Quinine in acetone | $-74.5^{\circ}$ (c 1, 96 % EtOH) (t = $25^{\circ}$)<br>on the Ac-derivative | AA 103 |
| $C_{10}H_{13}NO_2$ | $C_6H_5$-C($CH_2CH_3$)($NH_2$)COOH | N-Formyl | Quinine in EtOH-$H_2O$ | a) $+126^{\circ}$ (c 1.05, 1 N aq. NaOH) (t = $26^{\circ}$)<br>b) $-124^{\circ}$ (c 1.03, 1 N aq. NaOH) (t = $27^{\circ}$)<br>both on the N-formyl derivatives | AA 104 |

TABLE 13 (continued)

| Formula | Substance resolved | Derivative used | Resolving agent and solvent | Rotation $[\alpha]_D$ of resolved compound | Reference |
|---|---|---|---|---|---|
| $C_{10}H_{13}NO_2S$ | $CH_2CH_2CHCOOH$ with S–$C_6H_5$ on the first $CH_2$ and $NH_2$ on CH | Ac | Strychnine in $H_2O$ | a) - 10.1° (c 1, 95% EtOH) (t = 24°)<br>b) + 9.6° (c 1, 95% EtOH) (t = 25°)<br>both on the Ac-derivatives | AA 105 |
| $C_{10}H_{13}NO_3S$ | $CH_3S$–$C_6H_4$–CH(OH)–CH($NH_2$)COOH<br>erythro-β-(p-Methylmercapto-phenyl)-serine | Ac | (+)-threo-2-Amino-1-(p-nitro-phenyl)-1,3-propanediol in MeOH | a) + 52° (c 2%, DMF) (t = 20°)<br>b) - 52° (c 2%, DMF) (t = 20°)<br>both on the Ac-derivatives | AA 106 |
| $C_{10}H_{13}NO_4$ | See $C_9H_{11}NO_2$ Substituted β-phenylalanines<br>See $C_{10}H_{13}NO_2$ Substituted α-methylphenylalanines | | | | |
| $C_{10}H_{13}NO_4$ | See $C_{15}H_{18}N_2O_6$ below | | | | |
| $C_{10}H_{13}NO_5$ | HO–$C_6H_3$($OCH_3$)–CH(OH)–CH($NH_2$)COOH<br>threo-3-(4-Hydroxy-3-methoxy-phenyl)-serine | Z | Quinine | $\alpha_D^{26}$ - 32.5° (1 N HCl) | AA 107 |
| $C_{10}H_{13}N_5O_4$ | Theobromine ring system ($N$-$CH_3$ groups, C=O, R on N) bearing $CH_2CH(NH_2)COOH$ at C-8, R = H<br>β-(8-Theobrominyl)-α-alanine | Z | (+)-threo-2-Amino-1-(p-nitro-phenyl)-1,3-propanediol in EtOH-$H_2O$<br>(-)-Enantiomer | a) - 22° (c 5%, 5 N aq. HCl) (t = 18°)<br>b) + 24° (c 0.5%, 5 N aq. HCl) (t = 18°)<br>- 14° (c 1%, 2 N aq. HCl) (t = 25°) on the Z-derivative | AA 108<br>AA 108 |

| Substance resolved | Derivative used | Resolving agent and solvent | Rotation $[\alpha]_D$ of resolved compound | Reference |
|---|---|---|---|---|
| $C_{10}H_{14}N_2O_2$ (tetrahydroindazolyl-N-$CH(CH_3)COOH$) | 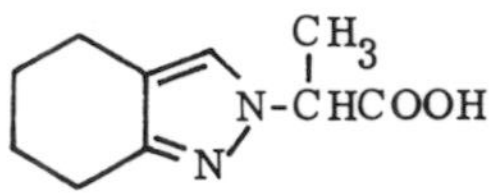 — | Cinchonidine in acetone-EtOAc (2:1) | a) + 36.4° (c 1.008, $CHCl_3$) (t = 25°)<br>b) - 35.9° (c 1.004, $CHCl_3$) (t = 25°) | AA 109 |
| $C_{10}H_{14}N_2O_4$ (3,4-(HO)$_2C_6H_3$-$CH_2C(CH_3)(NHNH_2)COOH$) | 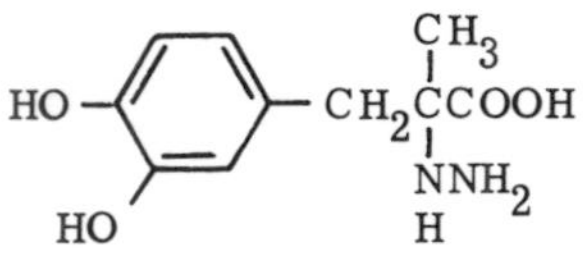 Nitrile | (-)-Menthoxyacetyl chloride; hydrazide Recr.: $CHCl_3$ | a) - 17.3° (MeOH)<br>b) + 17° (MeOH) | AA 110 |
| 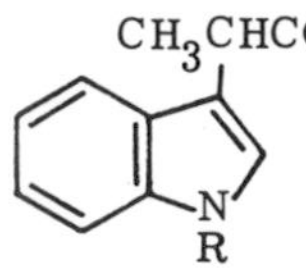 $C_{11}H_{11}NO_2$ (indol-3-yl-$CH(CH_3)COOH$), R = H | — | Brucine in EtOH-$H_2O$ | a) + 79.4 ± 0.5° (c 0.850, EtOH) (t = 25°) | AA 111 |
| | | Cinchonidine in EtOAc-MeOH | b) - 78.8 ± 0.5° (c 0.832, EtOH) (t = 25°) | AA 111 |
| 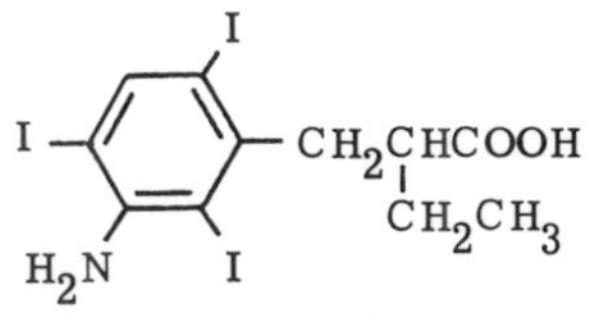 $C_{11}H_{12}NO_2I_3$ Iodopanoic acid | — | (-)-α-Methylbenzylamine<br>(+)-α-Methylbenzylamine<br>both in 95 % EtOH | a) - 5.2 ± 0.1° (c 2, EtOH) (t = 20°)<br>b) + 5.1 ± 0.1° (c 2, EtOH) (t = 20°) | AA 112 |
| 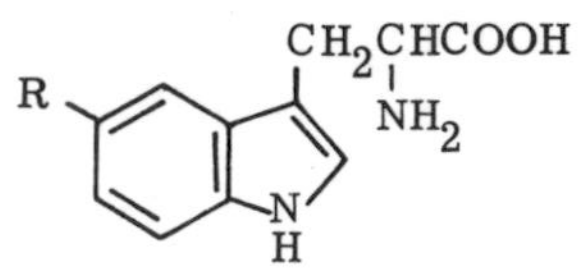 $C_{11}H_{12}N_2O_2$ Tryptophan R = H | Ac | (+)-Lysine in MeOH-$H_2O$ | + 21.2° (c 0.85, EtOH) (t = 29°) on the Ac-derivative | AA 113 |
| | Ac | Resolution with (-)-α-methylbenzylamine | | AA 114 |
| | Z | Quinine in acetone | a) + 15.55° (c 5, 1 equiv NaOH) (t = 25°)<br>b) - 10.11° (c 5, 1 equiv NaOH) (t = 25°)<br>both on the Z-derivatives | B 86 (p. 187) |
| | — | Also, resolutions with inosine (AA 115) and with (-)-pantolactone (AA 55). | | |

TABLE 13 (continued)

| Formula | Substance resolved | Derivative used | Resolving agent and solvent | Rotation $[\alpha]_D$ of resolved compound | Reference |
|---|---|---|---|---|---|
| $C_{11}H_{12}N_2O_3$ | (structure: indol-3-yl-$CH_2C(OH)(NH_2)COOH$) α-Hydroxy-tryptophan | Ac | Brucine in 95 % EtOH | a) - 29.0° (c 1.91%, 1 N aq. NaOH)<br>b) + 28.4° (c 1.99%, 0.1 N aq. NaOH)<br>both on the Ac-derivatives | AA 116 |
| $C_{11}H_{12}N_2O_3$ | See $C_{11}H_{12}N_2O_2$ R = OH | Z, O-Benzyl | Quinidine<br>Quinine<br>both in $C_6H_6$ | a) - 10.3° (c 1, EtOH) (t = 21°)<br>b) + 10.15° (c 1, EtOH) (t = 25°)<br>both on the O-benzyl Z-derivatives | AA 117 |
| $C_{11}H_{13}NO_4$ | See $C_{10}H_{13}NO_2$ | Substituted α-methylphenylalanines | | | |
| $C_{11}H_{15}NO_2$ | $CH_3CH(CH_3)-CH(NHC_6H_5)COOH$<br>N-Phenylvaline | — | Quinine in acetone-MeOH | + 83.8° (c 4.3%, EtOH) (t = 25°)<br>$[\alpha]_{D(max)}$ + 86.0° | AA 118 |
| $C_{11}H_{15}NO_4$ | See $C_9H_{11}NO_2$ | Substituted β-phenylalanines | | | |
| $C_{11}H_{15}N_5O_4$ | See $C_{10}H_{13}N_5O_4$ R = $CH_3$<br>β-(8-Caffeinyl)-α-alanine | Z | (+)-threo-2-Amino-1-(p-nitrophenyl)-1,3-propanediol in EtOH<br>(-)-Enantiomer | a) + 18° (c 1%, 45% EtOH) (t = 20°)<br>b) - 18° (c 1%, 45% EtOH) (t = 20°)<br>both on the Z-derivatives | AA 108<br>AA 108 |
| $C_{12}H_{13}NO_2$ | See $C_{11}H_{11}NO_2$ R = $CH_3$ | — | Cinchonine<br>Cinchonidine<br>both in EtOH-$H_2O$ | a) + 85.6° (c 2.061, acetone) (t = 25°)<br>b) - 85.6° (c 2.028, acetone) (t = 25°) | AA 119<br>AA 119 |
| $C_{12}H_{15}NO_2$ | (structure: 1,2,3,4-tetrahydronaphthalen-2-yl-$NHCH_2COOH$) | N, N-diethyl-amide | Tartaric acid in EtOH-$H_2O$ | a) + 35° (t = 25°)<br>b) - 36.8° (t = 25°)<br>both on the diethylamide derivatives | AA 120 |

TABLE 13 (continued)

| | Substance resolved | Derivative used | Resolving agent and solvent | Rotation $[\alpha]_D$ of resolved compound | Reference |
|---|---|---|---|---|---|
| $C_{12}H_{15}NO_5$ | See $C_9H_{11}NO_2$ | Substituted β-phenylalanines | | | |
| $C_{12}H_{17}NO_4$ | See $C_{10}H_{13}NO_2$ | Substituted α-methylphenylalanines | | | |
| $C_{12}H_{17}NO_5$ | See $C_9H_{11}NO_2$ | Substituted β-phenylalanines | | | |
| $C_{12}H_{18}N_2O_2$ | 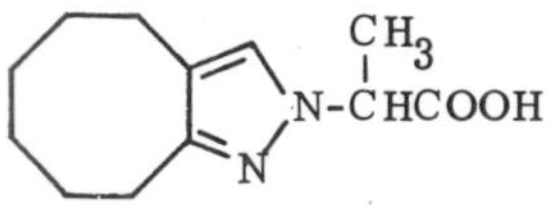 $CH_3$ N-CHCOOH, N | — | (-)-Ephedrine<br>(-)-α-(2-Naphthyl)-ethyl-amine; both in EtOAc | a) - 46.5° ($CHCl_3$) (t = 25°)<br>b) + 46.9° ($CHCl_3$) (t = 25°) | AA 121<br>AA 121 |
| $C_{14}H_{15}NO_2$ | 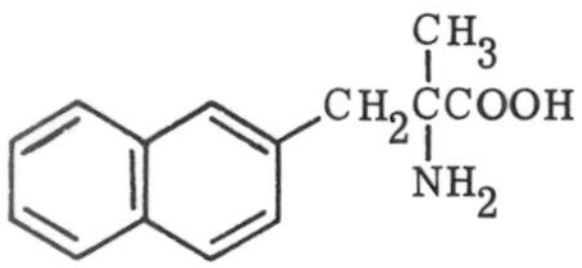 $CH_3$, $CH_2CCOOH$, $NH_2$ | Ac | (+)-α-Methylbenzylamine<br>(-)-α-Methylbenzylamine<br>both in acetone | a) - 160° (c 1%, MeOH) (t = 25°)<br>b) + 140° (c 1%, MeOH) (t = 25°)<br>both on the Ac-derivatives | AA 99 |
| $C_{14}H_{17}NO_6$ | See $C_{10}H_{13}NO_2$ | Substituted α-methylphenylalanines | | | |
| $C_{15}H_{15}NO_2$ | 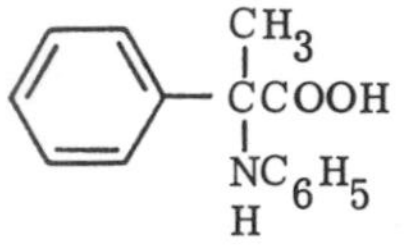 $CH_3$, CCOOH, $NC_6H_5$, H | — | Quinine<br>in MeOH | - 29.0 ± 2.0° (c 0.50, $CHCl_3$) (t = 28°) | AA 78 |
| $C_{15}H_{15}NO_2$ | 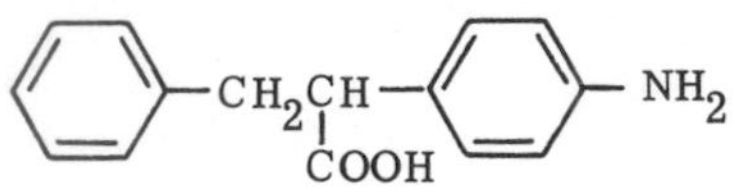 $CH_2CH$, COOH, $NH_2$ | — | (+)-Deoxyephedrine<br>in EtOAc | a) + 102° (c 5%, $H_2O$) (t = 25°)<br>b) - 103° (c 5%, $H_2O$) (t = 25°) | AA 122 |
| $C_{15}H_{18}N_2O_6$ | 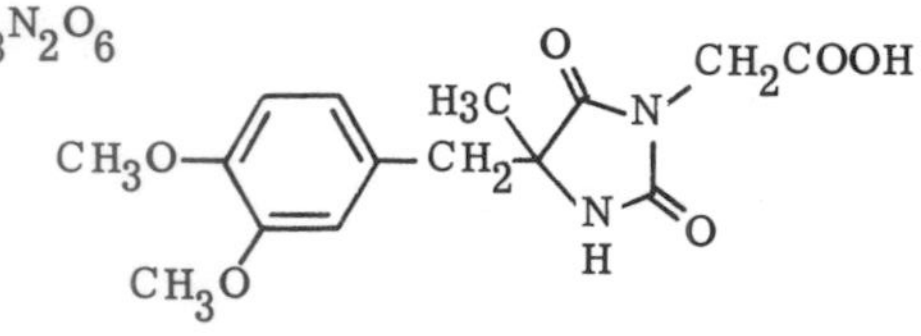 $CH_3O$, $CH_3O$, H3C, $CH_2$, O, N, $CH_2COOH$, N, H, O | — | (+)-threo-2-Amino-1-phenyl-1,3-propanediol<br>in BuOH-EtOH<br>(-)-Enantiomer<br>in BuOH | a) + 81.8° (c 1, MeOH) (t = 20°)<br>b) - 82° (c 1, MeOH) (t = 20°) | AA 123<br>AA 123 |

TABLE 13 (continued)

| Substance resolved | | Derivative used | Resolving agent and solvent | Rotation $[\alpha]_D$ of resolved compound | Reference |
|---|---|---|---|---|---|
| $C_{16}H_{16}NO_4BrS$ 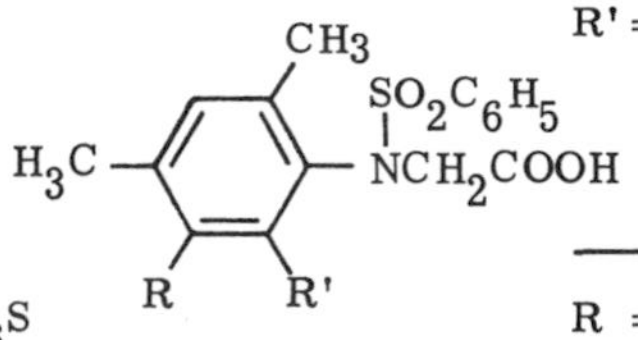 (structure: $H_3C$, $CH_3$, $SO_2C_6H_5$, $NCH_2COOH$, R, R') | R = H<br>R' = Br | — | Cinchonidine<br>in EtOAc-MeOH (9:1) | a) $-15^o$ (c 0.98, EtOH) (t = $28^o$)<br>b) $+15^o$ (c 1.02, EtOH) (t = $28^o$) | AA 124 |
| $C_{16}H_{16}N_2O_6S$ | R = H<br>R' = $NO_2$ | — | Cinchonine<br>in EtOAc-EtOH (2:5) | $+236^o$ (c 0.53, EtOH) (t = $30^o$) | AA 124 |
| $C_{17}H_{18}NO_4BrS$ | R = Br<br>R' = $CH_3$ | — | Cinchonidine<br>in EtOAc | a) $-6^o$ (c 1.02, EtOH) (t = $25^o$)<br>b) $+6^o$ (c 0.99, EtOH) (t = $25^o$) | AA 124 |
| $C_{19}H_{17}NO_4S$ 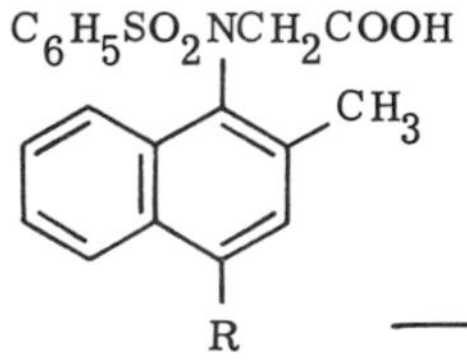 (structure: $C_6H_5SO_2NCH_2COOH$, $CH_3$, R) | R = H | — | Cinchonidine<br>in EtOAc-MeOH | a) $-82.4^o$ (DMF) (t = $30^o$) | AA 125 |
| | | — | Cinchonine<br>in $CHCl_3$ | b) $+91.0^o$ (c 2.62, DMF) (t = $28^o$) | AA 126 |
| | R = $NHCOCH_3$ | — | Brucine | | AA 126 |
| Also, | R = CN, $NO_2$, $C_6H_5S$, $C_6H_5SO_2$, $CH_3$, $CH_3O$, $NHCOC_6H_5$, OH | — | Cinchonidine | | AA 125, AA 126 |
| Also, | R = Br, Cl, I, $NO_2$, $NHSO_2C_6H_5$ | — | Cinchonine | | AA 126, AA 127 |
| $C_{20}H_{18}N_2O_4$ 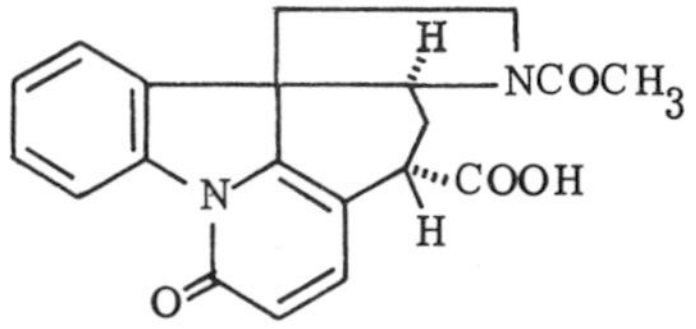 (structure: H, $NCOCH_3$, COOH, H, N, O) | | — | Quinidine<br>in $CHCl_3$-MeOH | $-285 \pm 5^o$ (c 1.25, $CHCl_3$) (t = $23^o$)<br>on the methyl ester | AA 128 |

TABLE 13 (continued)

| Formula | Substance resolved | Derivative used | Resolving agent and solvent | Rotation $[\alpha]_D$ of resolved compound | Reference |
|---|---|---|---|---|---|
| $C_{24}H_{25}NO_4S$ | $C_6H_5SO_2NCH_2COOH$ R = H (2,4,6-trimethyl-3-benzylphenyl with $H_3C$, $CH_3$, $CH_2$, $CH_3$ and R on the benzyl ring) | — | Cinchonidine in MeOH | $-53.5 \pm 1.0^o$ (DMF) ($t = 26^o$) | AA 129 |
| | Also, $R = CH_3$, Br, CN, $OCH_3$ | — | Cinchonidine in MeOH | | AA 129 |
| $C_{24}H_{25}NO_5$ | See $C_9H_{11}NO_2$ | Substituted β-phenylalanines | | | |

## References to Table 13
## Resolutions of Amino Acids, Amino Acid Derivatives and Related Compounds


(AA 1) M. Portelli and B. Soranzo, Farmaco, Ed. Sci., 17, 909 (1962) [Chem. Abstr., 63, 13235e (1965)].

(AA 2) Z. Gronka and B. Liberek, Tetrahedron, 27, 1783 (1971).

(AA 3) K. Vogler and P. Lanz, Helv. Chim. Acta, 49, 1348 (1966).

(AA 4) V.K. Burichenko, K.T. Poroshin and S.B. Davidyants, Dokl. Akad. Nauk Tadzh. SSR, 9, 21 (1966) [Chem. Abstr., 65, 2343a (1966)].

(AA 5) B. Halpern and J.W. Westley, Chem. Commun., 1965, 421.

(AA 6) V.M. Belikov, T.F. Savel'eva and E.N. Safonova, Izv. Akad. Nauk SSSR, Ser. Khim., 1971, 1461 [Chem. Abstr., 75, 88925n (1971)].

(AA 7) B. Liberek and S. Dziala, Zesz. Nauk. Wyzsz. Szk. Pedagog. Gdansku: Mat., Fiz., Chem., 1969, 165.

(AA 8) K. Oki, K. Suzuki, S. Tuchida, T. Saito and H. Kotake, Bull. Chem. Soc. Japan, 43, 2554 (1970).

(AA 9) T. Perlotto and M. Vignolo, Farmaco, Ed. Sci., 21, 30 (1966) [Chem. Abstr., 64, 19756g (1966)].

(AA 10) A. Stoll and T. Petrzilka, Helv. Chim. Acta, 35, 589 (1952).

(AA 11) A.S. Dutta and J.S. Morley, Chem. Commun., 1971, 883.

(AA 12) E. Wünsch and G. Fürst, Hoppe-Seyler's Z. Physiol. Chem., 329, 109 (1962).

(AA 13) D. Pitrè, S. Boveri and N. Buser, Farmaco, Ed. Sci., 23, 244 (1968) [Chem. Abstr., 69, 27757p (1968)].

(AA 14) D. Pitrè, S. Boveri and N. Buser, Farmaco, Ed. Sci., 23, 945 (1968) [Chem. Abstr., 70, 4563d (1969)].

(AA 15) I. Gimesi, S. Geri, G. Toth and S. Bajusz, Hung. Patent 155,047, Aug. 27, 1968 [Chem. Abstr., 70, 4605u (1969)].

(AA 16) T. Fujii and T. Fujita, Yakugaku Zasshi, 87, 95 (1967) [Chem. Abstr., 67, 64675q (1967)].

(AA 17) E. Felder, D. Pitrè and S. Boveri, Hoppe-Seyler's Z. Physiol. Chem., 351, 943 (1970).

(AA 18) W. Steglich, H.-U. Heininger, H. Dworschak and F. Weygand, Angew. Chem., 79, 822 (1967); Angew. Chem. Int. Ed. Engl., 6, 807 (1967).

(AA 19) R.M. Rodebaugh and N.H. Cromwell, J. Heterocycl. Chem., 6, 993 (1969).

(AA 20) K. Harada, Bull. Chem. Soc. Japan, 37, 1383 (1964).

(AA 21) H. Okai, N. Imamura and N. Izumyia, Bull. Chem. Soc. Japan, 40, 2154 (1967).

(AA 22) Y. Liwschitz, A. Singerman and Y. Wiesel, Israel J. Chem., 6, 647 (1968).

(AA 23) Y. Liwschitz, Y. Edlitz-Pfeffermann and A. Singer, J. Chem. Soc. C, 1967, 2104.

(AA 24) S.D. Mikhno, N.S. Kulachkina and V.M. Berezovskii, Zh. Org. Khim., 6, 81 (1970) [Engl. transl., p. 80].

(AA 25) G. Losse, Chem. Ber., 87, 1279 (1954).

(AA 26) D. Pitrè and S. Boveri, Farmaco, Ed. Sci, 26, 733 (1971) [Chem. Abstr., 75, 87990t (1971)].

(AA 27) M.D. Armstrong, J. Amer. Chem. Soc., 73, 4456 (1951).

(AA 28) K. Vogler and P. Lanz, Helv. Chim. Acta, 42, 209 (1959).

(AA 29) B.G. Christensen and W.J. Leanza, Ger. Offen., 1,816,103, July 24, 1969 [Chem. Abstr., 71, 91871q (1969)].

(AA 30) T. Inui and T. Kaneko, Nippon Kagaku Zasshi, 82, 1078 (1961) [Chem. Abstr., 59, 747a (1963)].

(AA 31) H. Arold, J. Prakt. Chem., 24, 23 (1964).

(AA 32) S. Lindstedt and G. Lindstedt, Ark. Kemi, 22, 93 (1964).

(AA 33) K. Okawa, K. Hori, K. Hirose and Y. Nakagawa, Bull. Chem. Soc. Japan, 42, 2720 (1969).

(AA 34) E. Iwasaki, Japan. Patent 69 29,459, Dec. 1, 1969 [Chem. Abstr., 72, 43647v (1970)].

(AA 35) A.V. Robertson and B. Witkop, J. Amer. Chem. Soc., 84, 1697 (1962).

(AA 36) S.G. Cohen and E. Khedouri, J. Amer. Chem. Soc., 83, 1093 (1961).

(AA 37) S. Terashima, K. Achiwa and S. Yamada, Chem. Pharm. Bull. (Tokyo), 13, 1399 (1965).

(AA 38) S. Akabori, T. Ikenaka and K. Matsumoto, Nippon Kagaku Zasshi, 73, 112 (1952); idem, Proc. Japan Academy, 27, 7 (1951).

(AA 39) K. Lenard and S. Bajusz, Hung. Patent 148,731, Dec. 15, 1961 [Chem. Abstr., 58, 10304a (1963)].

(AA 40) Brit. Patent 1,167,455, Oct. 15, 1969 [Chem. Abstr., 72, 3763y (1970)].

(AA 41) V. du Vigneaud and R.W. Holley, U.S. Patent 2,543,358, Feb. 27, 1951 [Chem. Abstr., 45, 7601g (1951)].

(AA 42) J. Oh-hashi and K. Harada, Bull. Chem. Soc. Japan, 39, 2287 (1966).

(AA 43) G.I. Tesser, R.J.F. Nivard and M. Gruber, Rec. Trav. Chim. Pays-Bas, 81, 713 (1962).

(AA 44) S. Gronowitz, I. Sjögren, L. Wernstedt and B. Sjöberg, Ark. Kemi, 23, 129 (1965).

(AA 45) H.M. Kolenbrander and C.P. Berg, Arch. Biochem. Biophys., 119, 435 (1967).

(AA 46) A.J. Bosch, J. Chem. Educ., 46, 691 (1969).

(AA 47) M. Claesen, G. Laridon and H. Vanderhaeghe, Bull. Soc. Chim. Belges, 77, 579 (1968).

(AA 48) L. Baláspiri, B. Denke, J. Petres and K. Kovács, Monatsh. Chem., 101, 1177 (1970).

(AA 49) H.C. Beyerman, Rec. Trav. Chim. Pays-Bas, 78, 134 (1959).

(AA 50) A. Shafi'ee and G. Hite, J. Med. Chem., 12, 266 (1969); R.L. Peck and A.R. Day, J. Heterocycl. Chem., 6, 181 (1969).

(AA 51) M. Claesen, A. Vlietinck and H. Vanderhaeghe, Bull. Soc. Chim. Belges, 77, 587 (1968).

(AA 52) M.S. Rabinovich, M.F. Shostakovskii and E.V. Preobrazhenskaya, Zh. Obshch. Khim., 30, 71 (1960) [Engl. transl., p. 73].

(AA 53) J.E. Nelemans, A.H. Pecasse, W. Pesch and U. Verstrijden, U.S. Patent 3,105,067, Sept. 24, 1963 [Chem. Abstr., 60, 2794 d (1964)].

(AA 54) J. Oonigi, K. Shibata, C. Hongo and M. Shibazaki, Japan. Patent 71 12,130, Mar. 29, 1971 [Chem. Abstr., 75, 36688t (1971)].

(AA 55) N. Okuda, I. Kuniyoshi and K. Saito, Japan. Patent 68 05,727, Mar. 2, 1968 [Chem. Abstr., 69, 97170r (1968)].

(AA 56) G.H. Suvercropp, Ger. Offen. 1,927,698, Dec. 4, 1969 [Chem. Abstr., 72, 55888n (1970)].

(AA 57) Belg. Patent 655,709, May 13, 1965 [Chem. Abstr., 65, 3964d (1966)].

(AA 58) S. Watanabe and K. Suga, Israel J. Chem., 7, 483 (1969).

(AA 59) K. Ueda, M. Iizuka, K. Shimada and H. Yamamoto, Ger. Offen., 2,024,062, Dec. 3, 1970 [Chem. Abstr., 74, 64393r (1971)].

(AA 60) R.D. Emmick, U.S. Patent 2,556,907, June 12, 1951 [Chem. Abstr., 46, 525h (1952)]; A.O. Rogers, U.S. Patent 2,657,230, Oct. 27, 1953 [Chem. Abstr., 48, 2767e (1954)].

(AA 61) M. Hara, K. Togo and T. Akashi, Japan. Patent 64 28,232, Dec. 7, 1964 [Chem. Abstr., 62, 11912a (1965)].

(AA 62) H. Kotake, T. Saito and K. Okubo, Bull. Chem. Soc. Japan, 42, 1367 (1969).

(AA 63) B.A. Ekstrom and B. Sjöberg, Ger. Offen., 1,912,904, Oct. 9, 1969 [Chem. Abstr., 72, 3484h (1970)].

(AA 64) A.L. Palomo-Coll, Afinidad, 28, 141 (1971).

(AA 65) B.O.H. Sjöberg and B.A. Ekstrom, Fr. Patent 1,351,144, Jan. 31, 1964 [Chem. Abstr., 60, 14441d (1964)].

(AA 66) A.A.W. Long, J.H.C. Nayler, H. Smith, T. Taylor and N. Ward, J. Chem. Soc. C, 1971, 1920.

(AA 67) J.C. Sheehan and K.R. Henery-Logan, J. Amer. Chem. Soc., 81, 3089 (1959).

(AA 68) J.H. Hunt and D. McHale, J. Chem. Soc., 1957, 2073.

(AA 69) S. Hase, R. Kiyoi and S. Sakakibara, Bull. Chem. Soc. Japan, 41, 1266 (1968).

(AA 70) A. Fredga and A. Lindgren, unpublished work; personal communication from A. Fredga.

(AA 71) A. Fredga and H. Gustafsson, unpublished work; personal communication from A. Fredga.

(AA 72) H. Kinoshita, M. Shintani, T. Saito and H. Kotake, Bull. Chem. Soc. Japan, 44, 286 (1971).

(AA 73) P.S. Portoghese, J. Pharm. Sci., 53, 228 (1964); idem., J. Med. Chem., 8, 147 (1965).

(AA 74) A. Fredga, Ark. Kemi, 11, 23 (1957).

(AA 75) D.J. Cram, L.K. Gaston and H. Jäger, J. Amer. Chem. Soc., 83, 2183 (1961).

(AA 76) H. Mizuno, S. Terashima, K. Achiwa and S. Yamada, Chem. Pharm. Bull. (Tokyo), 15, 1749 (1967).

(AA 77) Y. Sugi and S. Mitsui, Bull. Chem. Soc. Japan, 42, 2984 (1969).

(AA 78) H. Dahn, J.A. Garbino and C.O'Murchu, Helv. Chim. Acta, 53, 1370 (1970).

(AA 79) T. Sokolowska, Rocz. Chem., 40, 1895 (1966) [Chem. Abstr., 67, 64667p (1967)].

(AA 80) R.B. Fearing, Ph.D. Thesis, Iowa State University, 1951 and J.R. Mosser, M.S. Thesis, Iowa State University, 1952. Quoted in ref. B 58.

(AA 81) D. Pitrè and E.B. Grabitz, Hoppe-Seyler's Z. Physiol. Chem., 333, 105 (1963).

(AA 82) R.R. Sealock, M.E. Speeter and R.S. Schweet, J. Amer. Chem. Soc., 73, 5386 (1951).

(AA 83) S. Yamada, T. Fuji and T. Shioiri, Chem. Pharm. Bull. (Tokyo), 10, 680 (1962).

(AA 84) I. Hever and E. Villanyi, Hung. Teljes 383, June 6, 1970 [Chem. Abstr., 74, 64391p (1971)].

(AA 85) H. Nakamoto, M. Aburatami and M. Inagaku, Ger. Offen. 2,058,516, June 24, 1971 [Chem. Abstr., 75, 64270u (1971)].

(AA 86) A. Kaiser, M. Scheer, W. Haeuserman and L. Marti, Ger. Offen. 1,964,420, July 16, 1970 [Chem. Abstr., 74, 3864y (1971)].

(AA 87) A.M. Krubiner, Ger. Offen. 1,964,422, July 16, 1970 [Chem. Abstr., 73, 66897m (1970)].

(AA 88) E. Berenyi, Z. Budai, L. Pallos, P. Benko and L. Magdanyi, Hung. Teljes 859, Aug. 28, 1970 [Chem. Abstr., 74, 64390n (1971)].

(AA 89) (a) E. Berenyi, Z. Budai, L. Pallos, L. Magdanyi and P. Benko, Hung. Teljes 858, Aug. 28, 1970 [Chem. Abstr., 74, 64389u (1971)]; (b) idem, Ger. Offen. 2,052,995, May 6, 1971 [Chem. Abstr., 75, 49576e (1971)].

(AA 90) B.A. Berkowitz, S. Spector, A. Brossi, A. Focella and S. Teitel, Experientia, 26, 982 (1970).

(AA 91) B. Hegedues and P. Zeller, Can. Patent 866,850, Mar. 23, 1971 [Chem. Abstr., 75, 36685q (1971)].

(AA 92) C.G. Alberti, B. Camerino and A. Vercellone, Gazz. Chim. Ital., 83, 930 (1953).

(AA 93) Swiss Patent 285,135, Dec. 16, 1952 [Chem. Abstr., 48, 8257c (1954)].

(AA 94) G. Carrara, G.F. Cristiani, V. D'Amato, E. Pace and R. Pagani, Gazz. Chim. Ital., 82, 325 (1952).

(AA 95) W.M. McLamore, W.D. Celmer, V.V. Bogert, F.C. Pennington, B.A. Sobin and I.A. Solomons, J. Amer. Chem. Soc., 75, 105 (1953).

(AA 96) J.H. Ackerman, G.M. Laidlaw and G.A. Snyder, Tetrahedron Lett., 1969, 3879.

(AA 97) H. Gustafsson, Ark. Kemi, 31, 415 (1970).

(AA 98) S. Terashima, K. Achiwa and S. Yamada, Chem. Pharm. Bull. (Tokyo), 14, 1138 (1966).

(AA 99) H.R. Almond, Jr., D.T. Manning and C. Niemann, Biochemistry, 1, 243 (1962).

(AA 100) F.W. Bollinger, J. Med. Chem., 14, 373 (1971).

(AA 101) A. Hajós and P. Sohár, Acta Chim. Acad. Sci. Hung., 53, 295 (1967).

(AA 102) H. Dyrsting, Dan. Patent 112,738, Jan. 13, 1969 [Chem. Abstr., 71, 50518m (1969)].

(AA 103) E.W. Tristram, J. ten Broeke, D.F. Reinhold, M. Sletzinger and D.E. Williams, J. Org. Chem., 29, 2053 (1964).

(AA 104) H. Mizuno, S. Terashima and S. Yamada, Chem. Pharm. Bull. (Tokyo), 19, 227 (1971).

(AA 105) M.D. Armstrong and J.D. Lewis, J. Org. Chem., 17, 618 (1952).

(AA 106) M. Portelli, G. Renzi and B. Soranzo, Ann. Chim. (Rome), 60, 160 (1970).

(AA 107) B. Hegedues and P. Zeller, Ger. Offen. 2,004,301, Sept. 3, 1970 [Chem. Abstr., 73, 99206p (1970)].

(AA 108) R.G. Vdovina, A.V. Karpova and E.S. Chaman, Zh. Obshch. Khim., 37, 1007 (1967) [Engl. transl., p. 951].

(AA 109) H. Gustafsson, Ark. Kemi, 29, 587 (1968).

(AA 110) S. Karady, M.G. Ly, S.H. Pines and M. Sletzinger, J. Org. Chem., 36, 1946 (1971).

(AA 111) B. Sjöberg, Ark. Kemi, 12, 251 (1958).

(AA 112) D. Pitrè and S. Boveri, J. Med. Chem., 11, 406 (1968).

(AA 113) J.N. Coker, W.L. Kohlhase, M. Fields, A.O. Rogers and M.A. Stevens, J. Org. Chem., 27, 850 (1962).

(AA 114) L.R. Overby, U.S. Patent 3,167,566, Jan. 26, 1965 [Chem. Abstr., 62, 13232h (1965)].

(AA 115) Y. Suzuki, T. Akashi and T. Nakamura, Japan. Patent 69 20,983, Sept. 8, 1969 [Chem. Abstr., 71, 124919q (1969)].

(AA 116) M. Kotake, T. Sakan and T. Miwa, Chem. Ber., 85, 690 (1952).

(AA 117) H. Rinderknecht, Helv. Chim. Acta, 47, 2403 (1964).

(AA 118) J.F. Klebe and H. Finkbeiner, J. Amer. Chem. Soc., 90, 7255 (1968).

(AA 119) A. Fredga and L.-B. Agenäs, Ark. Kemi, 15, 327 (1960).

(AA 120) G.B. Marini-Bettòlo, H.A. Frediani and S. Chiavarelli, Rend. Ist. Super. Sanità, 15, 850 (1952) [Chem. Abstr., 48, 4489b (1954); Chem. Zentralbl., 124, 5496 (1953)].

(AA 121) H. Gustafsson and H. Ericsson, unpublished work; personal communication from A. Fredga.

(AA 122) T.R. Lewis, M.G. Pratt, E.D. Homiller, B.F. Tullar and S. Archer, J. Amer. Chem. Soc., 71, 3749 (1949).

(AA 123) Neth. Appl. 6,509,816, Feb. 1, 1966 [Chem. Abstr., 65, 3963f (1966)].

(AA 124) R. Adams and J.R. Gordon, J. Amer. Chem. Soc., 72, 2458 (1950).

(AA 125) R. Adams and H.H. Gibbs, J. Amer. Chem. Soc., 79, 170 (1957).

(AA 126) R. Adams and K.V.Y. Sundstrom, J. Amer. Chem. Soc., 76, 5474 (1954).

(AA 127) R. Adams and R.H. Mattson, J. Amer. Chem. Soc., 76, 4925 (1954).

(AA 128) R.B. Woodward, M.P. Cava, W.D. Ollis, A. Hunger, H.U. Daeniker and K. Schenker, Tetrahedron, 19, 247 (1963).

(AA 129) R. Adams and K.R. Brower, J. Amer. Chem. Soc., 78, 663 (1956).

TABLE 14 RESOLUTIONS OF HYDROCARBONS, HALOGEN COMPOUNDS, AND MISCELLANEOUS NEUTRAL COMPOUNDS*)

| Substance resolved | | Resolving agent and solvent | Rotation $[\alpha]_D$ of resolved compound | Reference |
|---|---|---|---|---|
| CHBrClF | Br–C(H)(Cl)(F): H–C–Cl with Br above and F below | Brucine *via* asymmetric destruction (Partial resolution) | $\alpha_{578}$ - 0.125 ± 0.005° (1 1 dm, neat) | H 1 |
| | | For the optical activation of CHBrClF from the ketone $C_3H_3OBrClF$, see Table 11 : $[\alpha]_D^{19}$ + 0.20° (c 125, cyclohexane) and $[\alpha]_D^{19}$ - 0.13° (c 124, cyclohexane). | | K 1 (p. 234) |
| $C_3H_6Br_2$ | $CH_3$–CH(Br)–$CH_2$(Br) | Brucine *via* asymmetric destruction (Partial resolution) | $\alpha_D^{25}$ + 0.81° (1 1 dm, neat) | H 2 |
| $C_4H_6N_2O_6Br_2$ | $O_2NO$–$CH_2$–C(Br)(H)–C(H)(Br)–$CH_2$–$ONO_2$ | Brucine in $CH_2Cl_2$ *via* asymmetric destruction (Partial resolution) | $\alpha_D^{25}$ + 6.26° (1 1 dm, neat) | H 3 |
| $C_4H_8Br_2$ | $CH_3$–C(Br)(H)–C(H)(Br)–$CH_3$ | Brucine *via* asymmetric destruction (Partial resolution) | $\alpha_D^{25}$ - 2.04° (1 1 dm, neat) | H 2 |
| $C_5H_8$ | ($CH_3$)(H)C=C=C(H)($CH_3$) | (+)-sym-Tetra-3-pinanyl-diborane in $Et_2O$-diglyme *via* asymmetric destruction (Partial resolution) | a) - 43.8° ($Et_2O$) | H 4 |
| | | (-)-Enantiomer | b) + 23° (hexane) (t = 25°) | H 5 |

*) References to Table 14 begin on p. 307.

TABLE 14 (continued)

| Formula | Substance resolved | Resolving agent and solvent | Rotation $[\alpha]_D$ of resolved compound | Reference |
|---|---|---|---|---|
| $C_6H_4Cl_6D_2$ | 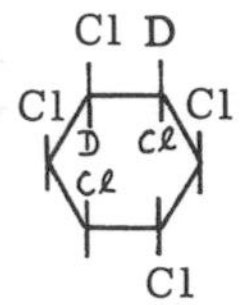 [structure: hexachlorocyclohexane with Cl, D substituents] | Brucine in acetone via asymmetric destruction (Partial resolution) | $\alpha^{20}_D$ + 0.046 ± 0.002° (1 1 dm, MeOAc) | H 6 |
| $C_6H_6Cl_6$ | 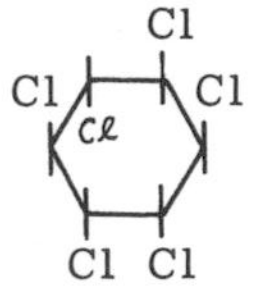 [structure: hexachlorocyclohexane] | Brucine in dioxane via asymmetric destruction | - 14.6° ($Et_2O$) | H 7 |
| $C_6H_{10}$ | 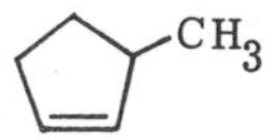 [structure: 3-methylcyclopentene, $CH_3$] | (+)-sym-Tetra-3-pinanyl-diborane in diglyme via asymmetric destruction (Partial resolution) | - 50.8° (t = 26°) | H 8 |
| $C_6H_{12}$ | $CH_3CH_2CH(CH_3)CH=CH_2$ | cis-Dichloro(ethylene)[(-)-α-methyl-benzylamine] platinum (II) (Partial resolution) | - 9.0° (t = 25°) | H 9 |
| $C_7H_{12}$ | 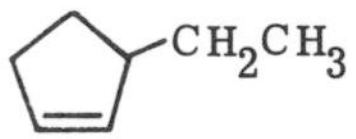 [structure: 3-ethylcyclopentene, $CH_2CH_3$] | (+)-sym-Tetra-3-pinanyl-diborane in diglyme via asymmetric destruction | - 45.2° (t = 26°) | H 8 |
| $C_7H_{12}$ | 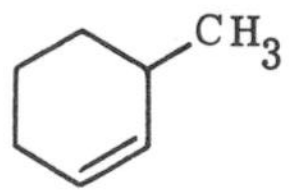 [structure: 3-methylcyclohexene, $CH_3$] | (+)-sym-Tetra-3-pinanyl-diborane in diglyme via asymmetric destruction | + 4.4° | H 10 |
| $C_7H_{12}$ | 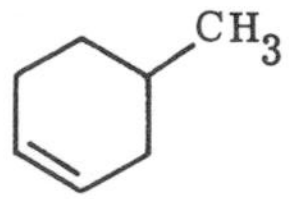 [structure: 4-methylcyclohexene, $CH_3$] | (-)-sym-Tetra-3-pinanyl-diborane in diglyme via asymmetric destruction (Partial resolution) | + 1.11 ± 0.5° (c 13.1, MeOH) (t = 26°) | H 11 |

TABLE 14 (continued)

| | Substance resolved | Resolving agent and solvent | Rotation $[\alpha]_D$ of resolved compound | Reference |
|---|---|---|---|---|
| $C_7H_{14}$ | $CH_3CH(CH_3)-CH(CH_3)CH=CH_2$ | cis-Dichloro(ethylene)[(-)-α-methylbenzylamine] platinum (II) (Partial resolution) | + 41.0° (t = 25°) | H 9 |
| $C_7H_{14}$ | $CH_3CH_2CH_2CH(CH_3)CH=CH_2$ | cis-Dichloro(ethylene)[(-)-α-methylbenzylamine] platinum (II) (Partial resolution) | + 10.4° (t = 25°) | H 9 |
| $C_8H_{12}$ | | (+)-trans-Dichloro(ethylene)-(α-methylbenzylamine)platinum (II) in $CH_2Cl_2$ | a) - 152° (c 0.87, $CH_2Cl_2$) (t = 0°) | H 12 |
| | | (-)-Enantiomer | b) + 109° (c 1.4, 1:1 $CH_2Cl_2$-$C_5H_{12}$) (t = 26°) | H 12 |
| $C_8H_{12}$ | | (+)-sym-Tetra-3-pinanyldiborane in diglyme via asymmetric destruction | - 6.51° (t = 26°) | H 8 |
| $C_8H_{12}$ | | (+)-sym-Tetra-3-pinanyldiborane in diglyme via asymmetric destruction | - 0.73° (t = 20°) | H 10 |
| $C_8H_{14}$ | | (+)-trans-Dichloro(ethylene)-(α-methylbenzylamine)platinum (II) in $CH_2Cl_2$ | a) - 458° (neat) (t = 25°) | H13, B10 (p.182) |
| | | (-)-Enantiomer | b) + 440° (neat) (t = 25°) | H 13, B 10 |
| | | Also, partial resolution via asymmetric destruction (H 14) | | |

TABLE 14 (continued)

| | Substance resolved | Resolving agent and solvent | Rotation $[\alpha]_D$ of resolved compound | Reference |
|---|---|---|---|---|
| $C_9H_{14}$ | $CH{=}C{=}CH$ / $(CH_2)_6$ (ring) | (+)-trans-Dichloro(ethylene)-(α-methylbenzylamine)platinum (II) in $CH_2Cl_2$ (Partial resolution) | $-71^{o}$ (neat) ($t = 24^{o}$) | B 10 (p. 182) |
| $C_9H_{16}$ | H, H | (+)-trans-Dichloro(ethylene)-(α-methylbenzylamine)platinum (II) in $CH_2Cl_2$ | $-119^{o}$ ($C_5H_{12}$) ($t = -20^{o}$) (578 nm) | H 15 |
| $C_9H_{20}$ | $CH_3CH_2CH(CH_3)(CH_2)_4CH_3$ | via crystallization of urea clathrate seeded with optically active hydrocarbon clathrate | a) $\alpha_D^{20}$ $-0.9074 \pm 0.003^{o}$ (l 1.9009, $C_6H_6$)<br>b) $\alpha_D^{20}$ $+0.055 \pm 0.003^{o}$ (l 1.9009, $C_6H_6$) | H 16 |
| $C_{10}H_{12}$ | | via formation and crystallization ($CCl_4$-cyclohexane) of the dimeric complex $[(-)\text{-}C_6H_5CH(CH_3)\text{-}NH_2]$ ($CH_3O$, Pt, Cl)$_2$<br>With the enantiomer | a) $+13.3^{o}$ (c 0.5, $CH_2Cl_2$) ($t = 25^{o}$)<br>b) $-13.3^{o}$ (c 0.5, $CH_2Cl_2$) ($t = 25^{o}$) | H 17 |
| $C_{10}H_{20}$ | $CH_3CH(CH_3)(CH_2)_3CH(CH_3)CH{=}CH_2$ | cis-Dichloro(ethylene)[(-)-α-methylbenzylamine]platinum (II) (Partial resolution) | $+6.9^{o}$ ($t = 25^{o}$) | H 9 |
| $C_{10}H_{22}$ | $CH_3CH_2CH(CH_3)(CH_2)_5CH_3$ | via crystallization of urea clathrate seeded with optically active hydrocarbon clathrate | a) $\alpha_D^{20}$ $-0.007 \pm 0.003^{o}$ (l 1.9009, $C_6H_6$)<br>b) $\alpha_D^{20}$ $+0.126 \pm 0.003^{o}$ (l 1.9009, $C_6H_6$) | H 16 |

TABLE 14 (continued)

| Substance resolved | | Resolving agent and solvent | Rotation $[\alpha]_D$ of resolved compound | Reference |
|---|---|---|---|---|
| $C_{12}H_8Br_2$ | X = Br | Brucine in $CHCl_3$ via asymmetric destruction (Partial resolution) | - 78° (c 3.65 %, $CHCl_3$) (t = 28°) | H 18, H 19 |
| $C_{12}H_8Cl_2$ | X = Cl | Brucine in dioxane via asymmetric destruction (Partial resolution) | - 27° (c 0.4, $Et_2O$) (t = 25°) | H 19 |
| $C_{14}H_{16}O$ | O-CH(CH₃)CH₂CH₃ | (-)-α-(2,4,5,7-Tetranitrofluorenylidene-aminoöxy)-propionic acid in HOAc<br>Also, chromatographic resolution (H 21) | a) - 7.8° (c 6.4, EtOAc) (t = 31°)<br>b) + 6.4° (c 11, EtOAc) (t = 32°) | H 20 |
| $C_{14}H_{22}$ | | (-)-sym-Tetra-3-pinanyl-diborane in diglyme via asymmetric destruction | - 58.2° (c 3.65, $CHCl_3$) (t = 23°) | H 22 |
| $C_{15}H_{12}$ | $C_6H_5(H)C{=}C{=}C(H)C_6H_5$ | (+)-sym-Tetra-3-pinanyl-diborane in $Et_2O$-diglyme via asymmetric destruction (Partial resolution)<br>Also, chromatographic resolution (H 23). See also ref. A 14 (p. 121). | - 180 ± 2° (c 0.4, $CHCl_3$) (t = 26°) | H 4 |
| $C_{15}H_{24}$ | $(CH_3)_3$ | (-)-sym-Tetra-3-pinanyl-diborane in $Et_2O$-diglyme via asymmetric destruction | - 15.7° (t = 20°) | H 24 |

TABLE 14 (continued)

| | Substance resolved | Resolving agent and solvent | Rotation $[\alpha]_D$ of resolved compound | Reference |
|---|---|---|---|---|
| $C_{15}H_{24}$ | 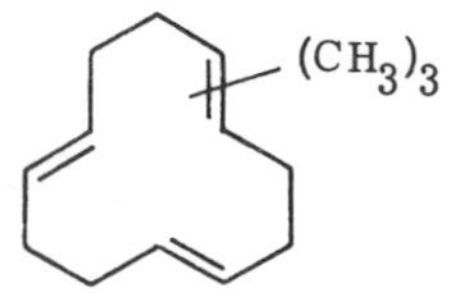 $(CH_3)_3$ | (-)-sym-Tetra-3-pinanyl-diborane in $Et_2O$-diglyme via asymmetric destruction | - 8.0° (t = 20°) | H 24 |
| $C_{16}H_{18}$ | $C_6H_5$-CH($CH_3$)CH($CH_3$)-$C_6H_5$ | See Table 9c, $C_{16}H_{20}N_2$ | | |
| $C_{18}H_{16}O_2$ | 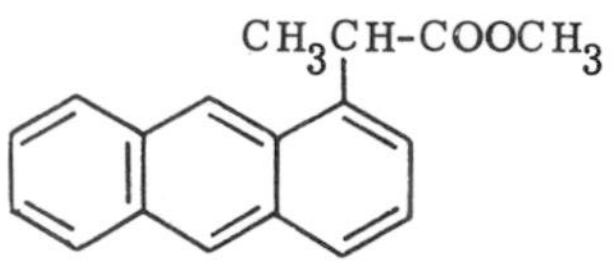 $CH_3CH$-$COOCH_3$ | (+)-α-(2,4,5,7-Tetranitrofluorenylidene-aminoöxy)-propionic acid in HOAc | a) + 101 ± 0.8° (c 1.32, dioxane) (t = 24°)<br>b) - 80.4 ± 0.8° (dioxane) (t = 19.5°) | H 20 |
| $C_{18}H_{18}$ | 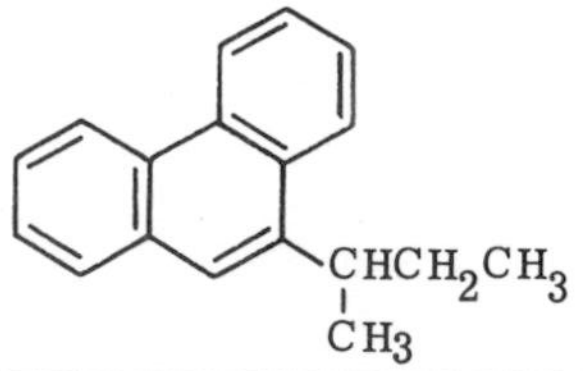 $CHCH_2CH_3$, $CH_3$ | via chromatographic resolution on silicic acid impregnated with (-)-α-(2,4,5,7-tetranitrofluorenylideneaminoöxy)-propionic acid (Partial resolution) | a) + 0.36° (cyclohexane) (t = 24°) (365 nm)<br>b) - 0.70° (cyclohexane) (t = 24°) (365 nm) | H 25 |
| $C_{18}H_{20}$ | 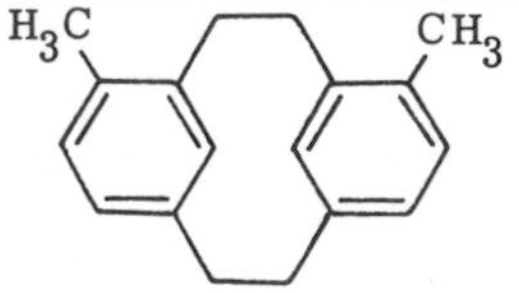 $H_3C$, $CH_3$ | (-)-α-(2,4,5,7-Tetranitrofluorenylideneaminoöxy)-propionic acid in $C_6H_6$ | a) - 3.4° (c 1.8, EtOH) (t = 26°)<br>b) + 4.2° (c 3.3, EtOH) (t = 26°)<br>both at 578 nm | A 291 (p.136) |
| $C_{19}H_{13}F$ | F $CH_3$ | (-)-α-(2,4,5,7-Tetranitrofluorenylideneaminoöxy)-propionic acid in $CHCl_3$ | a) - 580 ± 15° (c 0.5, $CHCl_3$) (t = 20°) | H 26 |
| | | Resolution in HOAc | b) + 573 ± 15° ($CHCl_3$) | H 26 |

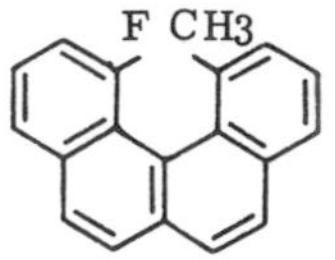

TABLE 14 (continued)

| Substance resolved | | Resolving agent and solvent | Rotation $[\alpha]_D$ of resolved compound | Reference |
|---|---|---|---|---|
| $C_{20}H_{14}$ | 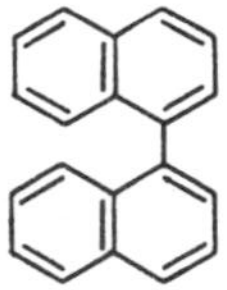 | See Table 8d, $C_{22}H_{14}O_4$, ref. A 298 (p. 137) and Table 9c, $C_{20}H_{16}N_2$, refs. B 141 and B 144 (p. 190). Also, spontaneous resolution by crystallization (H 27). | a) $+245^{\circ}$ ($C_6H_6$)<br>b) $-245^{\circ}$ ($C_6H_6$) | |
| $C_{20}H_{24}$ | 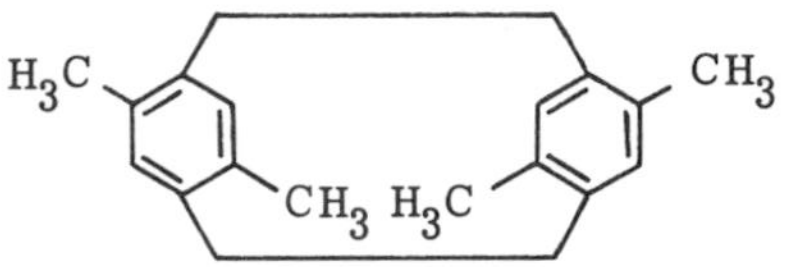 $H_3C$, $CH_3$, $CH_3$ $H_3C$ | (-)-$\alpha$-(2,4,5,7-Tetranitrofluorenylideneaminoöxy)-propionic acid in HOAc | a) $-3.2^{\circ}$ (c 1.52, $CHCl_3$) (t = $27^{\circ}$)<br>b) $+4.2^{\circ}$ (c 1.14, $CHCl_3$) (t = $27^{\circ}$) | H 28 |
| $C_{21}H_{16}$ | 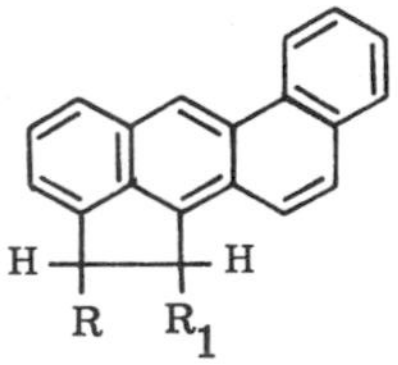 H, H, R, $R_1$<br>R = H<br>$R_1$ = $CH_3$ | (-)-$\alpha$-(2,4,5,7-Tetranitrofluorenylideneaminoöxy)-propionic acid in $C_6H_6$ | a) $-96 \pm 1^{\circ}$ (c 1, $C_6H_6$) (t = $25^{\circ}$) | H 29 |
| | | (+)-Enantiomer | b) ca. $+91^{\circ}$ ($C_6H_6$) (t = $20^{\circ}$) | H 29 |
| | R = $CH_3$<br>$R_1$ = H | (-)-Enantiomer | $-58.5 \pm 1^{\circ}$ (c 1, $C_6H_6$) (t = $20^{\circ}$) | H 29 |
| $C_{22}H_{14}$ | 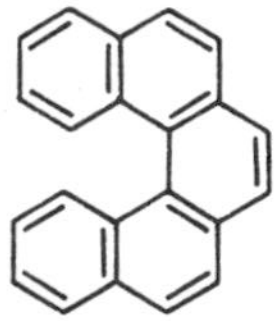 Pentahelicene | (+)-$\alpha$-(2,4,5,7-Tetranitrofluorenylideneaminoöxy)-propionic acid in HOAc | $-1670^{\circ}$ (isooctane) (t = $26^{\circ}$) (578 nm) | H 30 |
| $C_{22}H_{16}$ | 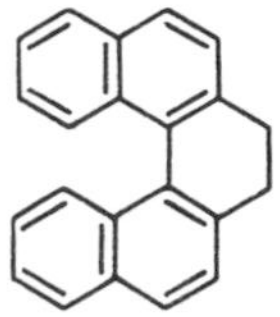 | *via* chromatographic resolution on silicic acid impregnated with (-)-$\alpha$-(2,4,5,7-tetranitrofluorenylideneaminoöxy)-propionic acid | a) $-1300^{\circ}$ (c 0.014, $C_6H_6$) (t = $24^{\circ}$) | H 21 |
| | | with (+)-enantiomer | b) $+574^{\circ}$ (t = $24^{\circ}$) | H 21 |

TABLE 14 (continued)

| | Substance resolved | Resolving agent and solvent | Rotation $[\alpha]_D$ of resolved compound | Reference |
|---|---|---|---|---|
| $C_{26}H_{16}$ | 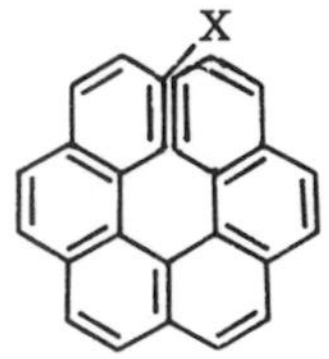 Hexahelicene X = H | (-)-α-(2,4,5,7-Tetranitrofluorenylideneaminoöxy)-propionic acid in $C_6H_6$-EtOH | a) $-3640 \pm 10^{\circ}$ (c 0.098, $CHCl_3$) ($t = 24^{\circ}$) | H 31 |
| | | (+)-Enantiomer | b) $+3707 \pm 12^{\circ}$ (c 0.082, $CHCl_3$) ($t = 25^{\circ}$) | H 31 |
| | | For asymmetric synthesis with circularly polarized light see ref. H 32 | | |
| $C_{26}H_{15}Br$ | X = Br | (-)-α-(2,4,5,7-Tetranitrofluorenylideneaminoöxy)-propionic acid | $-3560^{\circ}$ (c 0.066, $CHCl_3$) | H 33 |
| $C_{33}H_{34}O_6$ | 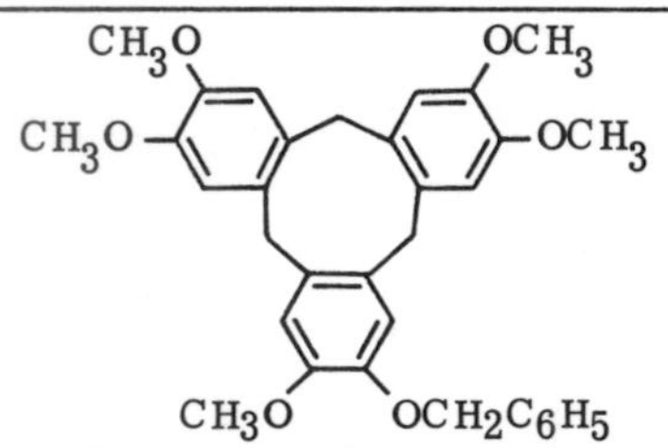 $CH_3O$, $CH_3O$, $OCH_3$, $OCH_3$, $CH_3O$, $OCH_2C_6H_5$ | via chromatography on cellulose 2,5-acetate (Partial resolution) | a) $+7.5^{\circ}$ (c 1%, $CHCl_3$) ($t = 22^{\circ}$)<br>b) $-5^{\circ}$ (c 1%, $CHCl_3$) ($t = 22^{\circ}$) | H 34 |

## References to Table 14
## Resolutions of Hydrocarbons, Halogen Compounds, and Miscellaneous Neutral Compounds

(H 1) S.H. Wilen, M.J. Wieder and C.M. Kasheres, unpublished results.

(H 2) H.J. Lucas and C.W. Gould, Jr., J. Amer. Chem. Soc., 64, 601 (1942).

(H 3) L. Fishbein and J.A. Gallaghan, J. Amer. Chem. Soc., 78, 1218 (1956).

(H 4) W.L. Waters, W.S. Linn and M.C. Caserio, J. Amer. Chem. Soc., 90, 6741 (1968); W.S. Linn, W.L. Waters and M.C. Caserio, ibid., 92, 4018 (1970).

(H 5) O. Rodriguez and H. Morrison, Chem. Commun., 1971, 679.

(H 6) R. Riemschneider, Chem. Ber., 89, 2713 (1956).

(H 7) S.J. Cristol, J. Amer. Chem. Soc., 71, 1894 (1949).

(H 8) H.C. Brown, N.R. Ayyangar and G. Zweifel, J. Amer. Chem. Soc., 86, 397 (1964).

(H 9) R. Lazzaroni, P. Salvadori and P. Pino, Tetrahedron Lett., 1968, 2507.

(H 10) H.C. Brown, N.R. Ayyangar and G. Zweifel, J. Amer. Chem. Soc., 86, 1071 (1964).

(H 11) S.I. Goldberg and F.-L. Lam, J. Org. Chem., 31, 240 (1966).

(H 12) A.C. Cope, J.K. Hecht, H.W. Johnson, Jr., H. Keller and H.J.S. Winkler, J. Amer. Chem. Soc., 88, 761 (1966).

(H 13) A.C. Cope, C.R. Ganellin, H.W. Johnson, Jr., T.V. Van Auken and H.J.S. Winkler, J. Amer. Chem. Soc., 85, 3276 (1963).

(H 14) W.L. Waters, J. Org. Chem., 36, 1569 (1971).

(H 15) A.C. Cope, K. Banholzer, H. Keller, B.A. Pawson, J.J. Whang and H.J.S. Winkler, J. Amer. Chem. Soc., 87, 3644 (1965).

(H 16) E.I. Klabunovskii, V.V. Patrikeev and A.A. Balandin, Izv. Akad. Nauk SSSR, 1960, 552; Engl. transl.: Bull. Acad. Sci. USSR, 1960, 521.

(H 17) A. Panunzi, A. De Renzi and G. Paiaro, Inorg. Chim. Acta, 1, 475 (1967).

(H 18) F.D. Greene, W.A. Remers and J.W. Wilson, J. Amer. Chem. Soc., 79, 1416 (1957).

(H 19) S.J. Cristol, F.R. Stermitz and P.S. Ramey, J. Amer. Chem. Soc., 78, 4939 (1956).

(H 20) M.S. Newman and W.B. Lutz, J. Amer. Chem. Soc., 78, 2469 (1956).

(H 21) L.H. Klemm and D. Reed, J. Chromatogr., 3, 364 (1960).

(H 22) P.S. Wharton and R.A. Kretchmer, J. Org. Chem., 33, 4258 (1968).

(H 23) T.L. Jacobs and D. Danker, J. Org. Chem., 22, 1424 (1957).

(H 24) J. Furukawa, T. Kazuken, H. Morikawa, R. Yamamoto and O. Okuno, Bull. Chem. Soc. Jap., 41, 155 (1968).

(H 25) L.H. Klemm, K.B. Desai and J.R. Spooner, Jr., J. Chromatogr., 14, 300 (1964).

(H 26) M.S. Newman, R.G. Mentzer and G. Slomp, J. Amer. Chem. Soc., 85, 4018 (1963).

(H 27) R.E. Pincock and K.R. Wilson, J. Amer. Chem. Soc., 93, 1291 (1971).

(H 28) D.T. Longone and M.T. Reetz, Chem. Commun., 1967, 46.

(H 29) M.S. Newman, R.W. Wotring, Jr., A. Pandit and P.M. Chakrabarti, J. Org. Chem., 31, 4293 (1966).

(H 30) Ch. Goedicke and H. Stegemeyer, Tetrahedron Lett., 1970, 937.

(H 31) M.S. Newman and D. Lednicer, J. Amer. Chem. Soc., 78, 4765 (1956).

(H 32) A. Moradpour, J.F. Nicoud, G. Balavoine, H. Kagan and G. Tsoucaris, J. Amer. Chem. Soc., 93, 2353 (1971).

(H 33) D.A. Lightner, D.T. Hefelfinger, G.W. Frank, T.W. Powers and K.N. Trueblood., Nature (London), Phys. Sci., 232, 124 (1971).

(H 34) A. Lüttringhaus and K.C. Peters, Angew. Chem., 78, 603 (1966); Angew. Chem. Int. Ed. Engl., 5, 593 (1966).